Molecular Research on Dental Materials and Biomaterials 2018

Molecular Research on Dental Materials and Biomaterials 2018

Editors

Ihtesham Ur Rehman
Mary Anne Melo

MDPI • Basel • Beijing • Wuhan • Barcelona • Belgrade • Manchester • Tokyo • Cluj • Tianjin

Editors
Ihtesham Ur Rehman
Lancaster University
UK

Mary Anne Melo
University of Maryland School of Dentistry
USA

Editorial Office
MDPI
St. Alban-Anlage 66
4052 Basel, Switzerland

This is a reprint of articles from the Special Issue published online in the open access journal *International Journal of Molecular Sciences* (ISSN 1422-0067) (available at: https://www.mdpi.com/journal/ijms/special_issues/dental_materials_biomaterials_2018).

For citation purposes, cite each article independently as indicated on the article page online and as indicated below:

LastName, A.A.; LastName, B.B.; LastName, C.C. Article Title. *Journal Name* **Year**, *Volume Number*, Page Range.

ISBN 978-3-0365-0086-7 (Hbk)
ISBN 978-3-0365-0087-4 (PDF)

Cover image courtesy of Mary Anne Melo.

© 2021 by the authors. Articles in this book are Open Access and distributed under the Creative Commons Attribution (CC BY) license, which allows users to download, copy and build upon published articles, as long as the author and publisher are properly credited, which ensures maximum dissemination and a wider impact of our publications.

The book as a whole is distributed by MDPI under the terms and conditions of the Creative Commons license CC BY-NC-ND.

Contents

About the Editors vii

Ihtesham Ur Rehman and Mary Anne Melo
Special Issue: Molecular Research on Dental Materials and Biomaterials 2018
Reprinted from: *Int. J. Mol. Sci.* **2020**, *21*, 9154, doi:10.3390/ijms21239154 1

Maria Salem Ibrahim, Ahmed S. Ibrahim, Abdulrahman A. Balhaddad, Michael D. Weir, Nancy J. Lin, Franklin R. Tay, Thomas W. Oates, Hockin H. K. Xu and Mary Anne S. Melo
A Novel Dental Sealant Containing Dimethylaminohexadecyl Methacrylate Suppresses the Cariogenic Pathogenicity of *Streptococcus mutans* Biofilms
Reprinted from: *Int. J. Mol. Sci.* **2019**, *20*, 3491, doi:10.3390/ijms20143491 5

Enrico Conserva, Alessandra Pisciotta, Laura Bertoni, Giulia Bertani, Aida Meto, Bruna Colombari, Elisabetta Blasi, Pierantonio Bellini, Anto de Pol, Ugo Consolo and Gianluca Carnevale
Evaluation of Biological Response of STRO-1/c-Kit Enriched Human Dental Pulp Stem Cells to Titanium Surfaces Treated with Two Different Cleaning Systems
Reprinted from: *Int. J. Mol. Sci.* **2019**, *20*, 1868, doi:10.3390/ijms20081868 19

Prakan Thanasrisuebwong, Rudee Surarit, Sompop Bencharit and Nisarat Ruangsawasdi
Influence of Fractionation Methods on Physical and Biological Properties of Injectable Platelet-Rich Fibrin: An Exploratory Study
Reprinted from: *Int. J. Mol. Sci.* **2019**, *20*, 1657, doi:10.3390/ijms20071657 35

Saturnino Marco Lupi, Arianna Rodriguez y Baena, Clara Cassinelli, Giorgio Iviglia, Marco Tallarico, Marco Morra and Ruggero Rodriguez y Baena
Covalently-Linked Hyaluronan versus Acid Etched Titanium Dental Implants: A Crossover RCT in Humans
Reprinted from: *Int. J. Mol. Sci.* **2019**, *20*, 763, doi:10.3390/ijms20030763 45

Takeo Karakida, Kazuo Onuma, Mari M. Saito, Ryuji Yamamoto, Toshie Chiba, Risako Chiba, Yukihiko Hidaka, Keiko Fujii-Abe, Hiroshi Kawahara and Yasuo Yamakoshi
Potential for Drug Repositioning of Midazolam for Dentin Regeneration
Reprinted from: *Int. J. Mol. Sci.* **2019**, *20*, 670, doi:10.3390/ijms20030670 65

Han Xie, Tong Cao, Alfredo Franco-Obregón and Vinicius Rosa
Graphene-Induced Osteogenic Differentiation Is Mediated by the Integrin/FAK Axis
Reprinted from: *Int. J. Mol. Sci.* **2019**, *20*, 574, doi:10.3390/ijms20030574 85

Sujatha Muthumariappan, Wei Cheng Ng, Christabella Adine, Kiaw Kiaw Ng, Pooya Davoodi, Chi-Hwa Wang and Joao N. Ferreira
Localized Delivery of Pilocarpine to Hypofunctional Salivary Glands through Electrospun Nanofiber Mats: An Ex Vivo and In Vivo Study
Reprinted from: *Int. J. Mol. Sci.* **2019**, *20*, 541, doi:10.3390/ijms20030541 99

Raphael Pilo, Sharon Agar-Zoizner, Shaul Gelbard and Shifra Levartovsky
The Retentive Strength of Laser-Sintered Cobalt-Chromium-Based Crowns after Pretreatment with a Desensitizing Paste Containing 8% Arginine and Calcium Carbonate
Reprinted from: *Int. J. Mol. Sci.* **2018**, *19*, 4082, doi:10.3390/ijms19124082 115

Susanne Jung, Lauren Bohner, Marcel Hanisch, Johannes Kleinheinz and Sonja Sielker
Influence of Implant Material and Surface on Differentiation and Proliferation of Human Adipose-Derived Stromal Cells
Reprinted from: *Int. J. Mol. Sci.* **2018**, *19*, 4033, doi:10.3390/ijms19124033 127

Izabela Barszczewska-Rybarek and Grzegorz Chladek
Studies on the Curing Efficiency and Mechanical Properties of Bis-GMA and TEGDMA Nanocomposites Containing Silver Nanoparticles
Reprinted from: *Int. J. Mol. Sci.* **2018**, *19*, 3937, doi:10.3390/ijms19123937 137

Mary Anne S. Melo, Michael D. Weir, Vanara F. Passos, Juliana P. M. Rolim, Christopher D. Lynch, Lidiany K. A. Rodrigues and Hockin H. K. Xu
Human In Situ Study of the effect of Bis(2-Methacryloyloxyethyl) Dimethylammonium Bromide Immobilized in Dental Composite on Controlling Mature Cariogenic Biofilm
Reprinted from: *Int. J. Mol. Sci.* **2018**, *19*, 3443, doi:10.3390/ijms19113443 155

Edina Lempel, Zsuzsanna Czibulya, Bálint Kovács, József Szalma, Ákos Tóth, Sándor Kunsági-Máté, Zoltán Varga and Katalin Böddi
Degree of Conversion and BisGMA, TEGDMA, UDMA Elution from Flowable Bulk Fill Composites
Reprinted from: *Int. J. Mol. Sci.* **2016**, *17*, 732, doi:10.3390/ijms17050732 169

Mohammad Zakir Hossain, Marina Mohd Bakri, Farhana Yahya, Hiroshi Ando, Shumpei Unno and Junichi Kitagawa
The Role of Transient Receptor Potential (TRP) Channels in the Transduction of Dental Pain
Reprinted from: *Int. J. Mol. Sci.* **2019**, *20*, 526, doi:10.3390/ijms20030526 183

Rafael Delgado-Ruiz and Georgios Romanos
Potential Causes of Titanium Particle and Ion Release in Implant Dentistry: A Systematic Review
Reprinted from: *Int. J. Mol. Sci.* **2018**, *19*, 3585, doi:10.3390/ijms19113585 215

Yu Hao, Xiaoyu Huang, Xuedong Zhou, Mingyun Li, Biao Ren, Xian Peng and Lei Cheng
Influence of Dental Prosthesis and Restorative Materials Interface on Oral Biofilms
Reprinted from: *Int. J. Mol. Sci.* **2018**, *19*, 3157, doi:10.3390/ijms19103157 251

About the Editors

Ihtesham Ur Rehman is a Professor of Bioengineering at Lancaster University. Prior to that, he was at The University of Sheffield, having previously moved from Queen Mary University of London. He has authored over 240 scientific articles and has four patents to his name. He has an h-index of 50 (Google Scholar). He has had a diverse research career, spanning over 30 years, and has led strong teams in research, teaching and management. He has given over 100 plenary and keynote lectures in the fields of biomaterials and bioengineering. In addition to authoring a number of book chapters, he has also co-authored a book (Rehman, I., Movasaghi, Z., Rehman, S. (2013). Vibrational Spectroscopy for Tissue Analysis. Boca Raton: CRC Press, https://doi.org/10.1201/b12949). Professor Rehman's research interests are of a multidisciplinary nature, including: (i) Chemical and structural evaluations of biological molecules, cells (cancer cells and stem cells) and proteins by spectroscopic means; (ii). Surface modification of polymers for the design and development of functional biomaterials; (iii). Creating bioactive functionalised materials with improved chemical, mechanical and biological properties; (iv). Antimicrobial resistance research (biofilms) using vibrational spectroscopy to study microbial interactions with blood, tissues and/or surfaces. Professor Rehman has a strong international research profile, holds visiting professorships at three international universities and is Executive Director and Founder of the Interdisciplinary Research Centre in Biomedical Materials (IRCBM) in Lahore, Pakistan (https://lahore.comsats.edu.pk/IRCBM/index.aspx). The IRCBM was set up in 2008 as a centre of excellence with a multidisciplinary approach to biomaterials science and engineering. The Centre works beyond subject boundaries with the aim of translating fundamental research into clinical care. He is also co-chairman and a founding member of UK-Pakistan Science and Innovation Global Network (http://chemb125.chem.ucl.ac.uk/UPSIGN/index.html), which aims to improve "Communication, Coordination and Cooperation" between Pakistan and UK institutions on all levels, including universities, public and private organisations, the business sector and non-government organisations (NGOs). Professor Rehman is also both a co-editor and editor for Europe for Applied Spectroscopy Reviews and a guest editor for the International Journal of Molecular Sciences.

Mary Anne Melo, DDS, MSc, Ph.D. FADM, is an Associate Professor and Division Director of Operative Dentistry at the School of Dentistry, University of Maryland, Baltimore, Maryland. Dr. Melo went to dental school at the University of Fortaleza, Brazil, completed a residency in operative dentistry, Master and Ph.D. at the Federal University of Ceara, and pursue fellowship training in dental materials at the University of Maryland before joining the operative faculty. Dr. Melo is a clinician, researcher, and educator. In her teaching work, she directs and participates in Restorative Dentistry predoctoral and postgraduate courses and mentors students in their master and Ph.D. research projects. Her main areas of research are focused primarily on 1) Oral biofilm control with nanotechnologies by investigating the anti-biofilm nanotechnologies for the oral health care field; 2) Biofilm modeling with the development of new in vitro caries/biofilm models; 3) evaluate antimicrobial properties of biomaterials and 4) the translational research for the development of clinically relevant therapeutic strategies for control of dental caries. She has authored over 100 original articles and serves as ad hoc reviewer and in the editorial board for several scientific journals in Dentistry, Medicine, and Biomaterials. Dr.Melo is a fellow of the Academy of Dental Materials and a current member of the Academy of Operative Dentistry, the International Association for Dental Research, the Society for Color and Appearance in Dentistry, and the American Academy of Cosmetic Dentistry. Dr. Melo's unique qualifications as a researcher with clinical, and materials science perspectives and preeminent research, have successfully aligned her role to advances to the dental field.

Editorial

Special Issue: Molecular Research on Dental Materials and Biomaterials 2018

Ihtesham Ur Rehman [1] and Mary Anne Melo [2,*]

[1] Engineering Department, Faculty of Science and Technology, Lancaster University, Gillow Avenue, Lancaster LA1 4YW, UK; i.u.rehman@lancaster.ac.uk
[2] Division of Operative Dentistry, Department of General Dentistry, University of Maryland School of Dentistry, 650 W. Baltimore St, Baltimore, MD 21201, USA
* Correspondence: mmelo@umaryland.edu

Received: 18 November 2020; Accepted: 26 November 2020; Published: 1 December 2020

Worldwide, populations of all ages suffer from oral diseases, disorders, pathological conditions of the oral cavity, and their impact on the human body. New dental materials and biomaterials, currently being introduced, under development, or envisioned, are expected to benefit the oral health status. These materials will provide a wide range of diverse functions, from promoting osteogenesis to bacterial biofilm formation inhibition.

Factually, dental biomaterials were intended to provide core functions, such as mechanical support to masticatory loads (e.g., dental crowns) or optical properties to display a pleasant and natural appearance (e.g., resin composites). This approach has led to the successful design of numerous clinically used materials over the years, such as sealants, orthodontic adhesives, luting cement, hybrids materials for computer-assisted design/computer-assisted manufacturing (CAD/CAM)-based restorative dentistry.

However, despite the ongoing development, unmet dental needs remain. The mouth is a dynamic environment, so dental materials regularly experience changes that alter performance, and this highlights the profound need for methods to allow tracking of its performance in physiologically complex especially simulating the intraoral environment. Both novel and cutting-edge preventive and therapeutic strategies are still in a claim.

In the special issue "Dental Materials and Biomaterials 2018", encouraging findings on progressing treatment of the most prevalent oral conditions were described, covering different aspects of technological development to treatment options, considering insights derived from a plethora of in vitro/ in situ models and clinical trials.

In efforts toward caries prevention, Ibrahim et al. [1], Zhou et al. [2], and Hao et al. [3] explored the application of dimethylaminohexadecyl methacrylate (DMAHDM) as an antibacterial strategy for resin-based materials. Barszczewska-Rybarek et al [4] also have explored the silver ions releasing approach intended for dental materials. Bioactive dimethacrylate composites filled with silver nanoparticles (AgNP) might be used in medical applications, such as dental restorations and bone types of cement. DMAHDM, an antibacterial monomer, acts primarily through direct contact killing. It has been previously hypothesized that a positively charged DMAHDM structure may interact with negatively charged bacteria to lead to cell membrane leakiness and rupture [1]. Various innovative methods have been applied to develop dental adhesives with particular functions to tackle these problems, such as incorporating matrix metalloproteinase inhibitors, antibacterial or remineralizing agents [3] into bonding systems, as well as improving the mechanical/chemical properties of adhesives, even combining these methods.

The targeted cariogenic biofilm is a densely packed community of oral microbial cells with predominantly acid-tolerant, acid-producing bacteria surrounded by an exopolysaccharide (EPS)-rich matrix. Within cariogenic dental plaque, Streptococcus mutans (S. mutans) has been strongly linked to cariogenic biofilm formation and carious lesion progression.

Delgado-Ruiz and Romanos [5] reviewed the current knowledge on titanium particles and ions released from metallic instruments used for implant bed preparation, from the implant surfaces during insertion implant-abutment interface during insertion and functional loading. Besides, the implant surfaces and restorations are exposed to the environment, saliva, bacteria, and chemicals that can potentially dissolve the titanium oxide layer. If these agents attack continuously, the implant surface can permanently lose its titanium oxide layer.

The formation of soluble compounds on the titanium surface will alter the implant surface chemistry and facilitate the dissolution and degradation of exposed bulk titanium, resulting in corrosion cycles. Implant maintenance procedures can potentially alter implant surfaces and produce titanium debris released into the peri-implant tissues. While this treatment is one of the most popular, side effects constrain its use.

Xie et al [6] have shown that graphene promotes osteogenesis via the activation of the mechanosensitive integrin/FAK axis future therapy to overcome this problem. Another possibility of improved osseointegration is discussed and reviewed by Marco Lupi et al. [7]. Biochemical modification of titanium surfaces (BMTiS) entails the immobilization of biomolecules to implant surfaces to induce specific host responses. This crossover randomized clinical trial assesses the clinical success and marginal bone resorption of dental implants bearing a surface molecular layer of covalently linked hyaluronan compared to control implants up to 36 months after loading.

In this special issue, insightful progress for dental material and biomaterials were described, and new avenues for application and translation were discussed. Particular emphasis was placed on dental materials—structure-property correlations in the context of both the clinical and non-clinical aspects. These papers provide readers with the necessary tools and principles of Dental Materials and biomaterials currently used in Clinical Dentistry and cover the underlying principles of bioactivity and biocompatibility.

Furthermore, Karakida et al. [8] studied drug repositioning is the process of discovering, validating, and marketing previously approved drugs for new indications. Due to the promise of reduced costs and expedited approval schedules, the research field of drug repositioning has been attracting attention. A standard database consisting of both approved and failed drugs has been developed as a web application to find candidates for drug repositioning. Without using the database, the present study demonstrated that MDZ enhances the differentiation of PPU-7 cells to odontoblast and promotes dentin-like hydroxyapatite formation. Further studies are required to elucidate the pharmacokinetics and pharmacological efficacy of MDZ in animal experiments. In the dental field, these findings support the repositioning of MDZ to promote dentin regeneration for endodontic treatments, such as pulp capping. Moreover, these findings support advancing research from pig to human experimental models using human DPSCs to discover MDZ's potential, not only for future dental treatments, but also for regenerative organ medicine. This special issue offers an exciting background of MS pathophysiology and disease development and prediction from animal models and clinical studies and suggests strategies for developing MS therapies.

Conflicts of Interest: The authors declare no conflict of interest.

References

1. Ibrahim, M.S.; Ibrahim, A.S.; Balhaddad, A.A.; Weir, M.D.; Lin, N.J.; Tay, F.R.; Oates, T.W.; Xu, H.H.K.; Melo, M.A.S. A Novel Dental Sealant Containing Dimethylaminohexadecyl Methacrylate Suppresses the Cariogenic Pathogenicity of Streptococcus mutans Biofilms. *Int. J. Mol. Sci.* **2019**, *20*, 3491. [CrossRef]
2. Zhou, W.; Liu, S.; Zhou, X.; Hannig, M.; Rupf, S.; Feng, J.; Peng, X.; Cheng, L. Modifying Adhesive Materials to Improve the Longevity of Resinous Restoration. *Int. J. Mol. Sci.* **2019**, *20*, 723. [CrossRef]
3. Hao, Y.; Huang, X.; Zhou, X.; Li, M.; Ren, B.; Peng, X.; Cheng, L. Influence of Dental Prosthesis and Restorative Materials Interface on Oral Biofilms. *Int. J. Mol. Sci.* **2018**, *19*, 3157. [CrossRef]

4. Barszczewska-Rybarek, I.; Chladek, G. Studies on the Curing Efficiency and Mechanical Properties of Bis-GMA and TEGDMA Nanocomposites Containing Silver Nanoparticles. *Int. J. Mol. Sci.* **2018**, *19*, 3937. [CrossRef]
5. Delgado-Ruiz, R.; Romanos, G. Potential Causes of Titanium Particle and Ion Release in Implant Dentistry: A Systematic Review. *Int. J. Mol. Sci.* **2018**, *19*, 3585. [CrossRef]
6. Xie, H.; Cao, T.; Franco-Obregón, A.; Rosa, V. Graphene-Induced Osteogenic Differentiation Is Mediated by the Integrin/FAK Axis. *Int. J. Mol. Sci.* **2019**, *20*, 574. [CrossRef] [PubMed]
7. Lupi, S.M.; Rodriguez y Baena, A.; Cassinelli, C.; Iviglia, G.; Tallarico, M.; Morra, M.; Rodriguez y Baena, R. Covalently-Linked Hyaluronan versus Acid Etched Titanium Dental Implants: A Crossover RCT in Humans. *Int. J. Mol. Sci.* **2019**, *20*, 763. [CrossRef]
8. Karakida, T.; Onuma, K.; Saito, M.M.; Yamamoto, R.; Chiba, T.; Chiba, R.; Hidaka, Y.; Fujii-Abe, K.; Kawahara, H.; Yamakoshi, Y. Potential for Drug Repositioning of Midazolam for Dentin Regeneration. *Int. J. Mol. Sci.* **2019**, *20*, 670. [CrossRef] [PubMed]

Publisher's Note: MDPI stays neutral with regard to jurisdictional claims in published maps and institutional affiliations.

© 2020 by the authors. Licensee MDPI, Basel, Switzerland. This article is an open access article distributed under the terms and conditions of the Creative Commons Attribution (CC BY) license (http://creativecommons.org/licenses/by/4.0/).

Article

A Novel Dental Sealant Containing Dimethylaminohexadecyl Methacrylate Suppresses the Cariogenic Pathogenicity of *Streptococcus mutans* Biofilms

Maria Salem Ibrahim [1,2], Ahmed S. Ibrahim [3], Abdulrahman A. Balhaddad [1,4], Michael D. Weir [1,5], Nancy J. Lin [6], Franklin R. Tay [7], Thomas W. Oates [5], Hockin H. K. Xu [1,5,8,9] and Mary Anne S. Melo [1,10,*]

1. Ph.D Program in Dental Biomedical Sciences, University of Maryland School of Dentistry, Baltimore, MD 21201, USA
2. Department of Preventive Dental Sciences, College of Dentistry, Imam Abdulrahman Bin Faisal University, Dammam 34212, Saudi Arabia
3. Medical Microbiology Department, Health Monitoring Centers, Ministry of Health, Jeddah 21176, Saudi Arabia
4. Department of Restorative Dental Sciences, College of Dentistry, Imam Abdulrahman Bin Faisal University, Dammam 34212, Saudi Arabia
5. Department of Advanced Oral Sciences and Therapeutics, Division of Biomaterials and Tissue Engineering, University of Maryland School of Dentistry, Baltimore, MD 21201, USA
6. Biosystems and Biomaterials Division, National Institute of Standards and Technology, Gaithersburg, MD 20899, USA
7. Department of Endodontics, The Dental College of Georgia, Augusta University, Augusta, GA 30912, USA
8. Center for Stem Cell Biology and Regenerative Medicine, University of Maryland School of Medicine, Baltimore, MD 21201, USA
9. University of Maryland Marlene and Stewart Greenebaum Cancer Center, University of Maryland School of Medicine, Baltimore, MD 21201, USA
10. Division of Operative Dentistry, Department of General Dentistry, University of Maryland School of Dentistry, Baltimore, MD 21201, USA
* Correspondence: mmelo@umaryland.edu; Tel.: +1-410-706-8705

Received: 11 April 2019; Accepted: 10 July 2019; Published: 16 July 2019

Abstract: Cariogenic oral biofilms are strongly linked to dental caries around dental sealants. Quaternary ammonium monomers copolymerized with dental resin systems have been increasingly explored for modulation of biofilm growth. Here, we investigated the effect of dimethylaminohexadecyl methacrylate (DMAHDM) on the cariogenic pathogenicity of *Streptococcus mutans* (*S. mutans*) biofilms. DMAHDM at 5 mass% was incorporated into a parental formulation containing 20 mass% nanoparticles of amorphous calcium phosphate (NACP). *S. mutans* biofilms were grown on the formulations, and biofilm inhibition and virulence properties were assessed. The tolerances to acid stress and hydrogen peroxide stress were also evaluated. Our findings suggest that incorporating 5% DMAHDM into 20% NACP-containing sealants (1) imparts a detrimental biological effect on *S. mutans* by reducing colony-forming unit counts, metabolic activity and exopolysaccharide synthesis; and (2) reduces overall acid production and tolerance to oxygen stress, two major virulence factors of this microorganism. These results provide a perspective on the value of integrating bioactive restorative materials with traditional caries management approaches in clinical practice. Contact-killing strategies via dental materials aiming to prevent or at least reduce high numbers of cariogenic bacteria may be a promising approach to decrease caries in patients at high risk.

Keywords: antibacterial agents; biofilm; dental caries; sealant; quaternary ammonium compounds; *Streptococcus mutans*

1. Introduction

Dental caries is a biofilm-triggered oral disease with an international pandemic distribution, primarily affecting school-age children [1]. The World Health Organization reports that dental caries is the most common oral condition included in the Global Burden of Disease Study [2]. Carious lesions of permanent teeth (2.3 billion people) ranks as the most prevalent oral condition, while carious lesions on deciduous teeth (560 million children) ranks 12th [2]. The pits and fissures on children's teeth present a topography with deep and narrow features, making the mechanical removal of plaque bacteria by brushing very challenging [3].

In efforts toward pediatric caries prevention, dental sealants were introduced for application on the occlusal surface. The sealants act as a physical barrier for microorganisms at the pits and fissures of the teeth. Evidence-based clinical recommendations, stated by the American Dental Association and the American Academy of Pediatric Dentistry, support sealants as a practical approach in preventing and arresting pit-and-fissure occlusal caries lesions [4]. Although sealants are a widely used preventive approach against occlusal carious lesions, the failure rate is high. Longitudinal data [5,6] show reasons for failure related to bacterial colonization underneath the sealed fissures with the progressive demineralization and development of cavitation.

Exploring the potential opportunities for new biotechnologies and biomaterials to improve dental outcomes requires an enhanced understanding of caries etiology and bacteria-material interactions. A material that can detrimentally affect the oral bacteria on or near its surface is a promising area of dental material development. For instance, bioactive or bioinductive materials can help reduce cariogenic biofilm formation at the sealant-tooth interface [7]. Likewise, contact-active dental materials that inhibit the growth of acid-producing bacteria can potentially disturb and even correct the localized dysbiosis (imbalance) of the oral microflora that may lead to caries around dental sealants [8,9]. Quaternary ammonium compounds are highly antimicrobial when applied in solution and also demonstrate antimicrobial activity when immobilized on a surface [10]. In the last decade, quaternary ammonium-based methacrylate monomers have been investigated to impart antibacterial surfaces to dental materials.

Recent works have demonstrated the application of dimethylaminohexadecyl methacrylate (DMAHDM) and amorphous calcium phosphate (NACP) as an antibacterial strategy for resin-based materials [8,11,12]. DMAHDM, an antibacterial monomer, acts primarily through direct contact killing [13]. It has been previously hypothesized that a positively charged DMAHDM structure may interact with negatively charged bacteria to lead to cell membrane leakiness and rupture [14,15]. Our previous report [16] has tuned sealant formulations to reach acceptable values for physical and mechanical properties using DMAHDM at 5 mass% and NACP at 20 mass%, respectively. However, the antibacterial performance of these formulations was not investigated.

The targeted cariogenic biofilm is a densely packed community of oral microbial cells with predominantly acid-tolerant, acid-producing bacteria surrounded by an exopolysaccharide (EPS)-rich matrix [13,17,18]. Within cariogenic dental plaque, *Streptococcus mutans* (*S. mutans*) has been strongly linked to cariogenic biofilm formation and carious lesion progression [19]. The main observable characteristics associated with *S. mutans* cariogenic pathogenicity include adhesion to tooth surfaces, synthesis of EPS via glucosyltransferases, biofilm formation, efficient use of sucrose to create an acidic environment, and aciduricity via acid tolerance response [13,20]. These virulence factors under a sucrose-rich environment contribute to the structural integrity and pathogenicity of the biofilm produced by *S. mutans* [17,18]. The remarkable ability of *S. mutans* to cope with significant and constant environmental variations, including changes in pH and oxygen tension, further leads it to be one of the primary cariogenic pathogens in this context [20].

The ability to reduce or modulate the virulence and viability of *S. mutans* biofilms using bioactive dental sealants could reduce caries associated with sealants. In the present study, we hypothesized that the formation and cariogenicity of *S. mutans* biofilms (acidogenicity, EPS production, aciduricity and tolerance to oxygen stress) would be reduced when the biofilms are in contact with sealant formulations containing 5% DMAHDM. Accordingly, this study aimed to investigate the effect of sealant formulations with 5% DMAHDM and 20% NACP on the virulence properties and viability of *S. mutans* biofilms.

2. Results

2.1. DMAHDM Reduced Biofilm Formation

Multiple properties of *S. mutans* were strongly reduced when the biofilms were cultured on the surfaces of sealants containing DMAHDM, relative to sealants without DMAHDM. Figure 1 shows the CFU counts for the 48 h *S. mutans* biofilms formed on the various sealants. The CFU values for groups containing 5% DMAHDM were significantly lower than those for control groups (commercial and experimental) and for groups containing 20% NACP ($p < 0.05$). A two-way ANOVA found a significant effect for 5% DMAHDM ($F_{(1,58)} = 5.453$; $p = 0.23$) but no significant effect for 20% NACP ($F_{(1,58)} = 0.884$; $p = 0.351$) or for the interaction between these factors ($p = 0.88$) on CFU counts.

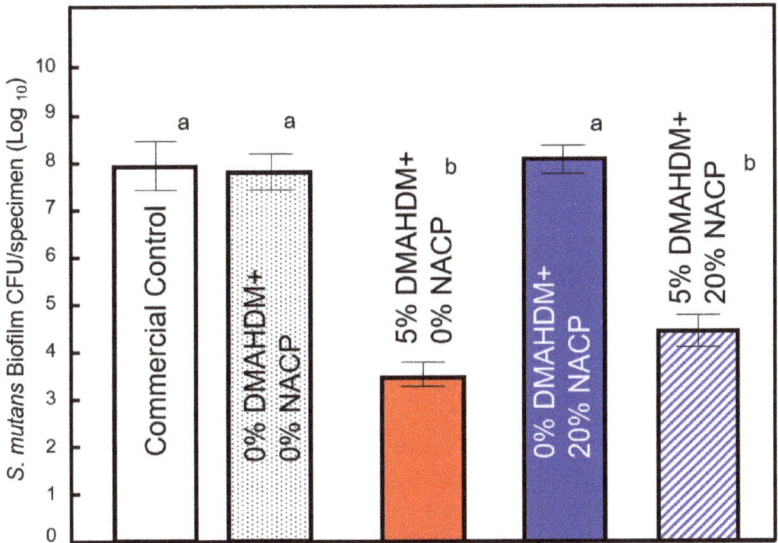

Figure 1. Colony-forming units per sealant specimen for *S. mutans* biofilms (mean ± SD of three independent experiments; $n = 6$ per group/experiment). Different lowercase letters denote a significant difference at a level of $\alpha = 0.05$ among groups.

Figure 2 shows the results from the MTT assay reflecting active metabolizing (respiring) *S. mutans* cells in the biofilms adherent to the sealant specimens. For the two formulations containing 5% DMAHDM, the MTT values were lower in comparison to the control ($p < 0.05$), representing approximately 82% and 87% reduction in biofilm metabolic activity for 5% DMAHDM + 0% NACP and 5% DMAHDM + 20% NACP, respectively. There is no main effect for the presence of 20% NACP ($F_{(1,64)} = 0.784$; $p = 0.467$).

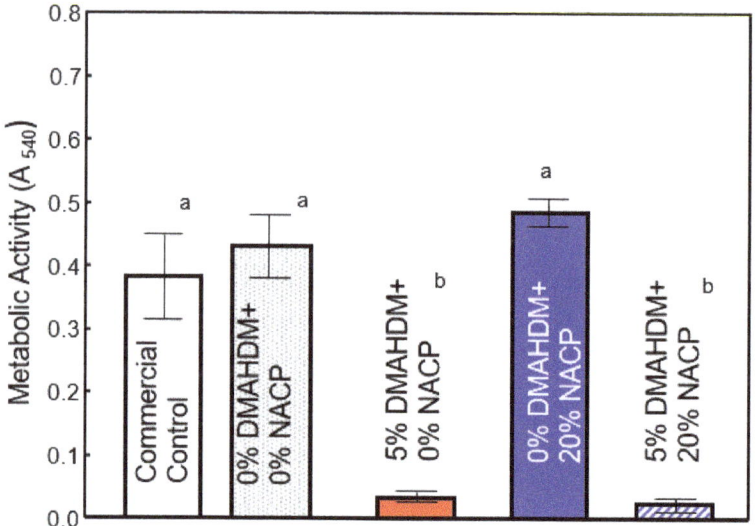

Figure 2. The metabolic activity of the 48 h *S. mutans* biofilms measured by MTT assay (mean ± SD of three independent experiments; $n = 6$ per group/experiment; non-normalized values). Different lowercase letters denote a significant difference at a level of $\alpha = 0.05$ among groups.

2.2. *DMAHDM Alone and in Combination with NACP Alters Acid and Oxygen Tolerance of S. mutans*

Figure 3A uses the change in CFU (normalized to time zero) to represent the *S. mutans* biofilm survival rate after exposure to pH 2.8 for up to 45 min. *S. mutans* biofilms grown on sealant specimens containing 5% DMAHDM displayed lower survival at 10 min (38% and 44%, respectively) compared with the control and NACP only groups (60% and 65%, respectively). This difference was not as pronounced at later time points.

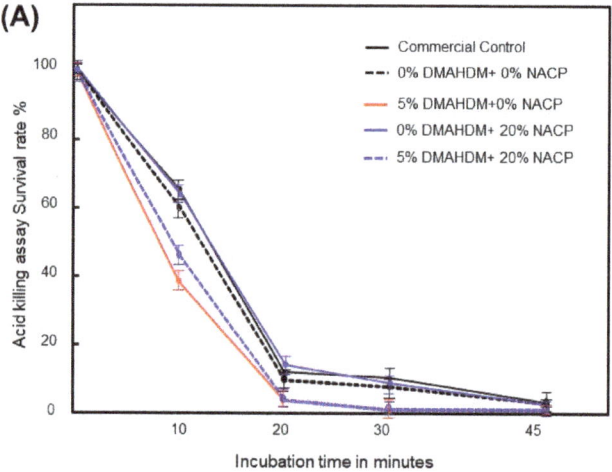

Figure 3. *Cont.*

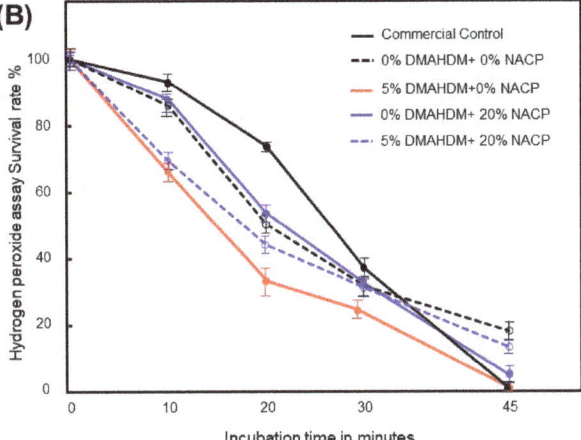

Figure 3. *S. mutans* biofilm survival expressed as the percent change in CFUs after exposure to (**A**) acid (pH 2.8) and (**B**) 0.2% H_2O_2 for up to 45 min (Mean ± SD of three independent experiments; n = 8 per group/experiment). Lines were drawn to improve readability.

For the hydrogen peroxide killing assay, Figure 3B depicts the survival rate of *S. mutans* biofilms exposed to an H_2O_2 concentration of 0.2% for 45 min. *S. mutans* biofilms formed on sealant specimens containing 5% DMAHDM exhibited lower survival than the control sealant specimens at all timepoints except at 30 min and 45 min.

2.3. Attenuation of Exopolysaccharide Produced by S. mutans

Measurements of EPS levels in the biofilms are plotted in Figure 4. Biofilms on the control sealant specimens had high exopolysaccharide production. Adding only 20% NACP to the formulation did not change the production of the polysaccharide ($F_{1,23}$ = 0.626, P = 0.436) substantially from the controls. Incorporation of DMAHDM significantly decreased the water-insoluble polysaccharide synthesis from *S. mutans* ($p < 0.05$). There was no significant interaction between DMAHDM and NACP content for polysaccharide production.

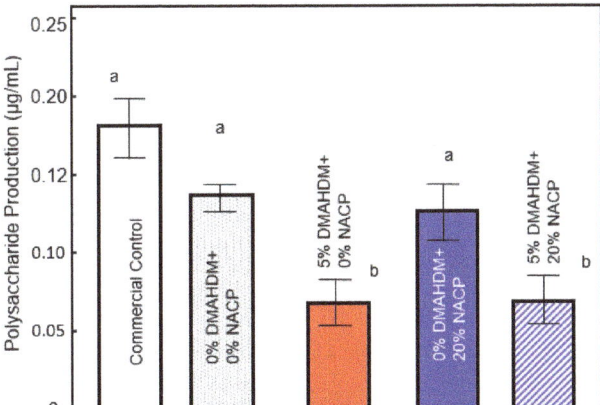

Figure 4. The polysaccharide production of 48 h *S. mutans* biofilms expressed as µg/mL (mean ± SD of three independent experiments; n = 6 per group/experimental). Different lowercase letters denote a significant difference at a level of α = 0.05 among groups.

Figure 5 plots the pH of the culture medium during biofilm formation. Biofilms in commercial and experimental control groups had the highest acid production, leading to a drop in pH sustained over 48 h. The addition of 20% NACP supported a near neutral pH over the 48-hour time course ($p < 0.05$). The addition of 5% DMAHDM had a minimal effect when NACP was present, but without NACP the initial pH drop within the first 8 h matched that of the controls. At 24 h and 48 h, however, the pH in the 5% DMAHDM only-sealant specimens increased relative to the controls, with pH values above 5.5.

The lactic acid production of biofilms on the sealants is presented in Figure 6. Biofilms on the control groups and the group containing 20% NACP only produced the most acid ($p < 0.05$), and 20% NACP had no significant effect on lactic acid production. Analysis of the lactic acid concentration indicates a significant effect of DMAHDM incorporation ($F_{1,40} = 4.611$, $P = 0.027$) on the biofilm lactic acid production. Low lactic acid concentrations at levels approximately 10% of those on the control groups were observed. There was no interaction for lactic acid production between DMAHDM content and NACP content.

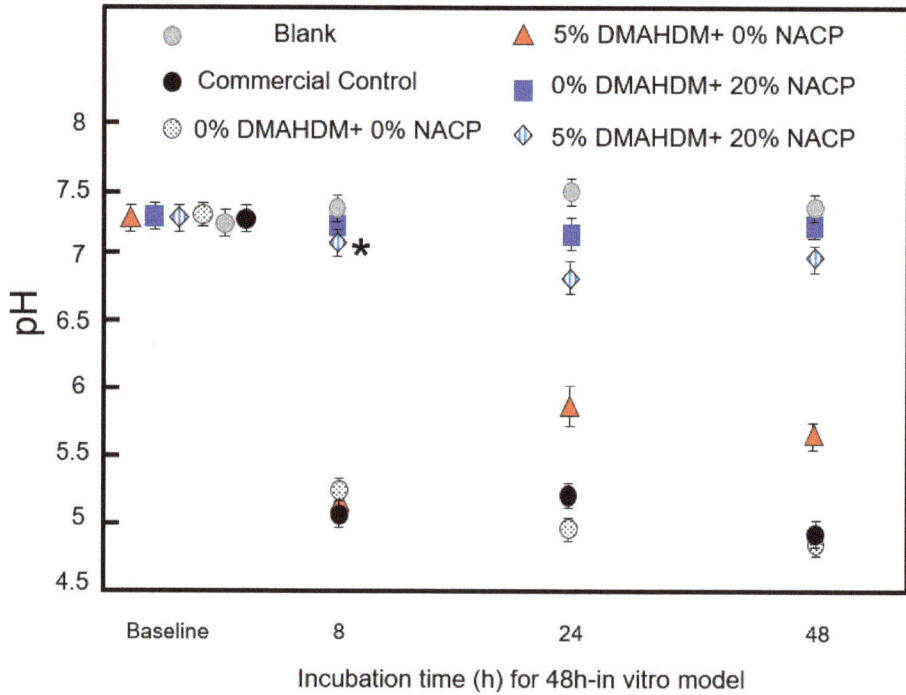

Figure 5. pH measurements (mean ± SD; $n = 3$) of growth medium from *S. mutans* biofilms grown on the various sealant specimens for 0 h, 8 h, 24 h, and 48 h. The blank represents growth medium incubated alongside tested groups. The asterisk represents a significant difference at a level of $\alpha = 0.05$ for 0% DMAHDM + 20% NACP and 5% DMAHDM + 20% NACP.

Imaging of green fluorescent protein (GFP)-expressing *S. mutans* was used to visualize the biofilm structures. The 48 h *S. mutans* biofilms formed on the 5% DMAHDM + 20% NACP sealant specimens (Figure 7A) were qualitatively reduced in visible biomass as compared to biofilms that grew on the control sealant specimens (Figure 7B). The biofilms on the control sealants (Figure 7C) appeared denser and thicker in comparison to biofilms on the DMAHDM + NACP sealant (Figure 7D).

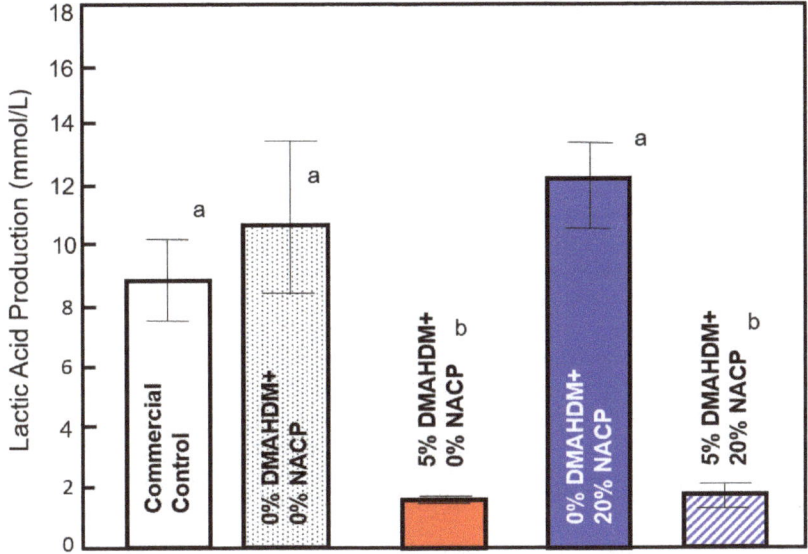

Figure 6. Lactic acid production of the 48 h *S. mutans* biofilms expressed as mmol/L (mean ± SD of three independent experiments; $n = 6$ per group/experimental). Different lowercase letters denote a significant difference at a level of $\alpha = 0.05$ among groups.

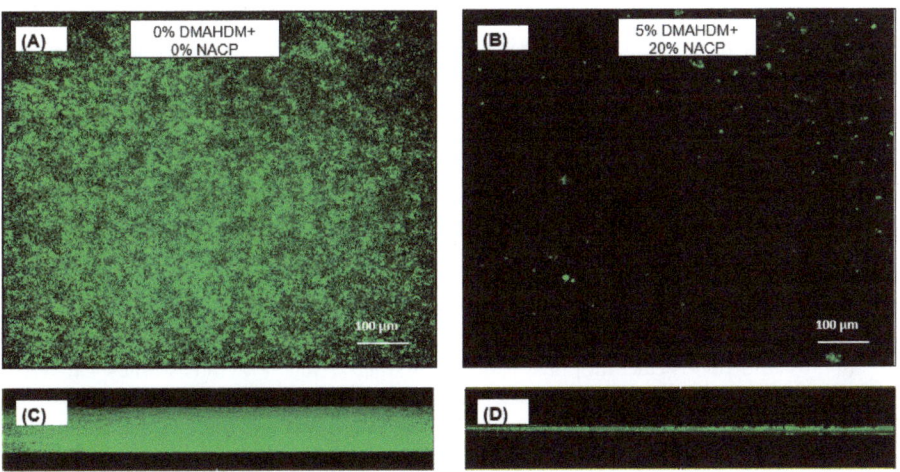

Figure 7. Scanning confocal laser microscope images of *S. mutans* biofilms formed on 0% DMAHDM + 0% NACP (control) and 5% DMAHDM + 20% NACP. (**A**,**B**) Random, representative images in the x-y plane (top view); (**C**,**D**) random, representative images in the x-z plane (side view).

3. Discussion

In this study, we explored the influence of an antibacterial contact killing approach for sealant formulations on virulence factors of *S. mutans*. The antibacterial strategy of incorporating 5% DMAHDM into a parental formulation containing 20% NACP resulted in an overall significant reduction of acid-producing biofilms. The data obtained from the CFU (Figure 1) and metabolic activity (Figure 2) assays showed that 5% DMAHDM, alone or combined with 20% NACP, can remarkably inhibit the activity of *S. mutans* biofilms, surpassing the clinically relevant greater-than-3-log (99.9%) reduction [21]. These results are consistent with previous reports for oral biofilms exposed to dental bonding formulations containing DMAHDM [22,23].

Subsequently, we examined whether the in vitro antibiofilm activity of DMAHDM would translate into changes in the ability of *S. mutans* to survive and resist high acid and oxygen stresses. *S. mutans* needs to present aciduricity to avoid falling victim to its own acidogenic metabolism. However, biofilms grown on DMAHDM-containing sealants were generally less tolerant of acidic and oxidative stresses (Figure 4). Upon exposure to acid, *S. mutans* promotes many changes at the transcriptional and physiological levels to respond to the threat of acid damage to sensitive and critical cytoplasmic molecules, such as metabolic organelles and DNA [24,25]. *S. mutans* protects itself from the damaging effects of acids through the maintenance of a cytoplasm that is more alkaline relative to the extracellular space [26]. This alkaline environment is mainly reached via plasma membrane mechanisms such as F1F0 ATPase for proton extrusion and changes in unsaturated fatty acid percentage [24]. The harmful interactions between DMAHDM and the *S. mutans* membrane can hypothetically damage a cell's ability to maintain the intracellular alkaline environment. Similarly, *S. mutans* makes changes to its membrane composition to limit the formation of hydrogen peroxide, promote its safe elimination, and repair damage [27]. If the membrane is dysfunctional, such as via perturbation by quaternary ammonium compounds like DMAHDM, the cell may have a reduced ability to survive in H_2O_2 environments, as seen here.

S. mutans relies on exopolysaccharide matrix synthesis to help it attach firmly to tooth surfaces. *S. mutans*' ability to synthesize exopolysaccharide is essential for a biofilm's structural stability and may limit antibacterial strategies [28]. According to the results of the EPS assay (Figure 5), the DMAHDM monomer lowers the level of extracellular polysaccharide production, which may make these biofilms less stable and may potentially be beneficial for the non-cariogenic microbial community. The representative confocal microscopy images (Figure 7) qualitatively support the results from the bioassays by showing reduced biofilm growth on the DMAHDM materials. The images suggest a reduced ability of *S. mutans* to form a dense biofilm architecture.

The presence of NACP alone (without DMAHDM) did not affect biofilm parameters including CFUs, metabolic activity, or polysaccharide production; however, an effect of NACP was observed on the pH values for the 48 h biofilm model where the pH stayed neutral for all time points tested when NACP was included in the formulation. NACP can affect pH through the release of supersaturating levels of calcium and phosphate ions. NACP is expected to enhance the remineralizing capacity relative to larger amorphous calcium phosphate particles due to the higher surface area-to-volume ratio of NACP. For instance, NACP has a relatively high specific surface area of 17.76 m^2/g, compared to about 0.5 m^2/g of traditional CaP particles [29]. The higher proportion of exposed surface area for NACP requires lower filler levels (by mass) to reach the same surface area and therefore expected outcome as compared to amorphous calcium phosphate on the micro scale. Previous studies showed that nanocomposites with NACP filler fraction of 20 mass% were able to reduce demineralization [30,31]. The importance of the calcium reservoir relies on the fact that Ca^{2+}, being one of the ions of hydroxyapatite, may reduce the driving force for tooth demineralization that occurs during a pH drop in dental biofilm. During demineralization, calcium release precedes phosphate release from enamel, dentin, and cementum, and calcium-deficient carbonated hydroxyapatite comprises the major substitution activity that takes place [32]. The presence of carbonates and other ionic substitutions significantly disrupts the crystal lattice in hydroxyapatite, weakening the hydroxyapatite and increasing its susceptibility to acid attack and solubility. Therefore, using high-calcium release materials to suppress the demineralization process could be beneficial.

Collectively, our data support the hypothesis that sealants with DMAHDM are able to disrupt *S. mutans* biofilm formation, thereby reducing their virulence potential. This outcome helps open the door for an important pathway to suppress dental caries via biofilm modulation and not only total biofilm eradication. A previous study conducted to investigate the antibacterial effect of DMAHDM on multispecies biofilms revealed that biofilms had a reduced relative amount of *S. mutans* when grown in contact with 3% DMAHDM as compared to the control, suggesting that DMAHDM may also be able to modulate the biofilm species composition toward a non-cariogenic tendency [33]. Shifting the biofilm composition from a cariogenic/acidogenic phenotype to a less virulent or even non-cariogenic biofilm is essential. However, the *S. mutans*/quaternary ammonium interactions involve various complex mechanisms where specific mechanisms and pathways are not clear and are still under debate [10,34]. While this study focused on the ability of sealant formulations with 5% DMAHDM and 20% NACP to alter the virulence-related traits of *S. mutans*, further research on the underlying mechanisms of bacterial/quaternary ammonium interactions is needed to optimize materials and achieve desired outcomes of improved oral health.

4. Materials and Methods

4.1. Development of Dental Resin Sealant Formulations

The parental resin matrix [16] consisted of (% by mass): 44.5% of pyromellitic glycerol dimethacrylate (PMGDM) (Hampford, Stratford, CT, USA); 39.5% of ethoxylated bisphenol a dimethacrylate (EBPADMA) (Sigma-Aldrich, St. Louis, MO, USA); 10% of 2-hydroxyethyl methacrylate (HEMA) (Esstech, Essington, PA, USA); 5% of bisphenol a glycidyl dimethacrylate (Esstech), and 1% of phenyl bis (2,4,6-trimethyl benzoyl)-phosphine oxide (BAPO) (Sigma-Aldrich) as a photo-initiator. For some formulations, DMAHDM at 5 mass% was added. The mass ratio of resin to filler was 1:1, where the filler fraction consisted of silanized barium boroaluminosilicate glass particles (average size of 1.4 µm (Caulk/Dentsply, Milford, DE, USA)) with or without the addition of NACP. DMAHDM and NACP were synthesized according to previous studies [16,22].

The tested dental sealant formulations were prepared as follows (% by mass):

1. Commercial High-viscosity Sealant/Flowable Composite control termed "Commercial Control" (Virtuoso, DenMat, Lompoc, CA, USA).
2. Experimental Control Sealant termed "0% DMAHDM + 0% NACP; Experimental Control" (50% PEHB + 0% DMAHDM + 50% Glass + 0% NACP).
3. Experimental Sealant termed "5% DMAHDM + 0% NACP" (45% PEHB + 5% DMAHDM + 50% Glass + 0% NACP).
4. Experimental Sealant termed "0% DMAHDM + 20% NACP" (50% PEHB + 0% DMAHDM + 30% Glass + 20% NACP).
5. Experimental Sealant termed "5% DMAHDM + 20% NACP" (45% PEHB + 5% DMAHDM + 30% Glass + 20% NACP).

4.2. Sample Preparation

Sealant specimens (diameter = 9 mm; thickness = 2 mm) were prepared using polytetrafluoroethylene molds covered with Mylar strips and glass slides [35]. Each sealant specimen was light-cured (1200 mW/cm^2; 60 s; Labolight DUO, GC America, Alsip, IL, USA) on each side. Specimens were sterilized using ethylene oxide (24-h cycle; Andersen Sterilizers, Inc., Haw River, NC, USA) and allowed to de-gas for at least seven days after sterilization.

4.3. S. mutans Biofilm Formation

An *S. mutans* biofilm model was used [36] with modifications. *S. mutans* UA159 from the American Type Culture Collection (ATCC, Manassas, VA, USA) was cultured overnight (≈18 h) in brain heart infusion (BHI) broth (Sigma-Aldrich) at 37 °C and 5% CO_2 (by volume). This *S. mutans* culture was adjusted to an optical density at 600 nm (OD_{600}) of 0.9 (≈1 × 10^8 CFU/mL according to growth curves monitored by spectrophotometry) and diluted in biofilm medium containing BHI broth supplemented with 2% (by mass) sucrose to prepare the inoculum.

The inoculum (1.5 mL) was added to wells of a 24-well culture plate containing the sterilized sealant specimens, and the well plates were incubated in an anaerobic chamber (37 °C; gas composition by volume: 10% H_2, 5% CO_2, 85% N_2; Whitley Workstation DG250; Microbiology International, Frederick, MD, USA) for 48 h without agitation to form a more established biofilm. The growth medium was changed at 8 h and 24 h. Control groups included wells with 1.5 mL of inoculum without sealants (Control inoculum) and wells with 1.5 mL of BHI + 2% sucrose without bacteria (Control BHI). Sealant specimens with the 48-hour biofilms were characterized as described in Sections 4.4–4.10.

4.4. Colony-Forming Unit Counts

After the sealant specimens ($n = 6$) had been incubated for 48 h, each sealant specimen with the attached biofilm was washed in 1 mL of phosphate-buffered saline (PBS), transported to a vial containing 2 mL of PBS, sonicated for 5 min (Branson 3510-DTH Ultrasonic Cleaner), and vortexed (5 s; maximum speed) to harvest the biofilm [37]. The resultant bacterial suspensions were serially diluted 10^1- through 10^6-fold, drop-plated onto BHI agar plates, and anaerobically incubated at 37 °C. After 48 h, the number of colony-forming units (CFU) was determined [37]. The results were calculated based on the number of CFU and the dilution factor; $\log_{10}$ transformed data are expressed as CFU/specimen.

4.5. Metabolic Activity

Metabolic activity was assessed via MTT (3-[4,5-dimethylthiazol-2-yl]-2,5-diphenyltetrazolium bromide) assay [37]. Briefly, each sealant specimen ($n = 6$) was washed in 1 mL of PBS and transported to a new 24-well plate containing 1 mL of tetrazolium dye (0.5 mg/mL) in PBS. Then, sealant specimens were incubated anaerobically for 1 h and transferred to a new 24-well plate. To solubilize the formazan crystals, 1 mL of dimethyl sulfoxide (DMSO) was added to each well, and the plate was incubated for 20 min at 37 °C. After that, 200 µL of the DMSO solution was collected from each specimen to measure the absorbance at 540 nm [37] (SpectraMax, Molecular Devices LLC, San Jose, CA, USA). The OD_{540} of DMSO defines the blanks for the spectrophotometric method. Values were not normalized prior to analysis.

4.6. Acid Stress and Oxygen Stress Tolerance

The differences in acid tolerance of *S. mutans* UA159 biofilms on the various formulations were evaluated using the method described by Kim et al. [38]. After 48 h of incubation, sealant specimens ($n = 8$) were washed once with 0.1 mol/L glycine (pH 7.0) and switched to 0.1 mol/L glycine at pH 2.8. Wells with biofilms but not treated with glycine served as controls. Specimens were incubated at room temperature, and 100 µL aliquots were collected at 0 min, 10 min, 20 min, 30 min, and 45 min. CFU was determined per Section 4.4.

The differences in oxidative stress tolerance of *S. mutans* UA159 biofilms were evaluated via hydrogen peroxide killing assay ($n = 8$) [38]. Sealant specimens were washed once with 0.1 mol/L glycine (pH 7.0) and placed in 5 mL of the same buffer. After that, hydrogen peroxide was added for a final concentration of 0.2%, and 100-µL aliquots were collected at 0 min, 10 min, 20 min, 30 min, and 45 min after the addition of hydrogen peroxide. Ten microliters of catalase (5 mg/mL) were added to each aliquot to inactivate the hydrogen peroxide immediately after aliquot collection. CFU was determined per Section 4.4.

4.7. Polysaccharide Production

The water-insoluble polysaccharide production in the 48 h *S. mutans* biofilms was measured using a phenol-sulfuric acid method [9] with modifications. Each sealant specimen ($n = 6$) was rinsed with 2 mL of PBS, transferred to a vial containing 1 mL of PBS, sonicated (Branson 3510-DTH Ultrasonic Cleaner) for 5 min, and vortexed (5 s; maximum speed) to harvest the biofilm. Centrifugation ($10,000 \times g$ (~10,000 r.c.f.) at 4 °C for 5 min) yielded a precipitate that was rinsed with PBS and resuspended in 1 mL of de-ionized water. Then, 1 mL of 6% (by volume) phenol solution was added to the vial, followed by 5 mL of 95–97% (by volume) sulfuric acid. After 30 min of incubation at 23 °C, 100 µL were transferred to a 96-well plate, and absorbance was measured at OD_{490} using a spectrophotometer. Standard glucose concentrations were used to convert OD readings to polysaccharide concentrations.

4.8. Acid-Neutralizing Activity

The pH of the biofilm growth medium ($n = 3$) was measured at 0 h, 8 h, 24 h, and 48 h using a digital pH meter (accuracy ± 5%; Accumet XL25, Thermo Fisher Scientific, Waltham, MA, USA). During the experiments, the pH meter was calibrated at regular intervals using commercial standard buffer solutions of pH 4, pH 7, and pH 10 at room temperature.

4.9. Lactic Acid Production

Lactate concentrations were determined using an enzymatic (lactate dehydrogenase) method [33]. After the sealant specimens had incubated for 48 h, each specimen ($n = 6$) with the attached biofilm was washed in 1.5 mL of cysteine peptone water (by mass: yeast extract 0.5%, peptone 0.1%, sodium chloride 0.85%, cysteine 0.005%), transferred to a new plate containing 1.5 mL of buffered peptone water (BPW) with 0.2% (by mass) sucrose, and incubated for 3 h at 37 °C in 5% CO_2 [37]. The lactic acid concentrations in the BPW solution were then measured based on the absorbance at 340 nm using a spectrophotometer. A standard curve was prepared using lactic acid standard solution (46937; Sigma) [39].

4.10. Confocal Laser Scanning Microscopy

Confocal laser scanning microscopy (CLSM) was performed for qualitative visualization of fully-hydrated, living cultures for two groups of sealants: the experimental control and 5% DMAHDM + 20% NACP. The qualitative approach was chosen to support the microbiological assay results. Biofilms were grown as described apart from the bacterial strain. For imaging, a green fluorescent protein (*gfp*)-expressing strain of *S. mutans* UA159 (*gfp*-UA159) was used in the biofilm model. *Gfp*-UA159 cells continually exhibit green fluorescence throughout growth.

The fluorescence of *gfp*-UA159 48 h biofilms was captured with a Yokogawa Spinning Disk Field Scanning Confocal System (CSU-W1) (Nikon Instruments Inc. Melville, NY, USA) on a Nikon Ti2 inverted microscope (Nikon Instruments Inc. Melville, NY, USA) using a 20× (numerical aperture, 1.00) water immersion objective. Biofilms were imaged four times on two sealant specimens. Biofilm slices in the x-y and x-z planes were collected at random locations on the biofilms.

4.11. Statistical Analysis

All experiments were performed with six to eight repetitions (sealant specimens) in each of three independent experiments. The average of the three experiments per specimen was considered as a statistical unit. The data satisfied the assumption of the equality of variances and normal distribution of errors. Factors included DMAHDM incorporation at two levels (0% and 5%) and NACP incorporation at two levels (0% and 20% NACP). Interactions between these two factors were also considered. To assess the *S. mutans* biofilm biological response, the following dependent variables were analyzed: Log CFU/specimen, metabolic activity, lactic acid production, and exopolysaccharide synthesis. The means and standard deviations (which serve as an estimate of the measurement uncertainty) for the dependable variables were analyzed by two-way ANOVA. Statistical significance was pre-set at $\alpha = 5\%$. All the statistical analyses were performed by SPSS statistics software (IBM version 26, Armonk, NY, USA).

4.12. Disclaimer

Certain commercial equipment, instruments, and materials are identified in this paper in order to specify the experimental procedure. Such identification does not imply recommendation or

5. Conclusions

S. mutans relies on its virulence-related traits to change oral plaque biofilms and promote the development and increase the severity of carious lesions in children. This study investigated the antibacterial strategy of incorporating 5% DMAHDM into dental sealants containing 20% NACP to reduce virulence and viability of *S. mutans* biofilms. Our findings suggest that sealants with 5% DMAHDM were able to impart a detrimental biological effect on *S. mutans* biofilms by significantly reducing the biofilm formation, metabolic activity, and EPS production. The antibacterial sealants were also able to reduce the acidicurity and the tolerance to oxygen stress. This less virulent phenotype is expected to reflect on the clinical pathogenicity of *S. mutans* and to play a key role in improving the longevity and integrity of dental sealants. Contact-killing strategies via dental materials aiming to prevent or modulate pathogenic biofilms are a promising approach in patients at high risk of caries.

Author Contributions: Conceptualization, M.A.S.M. and H.H.K.X.; Methodology, M.A.S.M., H.H.K.X., M.D.W., A.S.I. and N.J.L.; Investigation, M.S.I.; Analysis and Graphs—M.S.I., A.A.B. and M.A.S.M.; Writing—M.S.I., M.A.S.M. and H.H.K.X.; Writing—Review & Editing, H.H.K.X., N.J.L. and M.A.S.M.; Funding Acquisition, H.H.K.X., and M.A.S.M.; Resources, F.R.T.; M.D.W. and T.W.O.; Supervision, M.D.W. and T.W.O.; Project Administration, H.H.K.X. and M.A.S.M.

Funding: This work was supported by a University of Maryland Baltimore Seed grant (HX), and the University of Maryland School of Dentistry departmental fund (HX, MM).

Acknowledgments: The authors thank Mary Ann Jabra-Rizk for providing the *gfp*-expressing strain of *S. mutans* and Joseph Mauban for his assistance with the confocal microscopic images. M.S.I acknowledges the scholarship from the Imam AbdulRahman bin Faisal University, Dammam, Saudi Arabia, and the Saudi Arabia Cultural Mission (#428237).

Conflicts of Interest: The authors declare no conflict of interest. The funders had no role in the design of the study; in the collection, analyses, or interpretation of data; in the writing of the manuscript, or in the decision to publish the results.

References

1. Manton, D.J. Child dental caries—A global problem of inequality. *EClinicalMedicine* **2018**, *1*, 3–4. [CrossRef] [PubMed]
2. WHO. Sugars and Dental Caries. Available online: http://www.who.int/oral_health/publications/sugars-dental-caries-keyfacts/en/ (accessed on 24 February 2019).
3. Papageorgiou, S.N.; Dimitraki, D.; Kotsanos, N.; Bekes, K.; van Waes, H. Performance of pit and fissure sealants according to tooth characteristics: A systematic review and meta-analysis. *J. Dent.* **2017**, *66*, 8–17. [CrossRef] [PubMed]
4. Wright, J.T.; Crall, J.J.; Fontana, M.; Gillette, E.J.; Nový, B.B.; Dhar, V.; Donly, K.; Hewlett, E.R.; Quinonez, R.B.; Chaffin, J.; et al. Evidence-based clinical practice guideline for the use of pit-and-fissure sealants: A report of the American dental association and the American academy of pediatric dentistry. *J. Am. Dent. Assoc.* **2016**, *147*, 672–682. [CrossRef] [PubMed]
5. Alves, L.S.; Giongo, F.C.M.D.S.; Mua, B.; Martins, V.B.; Barbachan e Silva, B.; Qvist, V.; Maltz, M. A randomized clinical trial on the sealing of occlusal carious lesions: 3–4-year results. *Braz. Oral Res.* **2017**, *31*, e44. [CrossRef] [PubMed]
6. Fontana, M.; Platt, J.A.; Eckert, G.J.; González-Cabezas, C.; Yoder, K.; Zero, D.T.; Ando, M.; Soto-Rojas, A.E.; Peters, M.C. Monitoring of sound and carious surfaces under sealants over 44 months. *J. Dent. Res.* **2014**, *93*, 1070–1075. [CrossRef] [PubMed]
7. Cheng, Y.; Feng, G.; Moraru, C.I. Micro-and nanotopography sensitive bacterial attachment mechanisms: A review. *Front. Microbiol.* **2019**, *10*. [CrossRef] [PubMed]
8. Melo, M.A.S.; Weir, M.D.; Li, F.; Cheng, L.; Zhang, K.; Xu, H.H.K. Control of biofilm at the tooth-restoration bonding interface: A question for antibacterial monomers? A critical review. *Rev. Adhes. Adhes.* **2017**, *5*, 303–324. [CrossRef]

9. Wang, L.; Xie, X.; Imazato, S.; Weir, M.D.; Reynolds, M.A.; Xu, H.H.K. A protein-repellent and antibacterial nanocomposite for class-V restorations to inhibit periodontitis-related pathogens. *Mater. Sci. Eng. C* **2016**, *67*, 702–710. [CrossRef]
10. Inácio, Â.S.; Domingues, N.S.; Nunes, A.; Martins, P.T.; Moreno, M.J.; Estronca, L.M.; Fernandes, R.; Moreno, A.J.M.; Borrego, M.J.; Gomes, J.P.; et al. Quaternary ammonium surfactant structure determines selective toxicity towards bacteria: mechanisms of action and clinical implications in antibacterial prophylaxis. *J. Antimicrob. Chemother.* **2016**, *71*, 641–654. [CrossRef]
11. Melo, M.A.; Orrego, S.; Weir, M.D.; Xu, H.H.K.; Arola, D.D. Designing multiagent dental materials for enhanced resistance to biofilm damage at the bonded interface. *ACS Appl. Mater. Interfac.* **2016**, *8*, 11779–11787. [CrossRef]
12. Melo, M.A.S.; Guedes, S.F.F.; Xu, H.H.K.; Rodrigues, L.K.A. Nanotechnology-based restorative materials for dental caries management. *Trends Biotechnol.* **2013**, *31*, 459–467. [CrossRef] [PubMed]
13. Kuramitsu, H.K.; Wang, B.Y. The whole is greater than the sum of its parts: dental plaque bacterial interactions can affect the virulence properties of cariogenic Streptococcus mutans. *Am. J. Dent.* **2011**, *24*, 153–154. [PubMed]
14. Lin, W.; Yuan, D.; Deng, Z.; Niu, B.; Chen, Q. The cellular and molecular mechanism of glutaraldehyde-didecyldimethylammonium bromide as a disinfectant against Candida albicans. *J. Appl. Microbiol.* **2019**, *126*, 102–112. [CrossRef] [PubMed]
15. Li, J.; Zhao, L.; Wu, Y.; Rajoka, M.S.R. Insights on the ultra high antibacterial activity of positionally substituted 2′-O-hydroxypropyl trimethyl ammonium chloride chitosan: A joint interaction of -NH2 and -N + (CH3)3 with bacterial cell wall. *Colloids Surf. B Biointerf.* **2019**, *173*, 429–436. [CrossRef] [PubMed]
16. Ibrahim, M.S.; AlQarni, F.D.; Al-Dulaijan, Y.A.; Weir, M.D.; Oates, T.W.; Xu, H.H.K.; Melo, M.A.S. Tuning nano-amorphous calcium phosphate content in novel rechargeable Antibacterial dental sealant. *Materials* **2018**, *11*, 1544. [CrossRef] [PubMed]
17. Xiao, J.; Klein, M.I.; Falsetta, M.L.; Lu, B.; Delahunty, C.M.; Yates, J.R.; Heydorn, A.; Koo, H. The exopolysaccharide matrix modulates the interaction between 3D architecture and virulence of a mixed-species oral biofilm. *PLoS Pathog.* **2012**, *8*, e1002623. [CrossRef]
18. Klein, M.I.; Hwang, G.; Santos, P.H.S.; Campanella, O.H.; Koo, H. Streptococcus mutans-derived extracellular matrix in cariogenic oral biofilms. *Front. Cell. Infect. Microbiol.* **2015**, *5*, 10. [CrossRef]
19. Valdez, R.M.A.; Duque, C.; Caiaffa, K.S.; dos Santos, V.R.; de Aguiar Loesch, M.L.; Colombo, N.H.; Arthur, R.A.; de Negrini, T.C.; Boriollo, M.F.G.; Delbem, A.C.B. Genotypic diversity and phenotypic traits of *Streptococcus mutans* isolates and their relation to severity of early childhood caries. *BMC Oral Health* **2017**, *17*, 115. [CrossRef]
20. Krzyściak, W.; Jurczak, A.; Kościelniak, D.; Bystrowska, B.; Skalniak, A. The virulence of *Streptococcus mutans* and the ability to form biofilms. *Eur. J. Clin. Microbiol. Infect. Dis.* **2014**, *33*, 499–515. [CrossRef]
21. Michels, H.T.; Keevil, C.W.; Salgado, C.D.; Schmidt, M.G. From laboratory research to a clinical trial. *HERD* **2015**, *9*, 64–79. [CrossRef]
22. Wu, J.; Zhou, H.; Weir, M.D.; Melo, M.A.S.; Levine, E.D.; Xu, H.H.K. Effect of dimethylaminohexadecyl methacrylate mass fraction on fracture toughness and antibacterial properties of CaP nanocomposite. *J. Dent.* **2015**, *43*, 1539–1546. [CrossRef] [PubMed]
23. Zhou, H.; Li, F.; Weir, M.D.; Xu, H.H.K. Dental plaque microcosm response to bonding agents containing quaternary ammonium methacrylates with different chain lengths and charge densities. *J. Dent.* **2013**, *41*, 1122–1131. [CrossRef] [PubMed]
24. Baker, J.L.; Faustoferri, R.C.; Quivey, R.G. Acid-adaptive mechanisms of *Streptococcus mutans*-the more we know, the more we don't. *Mol. Oral Microbiol.* **2017**, *32*, 107–117. [CrossRef] [PubMed]
25. Len, A.C.L.; Harty, D.W.S.; Jacques, N.A. Stress-responsive proteins are upregulated in *Streptococcus mutans* during acid tolerance. *Microbiology* **2004**, *150*, 1339–1351. [CrossRef] [PubMed]
26. Kajfasz, J.K.; Rivera-Ramos, I.; Abranches, J.; Martinez, A.R.; Rosalen, P.L.; Derr, A.M.; Quivey, R.G.; Lemos, J.A. Two Spx proteins modulate stress tolerance, survival, and virulence in *Streptococcus mutans*. *J. Bacteriol.* **2010**, *192*, 2546–2556. [CrossRef] [PubMed]
27. Kajfasz, J.K.; Ganguly, T.; Hardin, E.L.; Abranches, J.; Lemos, J.A. Transcriptome responses of *Streptococcus mutans* to peroxide stress: Identification of novel antioxidant pathways regulated by Spx. *Sci. Rep.* **2017**, *7*, 16018. [CrossRef] [PubMed]

28. Hwang, G.; Klein, M.I.; Koo, H. Analysis of the mechanical stability and surface detachment of mature *Streptococcus mutans* biofilms by applying a range of external shear forces. *Biofouling* **2014**, *30*, 1079–1091. [CrossRef]
29. Sun, L.; Chow, L.C.; Frukhtbeyn, S.A.; Bonevich, J.E. Preparation and properties of nanoparticles of calcium phosphates with various Ca/P ratios. *J. Res. Natl. Inst. Stand. Technol.* **2010**, *115*, 243–255. [CrossRef]
30. Melo, M.A.S.; Weir, M.D.; Rodrigues, L.K.A.; Xu, H.H.K. Novel calcium phosphate nanocomposite with caries-inhibition in a human in situ model. *Dent. Mater.* **2013**, *29*, 231–240. [CrossRef]
31. Weir, M.D.; Chow, L.C.; Xu, H.H.K. Remineralization of demineralized enamel via calcium phosphate nanocomposite. *J. Dent. Res.* **2012**, *91*, 979–984. [CrossRef]
32. Abou Neel, E.A.; Aljabo, A.; Strange, A.; Ibrahim, S.; Coathup, M.; Young, A.M.; Bozec, L.; Mudera, V. Demineralization-remineralization dynamics in teeth and bone. *Int. J. Nanomed.* **2016**, *11*, 4743–4763. [CrossRef] [PubMed]
33. Wang, H.; Wang, S.; Cheng, L.; Jiang, Y.; Melo, M.A.S.; Weir, M.D.; Oates, T.W.; Zhou, X.; Xu, H.H.K. Novel dental composite with capability to suppress cariogenic species and promote non-cariogenic species in oral biofilms. *Mater. Sci. Eng. C* **2019**, *94*, 587–596. [CrossRef] [PubMed]
34. Wilkosz, N.; Jamróz, D.; Kopeć, W.; Nakai, K.; Yusa, S.I.; Wytrwal-Sarna, M.; Bednar, J.; Nowakowska, M.; Kepczynski, M. Effect of polycation structure on interaction with lipid membranes. *J. Phys. Chem. B* **2017**, *121*, 7318–7326. [CrossRef] [PubMed]
35. Al-Dulaijan, Y.A.; Cheng, L.; Weir, M.D.; Melo, M.A.S.; Liu, H.; Oates, T.W.; Wang, L.; Xu, H.H.K. Novel rechargeable calcium phosphate nanocomposite with antibacterial activity to suppress biofilm acids and dental caries. *J. Dent.* **2018**, *72*, 44–52. [CrossRef] [PubMed]
36. Wang, S.; Zhou, C.; Ren, B.; Li, X.; Weir, M.D.; Masri, R.M.; Oates, T.W.; Cheng, L.; Xu, H.K.H. Formation of persisters in *Streptococcus mutans* biofilms induced by antibacterial dental monomer. *J. Mater. Sci. Mater. Med.* **2017**, *28*, 178. [CrossRef] [PubMed]
37. Al-Qarni, F.D.; Tay, F.; Weir, M.D.; Melo, M.A.S.; Sun, J.; Oates, T.W.; Xie, X.; Xu, H.H.K. Protein-repelling adhesive resin containing calcium phosphate nanoparticles with repeated ion-recharge and re-releases. *J. Dent.* **2018**, *78*, 91–99. [CrossRef] [PubMed]
38. Kim, K.; An, J.S.; Lim, B.S.; Ahn, S.J. Effect of bisphenol a glycol methacrylate on virulent properties of *Streptococcus mutans* UA159. *Caries Res.* **2019**, *53*, 84–95. [CrossRef]
39. Melo, M.A.S.; Cheng, L.; Weir, M.D.; Hsia, R.C.; Rodrigues, L.K.A.; Xu, H.H.K. Novel dental adhesive containing antibacterial agents and calcium phosphate nanoparticles. *J. Biomed. Mater. Res. B Appl. Biomater.* **2013**, *101*, 620–629. [CrossRef]

© 2019 by the authors. Licensee MDPI, Basel, Switzerland. This article is an open access article distributed under the terms and conditions of the Creative Commons Attribution (CC BY) license (http://creativecommons.org/licenses/by/4.0/).

Article

Evaluation of Biological Response of STRO-1/c-Kit Enriched Human Dental Pulp Stem Cells to Titanium Surfaces Treated with Two Different Cleaning Systems

Enrico Conserva [1,2,†], Alessandra Pisciotta [1,†], Laura Bertoni [1], Giulia Bertani [1], Aida Meto [1], Bruna Colombari [1], Elisabetta Blasi [1], Pierantonio Bellini [1,2], Anto de Pol [1], Ugo Consolo [1,2] and Gianluca Carnevale [1,*]

1. Department of Surgery, Medicine, Dentistry and Morphological Sciences with interest in Transplant, Oncology and Regenerative Medicine, University of Modena and Reggio Emilia, 41125 Modena, Italy; enrico.conserva@unimore.it (E.C.); alessandra.pisciotta@unimore.it (A.P.); laura.bertoni@unimore.it (L.B.); giulia.bertani@unimore.it (G.B.); aida.meto@unimore.it (A.M.); bruna.colombari@unimore.it (B.C.); elisabetta.blasi@unimore.it (E.B.); pierantonio.bellini@unimore.it (P.B.); anto.depol@unimore.it (A.d.P.); ugo.consolo@unimore.it (U.C.)
2. Operative Unit of Dentistry and Maxillofacial Surgery, Department Integrated Activity-Specialist Surgeries, University-Hospital of Modena, 41125 Modena, Italy
* Correspondence: gianluca.carnevale@unimore.it
† These authors contributed equally to this work.

Received: 18 March 2019; Accepted: 10 April 2019; Published: 16 April 2019

Abstract: Peri-implantitis—an infection caused by bacterial deposition of biofilm—is a common complication in dentistry which may lead to implant loss. Several decontamination procedures have been investigated to identify the optimal approach being capable to remove the bacterial biofilm without modifying the implant surface properties. Our study evaluated whether two different systems—Ni-Ti Brushes (Brush) and Air-Polishing with 40 μm bicarbonate powder (Bic40)—might alter the physical/chemical features of two different titanium surfaces—machined (MCH) and Ca^{++} nanostructured (NCA)—and whether these decontamination systems may affect the biological properties of human STRO-1^+/c-Kit$^+$ dental pulp stem cells (hDPSCs) as well as the bacterial ability to produce biofilm. Cell morphology, proliferation and stemness markers were analysed in hDPSCs grown on both surfaces, before and after the decontamination treatments. Our findings highlighted that Bic40 treatment either maintained the surface characteristics of both implants and allowed hDPSCs to proliferate and preserve their stemness properties. Moreover, Bic40 treatment proved effective in removing bacterial biofilm from both titanium surfaces and consistently limited the biofilm re-growth. In conclusion, our data suggest that Bic40 treatment may operatively clean smooth and rough surfaces without altering their properties and, consequently, offer favourable conditions for reparative cells to hold their biological properties.

Keywords: human dental pulp stem cells; stemness properties; titanium surface properties

1. Introduction

The use of dental implants in daily clinical practice is currently widespread and in huge growth: the modern implant therapy allows in fact not only to offer a biological and functional advantage for many patients, compared with fixed or removable prosthetic solutions but also to obtain excellent long term results, as confirmed by previous studies reporting survival rates of 95.7% and 92.8% after 5 and 10 years, respectively [1]. However, despite these high survival rates, implant rehabilitation can fail. The most relevant troubles concerning osseointegrated implants are the peri-implant diseases, such as mucositis (reversible) and peri-implantitis (not reversible). Peri-implantitis has been defined as a disease with infectious pathogenesis characterized by a mucosal lesion often associated with bleeding, suppuration, increased probing depth always accompanied by marginal bone loss [2]. In this context, several factors affecting the systemic health status of the patients, that is, chronic diseases, autoimmune/inflammatory diseases, in combination with poor oral hygiene and smoking might influence the host-microbial interface very early in the healing phase following dental implant thus increasing the risk of peri-implant diseases [3].

The formation and organization of a bacterial biofilm on the implant portions exposed to the oral cavity is the initial cause of the peri-implant diseases and its removal is the goal to prevent or treat these diseases. Microbial biofilm commonly occurs in the oral cavity; it consists of a multispecies community, including Gram+ and Gram− bacteria as well as fungal cells, organized as sessile cells, tightly embedded in a matrix of polysaccharide origin. Biofilm formation and development is a process that begins with adhesion of initially planktonic microorganisms, followed by growth, extracellular polymeric matrix production, detachment and delocalization. Microbial biofilm occurs both in biotic and abiotic surfaces; once structured, biofilm-associated cells acquire enhanced resistance to cleaning treatment, antimicrobial drugs and host immune defences with respect to their planktonic counterparts [4]. To date, several methods have been developed with the aim of decontaminating implants: mechanical systems (Titanium brushes, Air-polishing systems, Ultra-Subsonic systems, Laser, Curettes) and chemical systems or antimicrobial solutions (Chlorhexidine in solution or in gel, Stannous fluoride, Tetracycline, Minocycline, Citric acid, Hydrogen peroxide and Gel etching with 35% phosphoric acid) [5–7]. Most of these decontaminating treatments might alter the chemical-physical properties of the implant surface [8–11]. Therefore, the ideal decontamination system is expected to be capable of breaking and removing the bacterial biofilm, without modifying the surface properties and thus maintaining the implant surface biologically favourable to adhesion and differentiation of the reparative cells grown onto the implant. It is well known indeed that the micro/nanotopography and the chemical composition of implant surfaces might influence the osseointegration process, by affecting and modifying the biological properties of the cells that interact with the surface [12–15]. Based on these considerations, among several decontamination techniques investigated, Nickel Titanium (Ni-Ti) brushes and Air Polishing systems have aroused interest. Ni-Ti is a resistant material with high flexibility when subjected to heating/cooling alternations. Moreover, the centrifugal force of the motor movement, the pressure exerted by the operator's hand and the heating/cooling alternations allow the brush to effectively reach even the most difficult spaces to clean. As a matter of fact, Ni-Ti brushes proved more effective in removing plaque than steel or plastic curettes, being, at the same time, less aggressive towards the implant surfaces [16–19]. Likewise, air-polishing devices based on the use of bicarbonate powder, which historically represents the first type of substance used in this approach, were shown to be highly effective in the mechanical removal of biofilm from different surface types, such as machined, SLA, sandblasted and TPS, with previous data showing that this approach provided better outcomes when compared to other instruments, such as plastic inserts of sonic devices [9,20,21].

Human DPSCs can be easily obtained from routine tooth extraction procedures and own self-renewal properties and a high regenerative potential [22]. Although dental pulp stem cells are a heterogeneous cell population, the immune selection against the stemness markers c-Kit and STRO-1 allows to isolate an adequate source of pre-osteoblast/odontoblast cells [23,24] which represent the ideal cell candidate in order to mimic in vitro the physiological processes of osseointegration and to investigate how different cleansing approaches might modify the cells-implants interactions.

The aim of this study was to investigate whether the use of two different mechanical decontamination systems, namely Ni-Ti Brushes and Air-Polishing System with 40 µm bicarbonate powder, might alter the roughness and chemical composition of two different implant surfaces (Machined and Ca^{++} Nano-incorporated) and whether the two decontamination systems may impact the biological and stemness features of DPSCs as well as the bacterial ability to produce biofilm.

2. Results

2.1. Titanium Surface Characterization

Scanning electron microscopy (SEM) analysis carried out on MCH surfaces from each experimental group is shown in Figure 1A. At lower magnifications, MCH control surface displayed concentric irregularities according to our previous findings [15]. The polishing treatment with Ni-Ti brush was characterized by grooves oriented in all the directions through the entire surface of the disks, as reported in either lower and higher magnifications. On the MCH titanium surfaces treated with Bic40, the presence of slight alterations of the whole surface was observed. Particularly, newly formed irregularities were detected on the entire treated disk (Figure 1A).

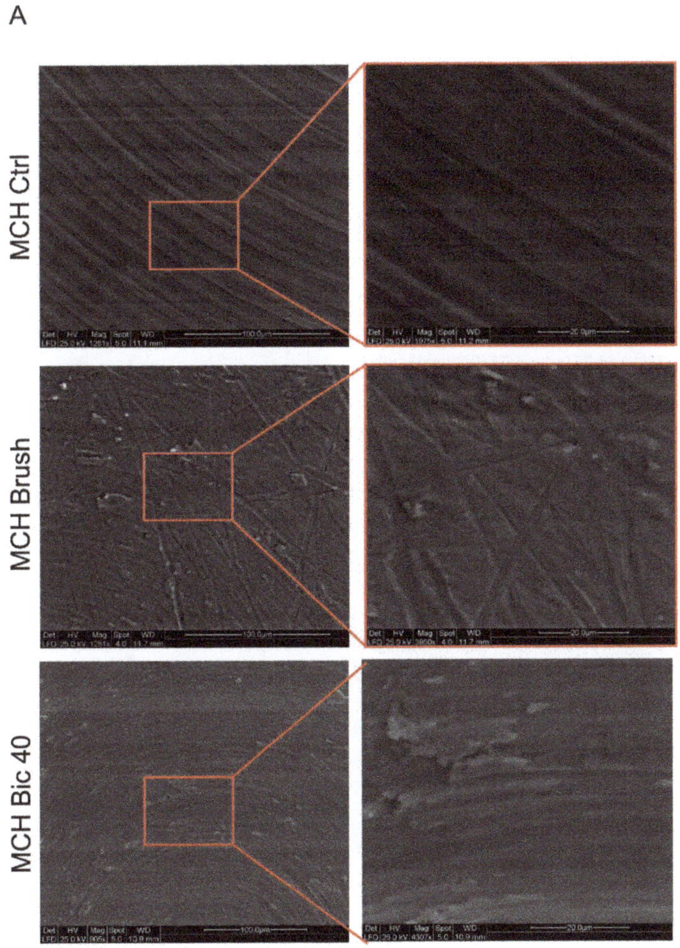

Figure 1. *Cont.*

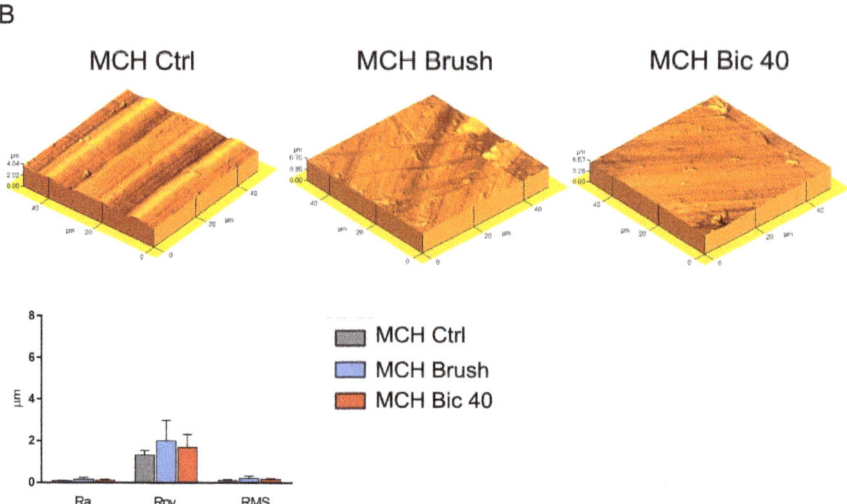

Figure 1. Machined (MCH) surfaces characterization after cleaning treatments. (**A**) Scanning electron microscopy (SEM) analysis at different magnifications was carried out on MCH titanium surfaces from the three experimental groups (Ctrl, Brush and Bic40, as indicated) in order to evaluate the surface topography. Scale bars: 100 μm (left) and 20 μm (magnifications on the right). (**B**) Atomic force microscopy (AFM) analysis of MCH surfaces. Histograms report the surface roughness expressed as Ra, Rpv and Rsm values. Values represent mean ± SD of three independent experiments; one-way ANOVA followed by Newman-Keuls post-hoc test.

Furthermore, atomic force microscopy (AFM) analysis was performed to evaluate the roughness of MCH disks following the different polishing treatments. In particular, as shown in Figure 1B, Ra, Rpv and Rsm were determined for each experimental group. With regard to Rpv parameter, higher values were recorded in MCH Brush group, in comparison to control MCH and MCH Bic40, although these differences were not statistically significant (Figure 1B). At the same time, SEM analysis was carried out on NCA surfaces from the three experimental groups. Data are reported in Figure 2A. According to previous data from our group [15], control NCA were characterized by homogeneous irregularities spread through the whole analysed area. The polishing treatment with Brush induced a notable modification of the surface: particularly, a flattening of the peaks typical of NCA surface was observed at lower and higher magnifications. Conversely, the air-polishing treatment with Bic40 did not induce any relevant alteration of the nanotopography of the surface. These observational data were not confirmed by AFM analyses, as a matter of fact, Ra, Rpv and Rms parameters did not differ among the three experimental groups. Likely, this evidence might be due to the AFM instrumental sensitivity (Figure 2B). Taken together, data on surface roughness of MCH and NCA surfaces treated with Bic40 did not show any significant difference, in terms of nanotopography, from both MCH and NCA controls.

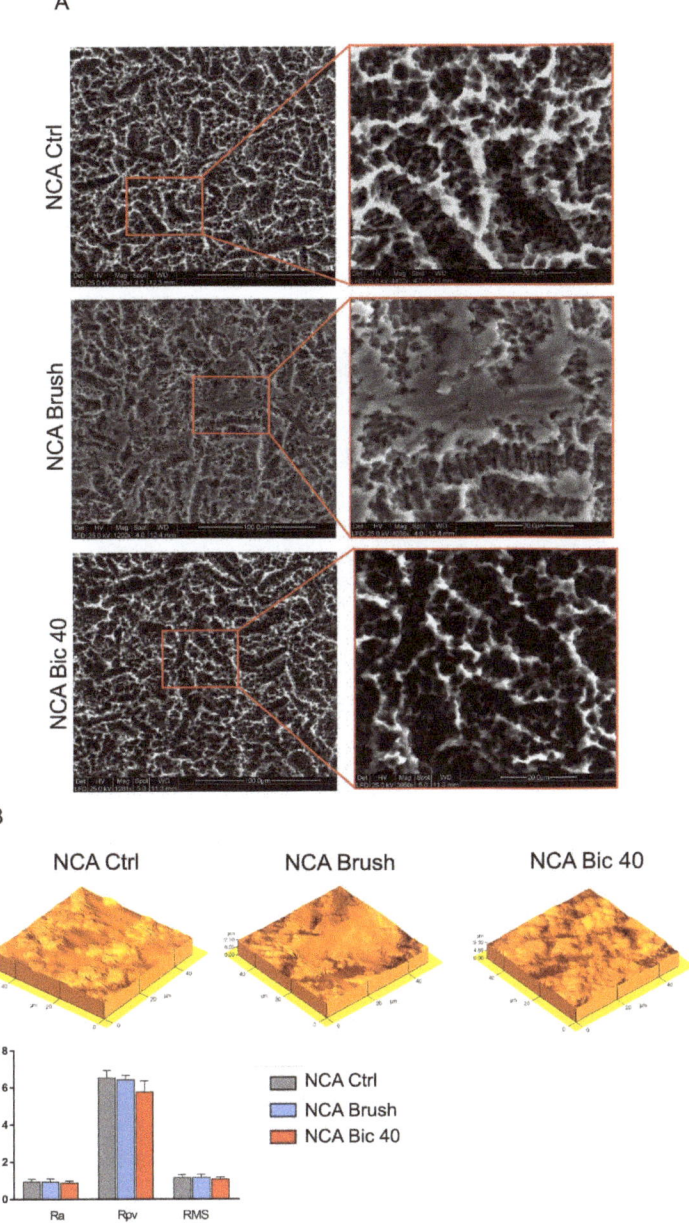

Figure 2. NCA surfaces characterization after cleansing treatments. (**A**) SEM analysis at different magnifications was carried out on NCA titanium surfaces from the three experimental groups (Ctrl, Brush and Bic40, as indicated) in order to evaluate the surface topography. Scale bars: 100 μm (left) and 20 μm (magnifications on the right). (**B**) AFM analysis of NCA surfaces. Histograms report the surface roughness expressed as Ra, Rpv and Rsm values. Values represent mean ± SD of three independent experiments. No statistically significant difference was detected among the groups; one-way ANOVA followed by the Newman-Keuls post-hoc test.

2.2. Stem Cells Morphology and Proliferation on Titanium Surfaces after Polishing Treatments

The hDPSCs morphology was evaluated by confocal microscopy, as shown in Figures 3 and 4. Cells were stained with phalloidin and DAPI. After 7 days of culture under standard conditions on MCH surfaces, hDPSCs displayed a fibroblast-like morphology with cells being arranged parallel to the surface grooves without showing significant differences following the Brush and Bic40 polishing treatments. With regard to the distribution of cells through the surface area, we noticed that hDPSCs cultured on MCH Brush oriented not only along the grooves due to the industrial fabrication but also along the scratches created after the Brush cleaning. No differences were observed in hDPSCs seeded on MCH Bic40 when compared to the control group (Figure 3A). Also, no differences in terms of proliferation rate were observed among the three experimental groups as indicated by histograms (Figure 3B).

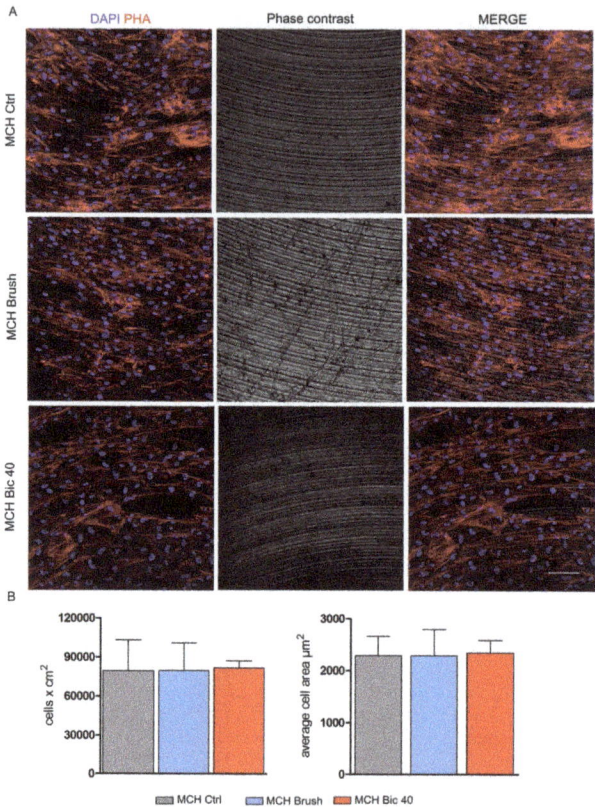

Figure 3. Evaluation of hDPSCs morphology and proliferation on MCH titanium surfaces. (**A**) hDPSCs morphology was assessed by confocal microscopy with phalloidin staining after 7 days of culture on MCH Ctrl, MCH Brush and MCH Bic40 surfaces. (**B**) Cell proliferation on titanium disks was measured by counting cell nuclei after DAPI staining. Histograms show cell numbers after 7 days of culture on titanium surfaces from the three experimental groups. Values represent mean ± SD. No statistically significant difference was detected among the groups; one-way ANOVA followed by the Newman-Keuls post-hoc test. Scale bar: 100 μm.

Figure 4 shows the morphology, distribution and proliferation of hDPSCs after 7 days of culture on NCA surfaces following Brush and Bic40 treatments. As formerly described, the culture on NCA disks induced a morphology alteration in hDPSCs. In particular, cells were homogeneously spread through the whole area and showed an irregular shape with reduction of the average cell area. The same features were observed in hDPSCs cultured on NCA Brush disks. Interestingly, when NCA disks were cleaned with Bic40 hDPSCs grew still homogeneously although recovering their typical fibroblast-like morphology, as reported when cultured on MCH disks. This shift in morphology was reflected also by an increased proliferation rate and by values of average cell area of hDPSCs cultured on NCA Bic40, with respect to NCA Ctrl and NCA Brush (* $p < 0.05$, Figure 4B).

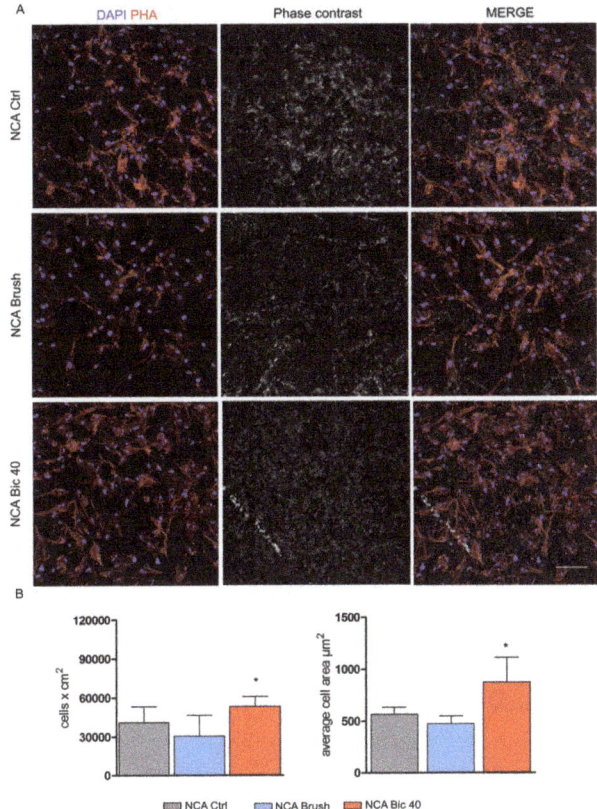

Figure 4. Evaluation of hDPSCs morphology and proliferation on NCA titanium surfaces. (**A**) hDPSCs morphology was assessed by confocal microscopy with phalloidin staining after 7 days of culture on NCA Ctrl, NCA Brush and NCA Bic40 surfaces. (**B**) Cell proliferation on titanium disks was measured by counting cell nuclei after DAPI staining. Histograms show cell numbers and average cell area after 7 days of culture on titanium surfaces from the three experimental groups. Values represent mean ± SD. * $p < 0.05$ NCA Bic 40 vs. NCA Brush, NCA Bic40 vs. NCA Ctrl; one-way ANOVA followed by the Newman-Keuls post-hoc test. Scale bar: 100 µm.

2.3. Expression of Stemness Markers

Human DPSCs were immune-selected against STRO-1 and c-Kit. After 7 days of culture on MCH and NCA surfaces, immunofluorescence analysis was performed in order to evaluate the maintenance of their biological properties. The expression of STRO-1 and c-Kit, two typical mesenchymal stem cells markers, were investigated in MCH and NCA after cleaning treatments.

As reported in Figure 5, hDPSCs cultured on MCH Ctrl showed the expression of both mesenchymal markers. These markers were still observed in hDPSCs cultured either on MCH Brush and MCH Bic40.

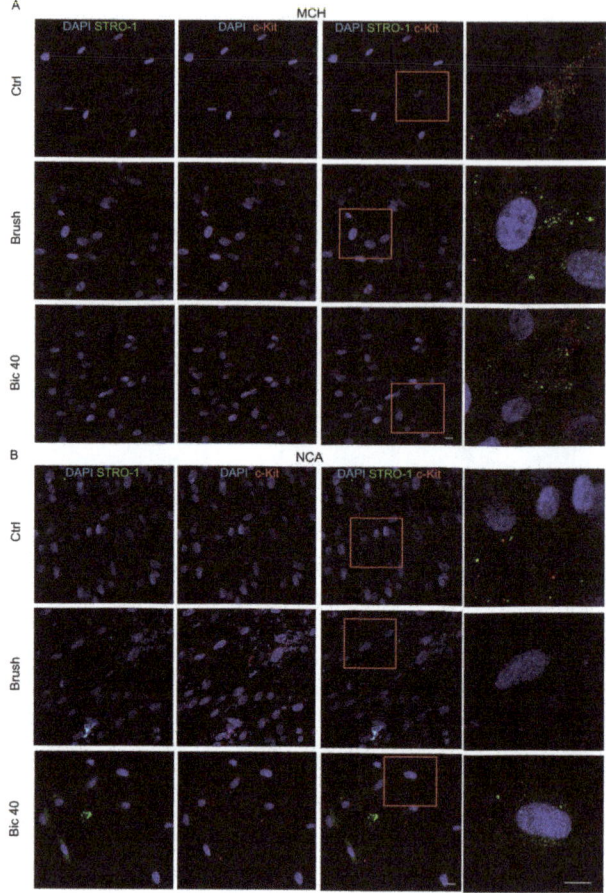

Figure 5. Evaluation of stemness markers of hDPSCs on titanium surfaces after cleaning treatments. Immunofluorescence analysis of STRO-1 and c-Kit in hDPSCs cultured for 7 days on MCH (**A**) and NCA (**B**) surfaces from the three experimental groups. Red squares indicate details reported at higher magnifications on the right. Nuclei were stained with DAPI. Scale bar: 10 μm.

On the contrary, we noticed a reduction in STRO-1 and c-Kit expression in hDPSCs grown on NCA Ctrl and NCA Brush after 7 days of culture. Conversely, the expression of these markers was evident in hDPSCs cultured on NCA Bic40. The morphology data in association with stemness evaluation indicate that after cleansing with Bic40, the NCA surface appeared more favourable/suitable to the growth of hDPSCs in their physiological microenvironment.

2.4. Microbial Biofilm Formation onto Titanium Disks

Based on the biological results concerning morphology, proliferation and stemness markers, it was then evaluated whether the polishing treatment with Bic40 on NCA surface may affect the microbial growth. Thus, *Pseudomonas aeruginosa* (10^6/mL) was seeded at time 0, in two sets of wells, containing or not the titanium disks; then, the plate was incubated at 35 °C for 24 h and microbial growth was assessed by bioluminescent analysis. As shown in Figure 6A, a superimposable trend was observed in the two groups of wells (control: no disks; sample: with disks); moreover, at 24 h, comparably high levels of total microbial load were achieved, being 2.6×10^8/mL and 3.4×10^8/mL the CFU/mL detected in control and sample groups, respectively. These data indicated that the presence of titanium disks did not affect microbial growth. To assess the occurrence of biofilm onto such disks, the latter were exposed to bacteria for 24 h, washed twice with PBS to eliminate the non-adherent microbial cells and then the residual bioluminescent signal was evaluated, as measure of formed biofilm. The obtained bioluminescent signal was converted in CFU and indicated that biofilm was produced at amounts as high as 1.95×10^8 CFU/mL (data not shown).

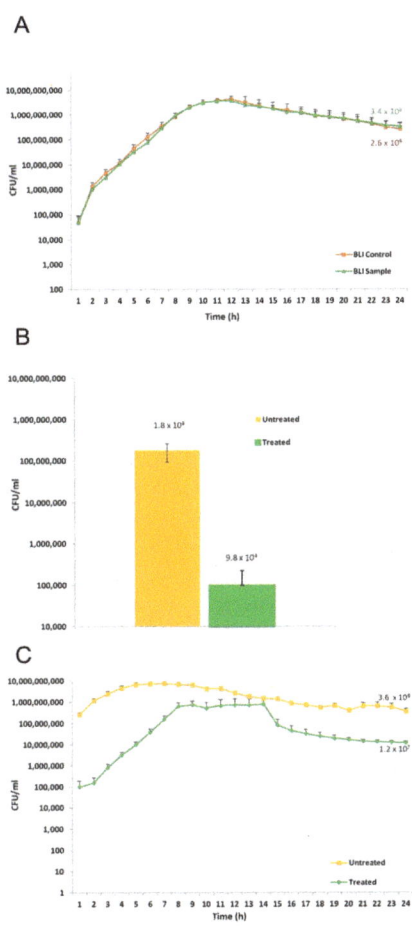

Figure 6. *P. aeruginosa* growth and biofilm formation on titanium disks, treated or not with Bic40 procedure, as assessed by a bioluminescent (BLI) bacterial strain. (**A**) *P. aeruginosa* growth was not affected by the presence of titanium disks. BLI-*Pseudomonas* (10^6/mL) in TSB plus 2% sucrose was seeded in 96 black well-plate and incubated at 35 °C in the presence (orange) or absence (green) of titanium disks (1/well). The plates were then incubated by Fluoroskan and the bioluminescent signal was recorded in real time (h), up to 24 h. By the calibration curve, the RLUs were converted in CFU/mL + SEM (Standard Error Mean), as indicated in the Y axis. (**B**) Effects of the cleaning procedure on titanium disks-associated biofilm. Microbial biofilm produced onto titanium disks (by 24 h incubation, as above) was exposed or not to the Bic40 cleaning procedure (treated vs. untreated group); the residual bioluminescent signal was measured and converted, by the calibration curve, in CFU/mL ± SEM, as indicated in the Y axis. (**C**) Real time monitoring of *P. aeruginosa* re-growth on treated and untreated titanium disks. The microbial re-growth, occurring in Bic40-treated and untreated disks, was evaluated in real time, for additional 24 h. Then, the bioluminescent signal was measured and converted in CFU, returning values of 3.6×10^8/mL and 1.2×10^7/mL, in untreated and treated groups, respectively.

2.5. Microbial Biofilm Removal from Titanium Disks

To assess the efficacy of the Bic40 cleaning procedure against a preformed biofilm, 4 disks containing a 24 h-old biofilm were treated with Bic40 procedure, as detailed above (treated group), while 3 disks, used as controls were not treated (untreated group); then, each disk was transferred in a new well with fresh medium and incubated for additional 24 h. As shown in Figure 6B, when compared to the untreated groups, the Bic40-treated disks displayed an about 3 logs drop in terms of CFU/mL.

2.6. Microbial Re-Growth on Treated and Untreated Titanium Disks

As detailed in Figure 6C, the microbial load in the untreated group remained almost constant during the subsequent 24 h range. In contrast, treated disks showed a delayed and time-related increase in CFU. In particular, the major differences between treated and untreated disks occurred within the first 5–6 h; then, microbial load on treated disks reached a plateau level, that consistently remained about 1 log below the values observed in the untreated disks until at 24 h.

3. Discussion

The peri-implant disease is due to the colonization of the implant surfaces by pathogens which constitute a biofilm [25]. Bacterial adhesion and biofilm formation play a fundamental role not only into the pathogenesis of peri-implantitis but also into the implant survival [12,13,26]. In a previous study [27] we observed that the biofilm occurred regardless of the degree of surface roughness. Therefore, the removal of the biofilm from the implant surface, indistinctly smooth or rough, is the primary objective. It is well known that the response of cells and tissues to a biomaterial depends on the properties of the material itself such as surface topography, chemical composition and capability to interact with body fluids [15,28]. Following bacterial contamination, the procedures used to decontaminate the implant surface can cause alterations of its topography and its chemical composition. To this regard and to the best of our knowledge the optimal cleansing approach is expected to effectively remove the bacteria biofilm without altering the chemical and physical properties of the implant and consequently the biological properties of cells involved in the osseointegration process. In this study we analysed two different titanium surfaces, before and after the treatment with two cleansing approaches, mimicking what physiologically occurs in terms of cells/implant interactions after decontamination procedures.

Qualitative analysis of surface morphology revealed how in both MCH and NCA groups the surfaces treated with Ni-Ti brushes are morphologically different from the untreated surfaces, in agreement with Park et al. [18]. In fact, SEM analysis showed the presence of deep grooves heterogeneously distributed over the MCH surface and flattened area over the NCA surface. Slight although not statistically significant differences were revealed by Ra, Rpv and Rsm physical parameters. Conversely, the treatment with Air-polishing system with Bic40 produced not significant alterations of both the MCH and NCA surfaces in terms of physical parameters. Subsequently, the aim of the study was to evaluate the biological features of stem cells that embryologically participate to the osseointegration process. The use of dental pulp stem cells represents a suitable choice in terms of stemness properties and ease of isolation. Although hDPSCs are a heterogeneous population, we used a stem cell population enriched for the expression of the stemness markers STRO-1 and c-Kit, which represent a strictly mesenchymal origin able to differentiate in bone, adipose and myogenic lineages. To the best of our knowledge, we noticed that on MCH titanium surface before and after both the cleansing treatments, hDPSCs maintained their fibroblast-like morphology without any alteration of cell proliferation. The only difference observed concerned the hDPSCs distribution pattern through the MCH Brush surface, in fact cells were spread along the grooves produced by Ni-Ti brush. On the contrary, hDPSCs grown on MCH Bic40 and MCH ctrl surfaces did not show any difference.

Regarding NCA surface, a change in cell morphology was observed in NCA ctrl and NCA brush, in accordance with our previous findings [15]. After the treatment with Bic40 hDPSCs showed their typical cell morphology and also demonstrated a statistically significant higher proliferation rate, when compared to NCA ctrl and NCA Brush surfaces: this phenomenon might be attributed to the interaction of calcium ions incorporated on the NCA surface with bicarbonate ions from the cleansing treatment with Bic40. These observations were confirmed by the evaluation of stemness markers on MCH and NCA surfaces before and after the cleansing treatments. In particular, whereas no differences were detected in any MCH group, we noticed that STRO-1 and c-Kit expression were maintained in NCA Bic40 group. As a matter of fact, the maintenance of stemness is a primary requirement to preserve the biological properties including self-renewal and differentiation capabilities and immunomodulatory properties and consequently to avoid cell senescence. Based on these results, Bic40 might represent the most suitable cleansing treatment. To further confirm the efficacy of Bic40, we also performed microbiology assays. Using a recently established model precious in assessing microbial biofilm formation onto medical devices [29], here we showed that BLI-*Pseudomonas* has the ability to adhere to the titanium disks and to form a consistent biofilm on their surface. Moreover, the cleaning treatment with BIC 40 μm is capable of reducing the biofilm of about 99% with respect to untreated control group (100% vs. 0.05%, respectively). In particular, the microbial load, evaluated as RLU and converted in CFU/mL, has been reduced of more than 3 logs in the treated groups as compared to their controls. Furthermore, microbial re-growth in treated disks remains consistently below the control values (difference of about 1 log). We may hypothesize that the combination of the physical treatment (dry spray) and the hypertonic condition (sodium bicarbonate accumulation onto titanium disks) negatively impact on microbial cell-viability. Furthermore, the enrichment of air-polishing powders with antimicrobial fillers such as Ciprofloxacin and/or mucosal defensive agents such as Zinc L-carnosine might improve the antibacterial action of the cleaning tool and its biocompatibility towards soft tissues [30]. In conclusion, we demonstrated that Bic40 provides a suitable cleansing approach either on smooth and rough surfaces.

4. Materials and Methods

4.1. Human DPSCs Isolation and Immune Selection

This study was carried out in compliance with the recommendations of Comitato Etico Provinciale—Azienda Ospedaliero-Universitaria di Modena (Modena, Italy), which provided the approval of the protocol (ref. number 3299/CE; 5 September 2017). Human DPSCs were isolated from third molars of adult subjects ($n = 3$; 30–35 years) undergoing routine dental extraction. All subjects gave written informed consent in accordance with the Declaration of Helsinki.

Cells were isolated from dental pulp as previously described [23]. Briefly, dental pulp was harvested from the teeth and underwent enzymatic digestion by using a digestive solution, (3 mg/mL type I collagenase plus 4 mg/mL dispase in α-MEM). Pulp was then filtered onto 100 μm Falcon Cell Strainers, in order to obtain a cell suspension. Cell suspension was then plated in 25 cm^2 culture flasks and expanded in standard culture medium [α-MEM supplemented with 10% heat inactivated foetal bovine serum (FBS), 2 mM L-glutamine, 100 U/mL penicillin, 100 μg/mL streptomycin] at 37 °C and 5% CO2. Following cell expansion, human DPSCs were immune-selected by using MACS® separation kit according to manufacturers' instructions. The immune-selections were performed by using primary antibodies: mouse IgM anti-STRO-1 and rabbit IgG anti-c-Kit (Santa Cruz, Dallas, TX, USA). The following magnetically labelled secondary antibodies were used: anti-mouse IgM and anti-rabbit IgG (Miltenyi Biotec, Bergisch Gladbach, Germany). The immune-selection resulted in the isolation of a homogeneous hDPSCs population expressing STRO-1 and c-Kit. All the experiments were performed using STRO-1$^+$/c-Kit$^+$ hDPSCs.

4.2. Titanium Surfaces Characterization

A total of 50 titanium disks (MegaGen Co. Ltd., Daegu, South Korea) measuring 13 mm in diameter and 3 mm in thickness were used in this study. Particularly, two different titanium surfaces were used: machined (MCH) and Ca^{2+} incorporated (NCA). The treatment processes are hold by the manufacturer. For this study, titanium surfaces were divided into 3 different experimental groups: (1) control surfaces (Ctrl), (2) surfaces cleaned with Ni-Ti brushes (Brush) and (3) surfaces cleaned by air-polishing with NaHCO$_3$ 40 µm (Bic40).

The cleansing of the disks was carried out using two mechanical methods: (1) Nickel-Titanium Brushes (I.C.T. Brush micro, Hans Korea Co. Ltd., Gyeonggi-do, Korea/ De Ore, Verona, Italy) and (2) Air-Polishing System (Combi-Touch, Mectron spa, Carasco, Genova, Italy) with 40 µm bicarbonate powder.

Briefly, the "I.C.T." (Implant Cleaning Technique) nickel-titanium brushes, made up of about 40 super elastic filaments with a diameter of 0.07–0.13 mm, were used at 400 rpm and 600 rpm, respectively, for two sequential rounds of 45 s each. The total duration of each surface treatment was 90 s and a 25 g pressure calibrated on an electronic scale was used, with 100 N of torque. All the treatments were performed by the same operator under irrigation with buffer saline solution (0.9% NaCl).

The "Combi-Touch" air polishing system with sodium bicarbonate particles (ø 40 µm) was used for 30 s at a distance of 5 mm. In particular, the operating principle of "Combi Touch" air-polishing system consists in the mechanical action of compressed air spreading an accelerated flow of particles onto the titanium surface. When the particles hit the surface, their kinetic energy is dissipated almost completely, thus producing a gentle although effectively cleansing action. The cleaning treatment is completed by a water jet that is arranged in the form of a bell around the main flow and that uses the pressure drop originated around the nozzle to prevent the powder cloud from bouncing and being dispelled and, at the same time, to dissolve the powder by washing the surface.

After the cleansing treatments, surface morphology for each experimental group was qualitatively evaluated through Scanning Electron Microscopy (EVO MA 10—Carl Zeiss, Oberkochen, Germany) working at 25 keV. Moreover, surface roughness was determined by Atomic Force Microscopy (Nanoscope IIIa, Veeco, Santa Barbara, CA, USA) and Ra, Rpv and Rms parameters were obtained. Ra (Roughness average) measures the average surface roughness considering the peaks and the valley means. The Rpv (peak to valley distance) describes the maximum observed range in a sample area and it is given by the distance between the highest peak and the lowest valley on a measured surface. Rms parameter describes the density of micropores on the surface.

4.3. Cell Morphology and Proliferation

Undifferentiated STRO-1$^+$/c-Kit$^+$ hDPSCs at passage 1 were seeded at a density of 2.5×10^3 cell/cm^2 on titanium disks in 12-multiwell units and expanded under standard culture conditions. After 7 days of culture, cells were fixed in ice-cold paraformaldehyde 4% for 15 min without dissociating them from the titanium disks. The cells were subsequently permeabilized with 0.1% Triton X-100 in PBS for 5 min, stained with AlexaFluor546 Phalloidin (Thermo Fisher Scientific) and rinsed with PBS 1%. Nuclei were stained with 1 µg/mL 4′,6-diamidino-2-phenylindole (DAPI) in PBS 1%. Titanium disks were mounted with DABCO anti-fading medium on cover glasses. Cell proliferation and morphology were assessed using confocal microscopy (Nikon A1), as formerly described by Bianchi et al. [24].

Cell proliferation was measured by counting the DAPI-positive nuclei on 10 randomly selected fields measuring 2.85×10^5 µm^2 for each disk by a blind operator. At the same time, average cell area was measured on hDPSCs from 10 randomly selected fields, measuring 2.85×10^5 µm^2, on 3 disks for each experimental group.

4.4. Evaluation of Stemness Markers in hDPSCs Cultured on Titanium Disks

After 1 week of culture on each disk, cells were fixed in 4% ice-cold paraformaldehyde in PBS for 15 min and then processed as previously described [31]. The following primary Abs diluted 1:100 were used: mouse IgM anti-STRO-1 and rabbit IgG anti-c-Kit (Santa Cruz, Dallas, TX, USA). Secondary Abs (goat anti-mouse IgM Alexa488, goat anti-rabbit Alexa546) were diluted 1:200 (Thermo Fisher Scientific, Waltham, MA, USA). Nuclei were stained with 1 µg/mL 4′,6-diamidino-2-phenylindole

(DAPI) in PBS 1%. The multi-labelling immunofluorescence experiments were carried out avoiding cross-reactions between primary and secondary Abs.

Confocal imaging was performed using a Nikon A1 confocal laser scanning microscope, as previously described [32]. The confocal serial sections were processed with ImageJ software to obtain 3-dimensional projections and image rendering was performed by Adobe Photoshop Software.

4.5. Microbial Strain

We used the bioluminescent *Pseudomonas aeruginosa* strain (P1242) (BLI-*Pseudomonas*) previously engineered in order to express the luciferase gene and substrate under the control of a constitutive P1 integron promoter 2 [33]; thus, live cells constitutively produce a detectable bioluminescent signal. To quantify the bioluminescence emission by BLI-Pseudomonas in the experimental groups, a calibration curve was created allowing to express such values in terms of colony forming units (CFU)/mL; in particular, serial dilutions (starting from 1×10^8/mL) of a bacterial suspension in Tryptic Soy Broth (TSB) (OXOID, Milan, Italy) with 2% sucrose were prepared and a volume of 100 µL of each dilution was seeded in a black-well microtiter plate. The plate was immediately read by using a Fluoroskan Luminescence reader (Thermo Fisher Scientific, Waltham, MA, USA).

4.6. Biofilm Formation onto Titanium Disks

In order to allow biofilm formation onto Ca-structured titanium disks, 180 µL of overnight cultures of BLI-Pseudomonas (10^6/mL) in TSB plus 2% sucrose were seeded in 96 black well-plate, containing 1 disc/well. In parallel, BLI-Pseudomonas was seeded in wells without the titanium disks. The plates were then incubated at 35 °C for 24 h, into the Fluoroskan reader and the bioluminescence was detected at every hour to evaluate in real time, the total microbial load. After incubation, the disks were washed twice with phosphate buffered saline (PBS) (EuroClone, Wetherby, UK) at room temperature (RT), transferred into new wells and the bioluminescence signal was again measured; the obtained values were referred to the amounts of biofilm formed onto disk surfaces. Through the calibration curve, the relative luminescent units (RLU) obtained in each experiment were converted in CFU/mL.

4.7. Biofilm Re-Growth onto Treated and Untreated Titanium Disks

Following biofilm formation, the disks were split in two groups and the cleaning treatment was performed, as detailed above; then, controls (untreated) and cleaned (exposed to Bic40 µm for 30 s/surface) were transferred into new wells containing fresh medium and further assessed for microbial residual load and time-related re-growth. Briefly, treated and untreated disks were analysed by Fluoroskan reader, immediately (residual biofilm) and at any hour during a further 24 h incubation at 35 °C. Through the calibration curve, the RLU were converted in CFU/mL.

4.8. Statistical Analysis

All experiments were performed in triplicate. Data were expressed as mean ± standard deviation (SD). Differences between two experimental conditions were analysed by paired, Student's *t*-test. Differences among three or more experimental samples were analysed by ANOVA followed by Newman–Keuls post hoc test (GraphPad Prism Software version 5 Inc., San Diego, CA, USA). In any case, p value < 0.05 was considered statistically significant.

Author Contributions: Conceptualization, E.C., A.P., U.C. and G.C.; Methodology, E.C., A.P., G.C. and E.B.; Validation, A.P., G.C., A.M., B.C. and E.B.; Formal Analysis, A.P., E.B., A.d.P., P.B., U.C. and G.C.; Investigation, E.C., A.P., G.B., A.M., B.C. and E.B.; Data Curation, E.C., A.P., G.B., L.B. and E.B.; Writing—Original Draft Preparation, E.C., A.P., E.B. and G.C.; Writing—Review & Editing, E.C., A.P., L.B., G.B., A.M., B.C., E.B., A.d.P., P.B., U.C. and G.C.; Supervision, A.d.P., U.C. and G.C.; Funding Acquisition, U.C.

Funding: Ugo Consolo was funded by Mectron spa, Carasco, Genova, Italy.

Conflicts of Interest: The authors declare no conflict of interest.

References

1. Albrektsson, T.; Donos, N. Implant survival and complications. The Third EAO consensus conference 2012. *Clin. Oral Implant. Res.* **2012**, *23*, 63–65. [CrossRef] [PubMed]
2. Lindhe, J.; Meyle, J.; on behalf of Group D of the European Workshop on Periodontology. Peri-implant diseases: Consensus Report of the Sixth European Workshop on Periodontology. *J. Clin. Periodontol.* **2008**, *35*, 282–285. [CrossRef]
3. Kumar, P.S. Systemic Risk Factors for the Development of Periimplant Diseases. *Implant. Dent.* **2019**, *28*, 115–119. [CrossRef] [PubMed]
4. Derks, J.; Tomasi, C. Peri-implant health and disease. A systematic review of current epidemiology. *J. Clin. Periodontol.* **2015**, *42*, 158. [CrossRef] [PubMed]
5. Dennison, D.K.; Huerzeler, M.B.; Quinones, C.; Caffesse, R.G. Contaminated Implant Surfaces: An In Vitro Comparison of Implant Surface Coating and Treatment Modalities for Decontamination. *J. Periodontol.* **1994**, *65*, 942–948. [CrossRef]
6. Toma, S.; Lasserre, J.F.; Taïeb, J.; Brecx, M.C. Evaluation of an air-abrasive device with amino acid glycine-powder during surgical treatment of peri-implantitis. *Quintessence Int* **2014**, *45*, 209–219.
7. Sahrmann, P.; Ronay, V.; Hofer, D.; Attin, T.; Jung, R.E.; Schmidlin, P.R. In vitro cleaning potential of three different implant debridement methods. *Clin. Oral Implant. Res.* **2015**, *26*, 314–319. [CrossRef] [PubMed]
8. Schwarz, F.; Ferrari, D.; Popovski, K.; Hartig, B.; Becker, J. Influence of different air-abrasive powders on cell viability at biologically contaminated titanium dental implants surfaces. *J. Biomed. Mater. Res. B Appl. Biomater.* **2009**, *88*, 83–91. [CrossRef] [PubMed]
9. Tastepe, C.S.; Liu, Y.; Visscher, C.M.; Wismeijer, D. Cleaning and modification of intraorally contaminated titanium disks with calcium phosphate powder abrasive treatment. *Clin. Oral Implant. Res.* **2013**, *24*, 1238–1246.
10. Wei, M.C.; Tran, C.; Meredith, N.; Walsh, L.J. Effectiveness of implant surface debridement using particle beams at differing air pressures. *Clin. Exp. Dent. Res.* **2017**, *3*, 148–153. [CrossRef]
11. Louropoulou, A.; Slot, D.E.; Van der Weijden, F. Influence of mechanical instruments on the biocompatibility of titanium dental implants surfaces: A systematic review. *Clin. Oral Implant. Res.* **2015**, *26*, 841–850. [CrossRef]
12. Mavrogenis, A.F.; Dimitriou, R.; Parvizi, J.; Babis, G.C. Biology of implant osseointegration. *J. Musculoskelet. Neuronal Interact.* **2009**, *9*, 61–71.
13. Wennerberg, A.; Albrektsson, T. Effects of titanium surface topography on bone integration: A systematic review. *Clin. Oral Implant. Res.* **2009**, *20*, 172–184. [CrossRef]
14. Feller, L.; Jadwat, Y.; Khammissa, R.A.G.; Meyerov, R.; Schechter, I.; Lemmer, J. Cellular Responses Evoked by Different Surface Characteristics of Intraosseous Titanium Implants. *Biomed Res. Int.* **2015**, *2015*, 1–8. [CrossRef]
15. Conserva, E.; Pisciotta, A.; Borghi, F.; Nasi, M.; Pecorini, S.; Bertoni, L.; de Pol, A.; Consolo, U.; Carnevale, G. Titanium Surface Properties Influence the Biological Activity and FasL Expression of Craniofacial Stromal Cells. *Stem Cells Int.* **2019**, *2019*, 4670560. [CrossRef]
16. John, G.; Becker, J.; Schwarz, F. Rotating titanium brush for plaque removal from rough titanium surfaces—An in vitro study. *Clin. Oral Implant. Res.* **2014**, *25*, 838–842. [CrossRef] [PubMed]
17. Duddeck, D.; Karapetian, V.; Grandoch, A. Time-saving debridment of implants with rotating titanium brushes. *Implants* **2012**, *3*, 20–22.
18. Park, J.-B.; Jeon, Y.; Ko, Y. Effects of titanium brush on machined and sand-blasted/acid-etched titanium disc using confocal microscopy and contact profilometry. *Clin. Oral Implant. Res.* **2015**, *26*, 130–136. [CrossRef] [PubMed]
19. Louropoulou, A.; Slot, D.E.; Van der Weijden, F. The effects of mechanical instruments on contaminated titanium dental implant surfaces: A systematic review. *Clin. Oral Implant. Res.* **2014**, *25*, 1149–1160. [CrossRef]
20. Tastepe, C.S.; Van Waas, R.; Liu, Y.; Wismeijer, D. Air powder abrasive treatment as an implant surface cleaning method: A literature review. *Int. J. Oral Maxillofac. Implant.* **2012**, *27*, 1461–1473.
21. Nemer Vieira, L.F.; Lopes de Chaves e Mello Dias, E.C.; Cardoso, E.S.; Machado, S.J.; Pereira da Silva, C.; Vidigal, G.M. Effectiveness of implant surface decontamination using a high-pressure sodium bicarbonate protocol: An in vitro study. *Implant. Dent.* **2012**, *21*, 390–393. [CrossRef]

22. Pisciotta, A.; Bertoni, L.; Riccio, M.; Mapelli, J.; Bigiani, A.; La Noce, M.; Orciani, M.; De Pol, A.; Carnevale, G. Use of a 3D Floating Sphere Culture System to Maintain the Neural Crest-Related Properties of Human Dental Pulp Stem Cells. *Front. Physiol.* **2018**, *9*, 547. [CrossRef]
23. Pisciotta, A.; Carnevale, G.; Meloni, S.; Riccio, M.; De Biasi, S.; Gibellini, L.; Ferrari, A.; Bruzzesi, G.; De Pol, A. Human Dental pulp stem cells (hDPSCs): Isolation, enrichment and comparative differentiation of two sub-populations. *Bmc Dev. Boil.* **2015**, *15*, 14. [CrossRef]
24. Bianchi, M.; Pisciotta, A.; Bertoni, L.; Berni, M.; Gambardella, A.; Visani, A.; Russo, A.; De Pol, A.; Carnevale, G. Osteogenic Differentiation of hDPSCs on Biogenic Bone Apatite Thin Films. *Stem Cells Int.* **2017**, *2017*, 1–10.
25. Heitz-Mayfield, L.J.A.; Lang, N.P. Comparative biology of chronic and aggressive periodontitis vs. peri-implantitis. *Periodontology* **2010**, *53*, 167–181. [CrossRef]
26. Veerachamy, S.; Yarlagadda, T.; Manivasagam, G.; Yarlagadda, P.K. Bacterial adherence and biofilm formation on medical implants: A review. *Proc. Inst. Mech. Eng. H J. Eng. Med.* **2014**, *228*, 1083–1099. [CrossRef]
27. Conserva, E.; Generali, L.; Bandieri, A.; Cavani, F.; Borghi, F.; Consolo, U. Plaque accumulation on titanium disks with different surface treatments: An in vivo investigation. *Odontology* **2018**, *106*, 145–153. [CrossRef]
28. Anselme, K.; Ponche, A.; Bigerelle, M. Relative influence of surface topography and surface chemistry on cell response to bone implant materials. Part 2: Biological aspects. *Proc. Inst. Mech. Eng. H* **2010**, *224*, 1487–1507. [CrossRef]
29. Pericolini, E.; Colombari, B.; Ferretti, G.; Iseppi, R.; Ardizzoni, A.; Girardis, M.; Sala, A.; Peppoloni, S.; Blasi, E. Real-time monitoring of Pseudomonas aeruginosa biofilm formation on endotracheal tubes in vitro. *Bmc Microbiol.* **2018**, *18*, 84. [CrossRef]
30. Pagano, S.; Chieruzzi, M.; Balloni, S.; Lombardo, G.; Torre, L.; Bodo, M.; Cianetti, S.; Marinucci, L. Biological, thermal and mechanical characterization of modified glass ionomer cements: The role of nanohydroxyapatite, ciprofloxacin and zinc l-carnosine. *Mater. Sci. Eng. C Mater. Biol. Appl.* **2019**, *94*, 76–8516. [CrossRef]
31. Carnevale, G.; Riccio, M.; Pisciotta, A.; Beretti, F.; Maraldi, T.; Zavatti, M.; Cavallini, G.M.; La Sala, G.B.; Ferrari, A.; De Pol, A. In vitro differentiation into insulin-producing β-cells of stem cells isolated from human amniotic fluid and dental pulp. *Dig. Liver Dis.* **2013**, *45*, 669–676.
32. Carnevale, G.; Pisciotta, A.; Riccio, M.; Bertoni, L.; De Biasi, S.; Gibellini, L.; Zordani, A.; Cavallini, GM.; La Sala, GB.; Bruzzesi, G.; et al. Human dental pulp stem cells expressing STRO-1, c-kit and CD34 markers in peripheral nerve regeneration. *J. Tissue Eng. Regen. Med.* **2018**, *12*, e774–e785. [CrossRef]
33. Choi, K.-H.; Schweizer, H.P. mini-Tn7 insertion in bacteria with single attTn7 sites: Example Pseudomonas aeruginosa. *Nat. Protoc.* **2006**, *1*, 153–161. [CrossRef]

© 2019 by the authors. Licensee MDPI, Basel, Switzerland. This article is an open access article distributed under the terms and conditions of the Creative Commons Attribution (CC BY) license (http://creativecommons.org/licenses/by/4.0/).

Article

Influence of Fractionation Methods on Physical and Biological Properties of Injectable Platelet-Rich Fibrin: An Exploratory Study

Prakan Thanasrisuebwong [1,2], Rudee Surarit [3], Sompop Bencharit [4,5] and Nisarat Ruangsawasdi [6,*]

1. Department of Oral and Maxillofacial Surgery, Faculty of Dentistry, Mahidol University, Bangkok 10400, Thailand; ae13ea@gmail.com
2. Ph.D. program in Oral Biology, Faculty of Dentistry, Mahidol University, Bangkok 10400, Thailand
3. Department of Oral Biology, Faculty of Dentistry, Mahidol University, Bangkok 10400, Thailand; rudee.sur@mahidol.ac.th
4. Department of General Dentistry, School of Dentistry, Virginia Commonwealth University, Richmond, VA 23298, USA; sbencharit@vcu.edu
5. Department of Biomedical Engineering, College of Engineering, Virginia Commonwealth University, Richmond, VA 23298, USA
6. Department of Pharmacology, Faculty of Dentistry, Mahidol University, Bangkok 10400, Thailand
* Correspondence: nisarat.rua@mahidol.ac.th

Received: 27 February 2019; Accepted: 30 March 2019; Published: 3 April 2019

Abstract: Injectable platelet-rich fibrin (i-PRF) has been used as an autografting material to enhance bone regeneration through intrinsic growth factors. However, fractionation protocols used to prepare i-PRF can be varied and the effects of different fractionation protocols are not known. In this study, we investigated the influence of different fractions of i-PRF on the physical and biological properties derived from variations in i-PRF fractionation preparation. The i-PRF samples, obtained from the blood samples of 10 donors, were used to harvest i-PRF and were fractioned into two types. The yellow i-PRF fractionation was harvested from the upper yellow zone, while the red i-PRF fractionation was collected from both the yellow and red zone of the buffy coat. The viscoelastic property measurements, including the clot formation time, α-angle, and maximum clot firmness, were performed by rotational thromboelastometry. The fibrin network was examined using a scanning electron microscope. Furthermore, the concentration of growth factors released, including VEGF, TGF-β1, and PDGF, were quantified using ELISA. A paired t-test with a 95% confidence interval was used. All three viscoelastic properties were statistically significantly higher in the yellow i-PRF compared to the red i-PRF. The scanning electron microscope reviewed more cellular components in the red i-PRF compared to the yellow i-PRF. In addition, the fibrin network of the yellow i-PRF showed a higher density than that in the red i-PRF. There was no statistically significant difference between the concentration of VEGF and TGF-β1. However, at Day 7 and Day 14 PDGF concentrations were statistically significantly higher in the red i-PRF compared to the yellow group. In conclusion, these results showed that the red i-PRF provided better biological properties through the release of growth factors. On the other hand, the yellow i-PRF had greater viscoelastic physical properties. Further investigations into the appropriate i-PRF fractionation for certain surgical procedures are therefore necessary to clarify the suitability for each fraction for different types of regenerative therapy.

Keywords: bone regeneration; fractionation; growth factors; i-PRF; PRF; ROTEM®

1. Introduction

Platelet-rich fibrin (PRF) derived from the centrifugation of whole blood has long been used successfully in regenerative dentistry and medicine [1–5]. A centrifugation speed of 2700 rpm for 12 min in PRF with a glass or glass-coated tubes permits the use of PRF material without any additional anticoagulants. The fibrin matrix scaffold network in PRF appears to be beneficial in tissue regeneration [6]. PRF has shown to enhance regeneration of bone [1,7], cartilage [8], soft tissue [5,9–11], and nerve fiber [12]. However, the application of solid PRF, resulting from a blood sample preparation at 400 g relative centrifugal force (RCF) or 2700 round per minute (rpm) for 12 min, has its limitations. For instance, it cannot be mixed with a particulate bone substitute. To mix with particulate bone, the biomaterials in a liquid form are more suitable. Both injectable platelet-rich fibrin (i-PRF) and PRF-predecessor platelet-rich plasma (PRP) are the liquid formula within the platelet concentrated group that can be used efficiently for mixing with the particulate bone graft. PRP is traditionally prepared from whole blood using a higher speed centrifugation compared to PRF [1–5]. Unfortunately, PRP has some drawbacks because it requires either synthetic anticoagulants, such as citrate dextrose-A (ACD-A) and citrate phosphate dextrose (CPD), or other activating agents, such as thrombin. Although those additive agents allow for appropriate clinical working time and manipulation, they can inhibit healing and regeneration due to the lack of biocompatibility [9–11,13–15].

The concept of low speed centrifugation is central to the preparation of the liquid form of i-PRF as well as to create a material rich in leukocytes, platelets, and growth factors, including the vascular endothelial growth factor (VEGF), transforming growth factor-beta 1 (TGF-β1), and platelet-derived growth factor (PDGF), compared to the first generation platelet-derived PRP [9–11,16]. According to the low speed centrifugation concept, i-PRF can be obtained by using a centrifugation speed at 60 g RCF or 700 rpm for 3 min [10,16,17]. Blood harvesting is similar to the PRF membrane method, however, the non-coating plastic tube was recommended in order to prevent the formation of early clotting in the tube. After centrifugation, whole blood is separated into three main parts based on the buffy coat layer: A yellow upper part, a buffy coat middle part, and a red blood cell containing lower part. A small syringe with an 18G hypodermic needle is used for collecting the i-PRF to be used. The harvesting method of i-PRF after centrifugation has been described as harvesting only the whole upper layer above the buffy coat [16,17], however, the amount of the upper layer harvested can vary among individuals and the position of needle tips during harvesting can also be varied based on different clinical practice. To the best of our knowledge, there is no study examining the different properties of yellow and red i-PRF.

The aim of this study was to investigate the influence of different separation techniques of i-PRF on its mechanical and biological properties. We further hypothesize that fractionation of the centrifuged plasma for i-PRF is central to the i-PRF's physical properties and biological components. We proposed to examine two fractionation protocols producing yellow i-PRF and red i-PRF based on the fraction of centrifuged plasma above and within the buffy coat, respectively. We expected that the yellow i-PRF would have less cellular components, a denser fibrin network, and therefore have better physical properties than the red i-PRF. On the contrary, we expected that the red i-PRF would have more cellular components and therefore have better biological properties reflecting the greater release of known PRF-related growth factors, including VEGF, TGF-β1, and PDGF.

2. Results

Rotational thromboelastometry (ROTEM®, Tem International GmbH, Munich, Germany) [18] was applied to examine the viscoelastic properties of red and yellow i-PRF. The means and standard deviations for the clot formation time (CFT), α-angle, and maximum clot firmness (MCF) were 52.8 $\pm$ 13.14 s, 81.2 $\pm$ 1.81 degree, and 85.3 $\pm$ 3.56 mm, respectively, for the yellow i-PRF and 68.2 $\pm$ 12.31 s, 77.8 $\pm$ 2.82 degree, and 81.6 $\pm$ 4.50 mm, respectively, for the red i-PRF as shown in Figure 1. The results showed statistically significant differences (paired t-test) between the two types of i-PRF in all three properties: CFT, α-angle, and MCF, with p-value = 0.008, 0.004, and 0.001, respectively, as shown in

Figure 1. The CFT for the yellow i-PRF was shorter than the red i-PRF. The α-angle was higher in the yellow i-PRF compared to the red i-PRF. The MCF was also higher in the yellow i-PRF compared to the red i-PRF.

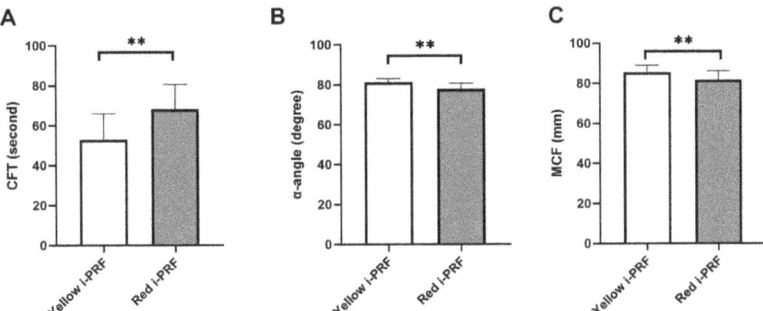

Figure 1. Rotational thromboelastometry (ROTEM®) analysis demonstrating (**A**) clot formation time (CFT), (**B**) α-angle, and (**C**) maximum clot firmness (MCF) of the yellow injectable platelet-rich fibrin (i-PRF) and the red i-PRF. ** $p < 0.01$.

The examination of the i-PRF by SEM showed the fibrin network architecture and cellular components. In the yellow i-PRF, a dense fibrin network with less cellular components than the red i-PRF was observed (Figure 2). On the contrary, SEM showed that more cellular components, such as leukocyte, platelets, and erythrocytes, were in the red i-PRF compared to the yellow i-PRF. In addition, the numerous erythrocytes were enmeshed in the fibrin network. The shapes of these erythrocytes were normal, but the fibrin network appeared at a lower density and was less organized compared to the yellow i-PRF (Figure 2).

Figure 2. Scanning electron microscope images of i-PRF using different fraction methods. A different position of the harvesting needle led to two types of i-PRF: The yellow and red i-PRF. (**A**) The yellow i-PRF showed a dense and highly organized fibrin network. (**B**) On the other hand, the red i-PRF demonstrated more cellular components of leukocyte, platelets, and erythrocytes than the yellow i-PRF.

The growth factor concentrations were quantified at 3 h, 24 h, 72 h (3 days), 168 h (7 days), and 336 h (14 days) after preparation. There was a distinct difference between the releasing patterns for VEGF, TGF-β1, and PDGF (Figure 3). At 3 h, 24 h, and 3 days, the VEGF concentrations were increased. Three hours after preparation, the VEGF concentration of both i-PRFs was at a similar level. This level presented in the yellow i-PRF was similar at 24 h again while the red i-PRF dramatically increased. The concentrations of VEGF peaked at Day 3 for both the yellow and red i-PRF. There was a slightly higher concentration of VEGF in the red i-PRF compared to the yellow i-PRF but the difference was not statistically significant (Figure 3A). Different results were detected at Day 7 and 14 after clotting. At this time point, both i-PRFs showed a decreasing VEGF concentration. In conclusion, the released accumulation of VEGF showed a higher concentration in the red i-PRF than the yellow i-PRF (Figure 3B).

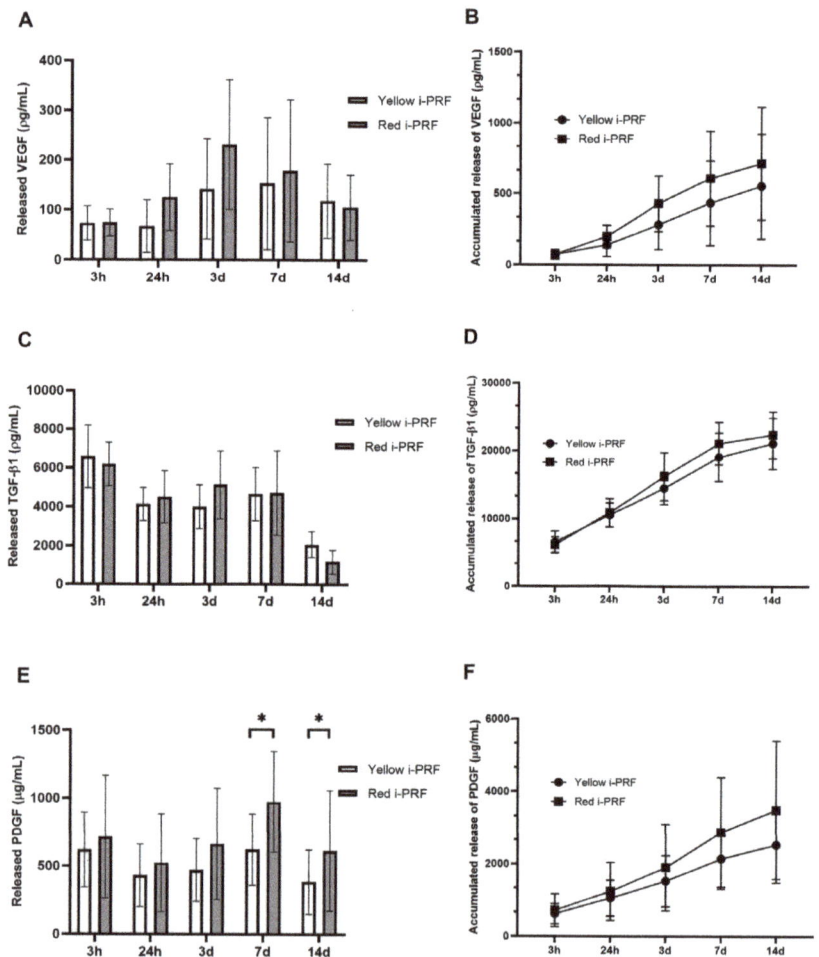

Figure 3. Time point releases and accumulated releases of three growth factors: (**A,B**) vascular endothelial growth factor (VEGF), (**C,D**) transforming growth factor-β1 (TGF-β1), and (**E,F**) platelet-derived growth factor (PDGF). * $p < 0.05$.

The growth factor concentration for the human TGF-β1 showed a general trend at all time points. The highest concentration was recorded at 3 h for the yellow and red i-PRF then decreased and was stable at 24 h, 3 days, and 7 days. There was a dramatic decrease of the TGF-β1 concentration for both i-PRFs at Day 14 (Figure 3C). Note that the released accumulation of TGF-β1 presented a similar trend and value for the yellow and red i-PRFs (Figure 3D). Using a paired t-test, there was however no statistically significant difference between the concentration of VEGF and TGF-β1 (Figure 3).

The release of PDGF was more constant throughout the experimental timeline. Analysis of the PDGF release (Figure 3E) revealed that more PDGF was released from the red i-PRF at 7 and 14 days (975.47 ± 371.24 µg/mL and 615.98 ± 443.59 µg/mL) compared with the yellow i-PRF (623.46 ± 263.30 µg/mL and 385.59 ± 237.53 µg/mL). These data were statistically significant for Day 7 ($p = 0.02$) and Day 14 ($p = 0.03$). Cumulatively, PDGF continued to release at all time points and a stronger releasing trend was observed for the red i-PRF over the yellow i-PRF (Figure 3F).

Analysis of the kinetics of the growth factors released showed that VEGF, TGF-β1, and PDGF reached peak release at Day 3, 3 h, and Day 7, respectively. On top of that, TGF-β1 was almost completely released within the first 7 days while VEGF and PDGF showed a longer releasing over 14 days.

3. Discussion

The main difference of i-PRF from solid PRF is the lower speed and time in centrifugation for i-PRF [10,19]. The idea of a slower centrifugation is to allow some cellular and growth factor components to stay in the final product [6,9–11,17,20–24]. The i-PRF can be used in combination with particulate bone allograft or autograft materials [25]. This advantage of a liquid formula of i-PRF allows a more efficient incorporation of the bone graft material and increases the signaling molecules throughout the whole graft.

Little information is known about the physical properties of i-PRF, which are important either when being used alone or in combination with grafting materials. This study is one of the first to apply ROTEM® technology, which is commonly used in hematology research [18,26–31], to examine i-PRF. We hypothesized that compared to the lower layer—red i-PRF, the upper layer—yellow i-PRF would have superior physical properties because of its lack of cellular components. All physical properties confirmed our hypothesis. Although the yellow i-PRF was statistically superior compared to the red i-PRF in viscoelastic properties, the value of the CFT, α-angle, and MCF of both i-PRFs were similar to the reference value or slightly better compared to the standard whole blood that was measured using ROTEM® [32]. In addition, the value of the CFT, α-angle, and MCF showed only a minor difference between both i-PRFs. The SEM analysis demonstrated a dense, organized, acellular fibrin network as a reason for the superior physical properties in the yellow i-PRF compared to the red i-PRF. This clinically suggests that if i-PRF is to be used mainly as filler material or enhancement stability and handling of grafting material, the yellow and red i-PRF may achieve a similar outcome.

A previous SEM analysis of leukocyte-PRF showed multi-different plasma layers that constituted of a fibrin-rich layer at the uppermost layer, followed by an enriched platelet layer, and the buffy coat layer with numerous leukocytes before the base layer of erythrocytes [33]. The SEM in our analysis demonstrated a dense, organized, acellular fibrin network as in the fibrin-rich layer at the plasma top layer of leukocyte-PRF. This characteristic might be a reason for the superior physical properties in the yellow i-PRF compared to the red i-PRF. Meanwhile, the red i-PRF showed a greater number of cells and platelets attached to the fibrin network similar to a combination of all the middle layers and the erythrocyte base layer of leukocyte-PRF together. This additional cell and platelet content in the red i-PRF might lead to better biological properties as we could observe a greater release of the growth factors.

Contrary to the physical properties, we hypothesized that the red i-PRF, which has more cellular components, would have higher release of growth factors. While there was no statistically significant difference in the releases of VEGF and TGF-β1, PDGF at later time points were significantly higher in the red i-PRF compared to the yellow i-PRF. This may be a result of the remaining platelet and cellular components shown in the SEM in the red i-PRF. Note also that the releasing level of all three growth factors was similar to previously reported levels [10]. In addition, we found that the growth factors continue to release even at Day 14. This result is similar to other PRF studies [9–11,17,20–24]. Self-cooperation of the fibrin network slowly forms the high fibrillar aggregation in PRF, which also entraps proteins and growth factors to the binding domains of fibrin molecules. In contrast, PRP is rapidly activated to gelation from a load of thrombin. Therefore, an unstable fibrin network of PRP is formed, which results in lower growth factor retention and incompatible cell homing. PRP therefore only releases growth factors in the early stage. Unlike the PRP, the kinetic release of growth factors from i-PRF is longer, up to 14 days, because of its preserved valuable components, including platelets and leukocytes. The growth factor release from this component can be sustained over a longer period of time from the cellular and acellular component trapped in a naturally progressive polymerization of the fibrin matrix [34]. On top of this, i-PRF potentially allows more growth factor attachment to the heparin-binding domain, which is the high-affinity growth factor-binding site of the fibrinogen that causes prolonged retention of the growth factor within fibrin [35]. This result suggests that when i-PRF is used in combination with particulate bone grafts, the signaling molecules for bone tissue engineering can be achieved. The red i-PRF may therefore have more benefits than the yellow i-PRF because of the greater release of the growth factor shown. In such cases, clinicians should harvest the red i-PRF just with the buffy coat.

This study demonstrated for the first time that minimal changes in the fractionation protocol for i-PRF could alter the physical and biological properties of the final product. This in turn may change the clinical outcomes/applications. These techniques need to be examined further through cell culture analysis and human clinical studies. In any case clinicians should pay particular attention to centrifugation and fractionation protocols. Deviations from the published protocols can result in variations of physical properties and biological components as seen in this study and in others [17]. In addition, further research to develop a novel biomaterial product from i-PRF is an interesting topic, where the freeze-dried or lyophilized technique may be used for sterilization [36,37].

4. Materials and Methods

4.1. Sample Collection and Preparation

The study protocol was approved by the Institutional Review Board of Human Ethic Committee of the Faculty of Dentistry, Mahidol University (COA. No. MU-DT/PY-IRB 2017/061.221). Blood samples were harvested from 10 healthy volunteers (age range 30–35, gender M/F = 2/8). The donors had no history of using antiplatelet or anticoagulant medications. All donors gave written consent for blood collection and for sample preparation and experiments (Figure 4).

Sample preparation was adopted from a previous published protocol, Miron et al. 2017 [10]. Immediately after the blood sample was collected, the sample was centrifuged at 60 g RCF for 3 min using a Duo Centrifuge (Process for PRF, Nice, France) at room temperature; 60 g RCF is equivalent to 700 rpm for this device, which has a 110 mm radius. After centrifugation, two types of i-PRF samples were harvested, yellow and red i-PRF. According to Wang et al. 2017 [24], about 1 mL of sample, either the yellow or red i-PRF, was collected. The yellow i-PRF referred to the sample harvested only in the upper liquid yellow zone above the buffy coat. The red i-PRF referred to the sample harvested from the zone, red and yellow, with the buffy coat. The bevel edge of the harvesting needle was used as a reference point (Figure 4). Subsequently, both types of i-PRF were used to observe the physical properties, including viscoelasticity and morphology, and the biological properties, which is the release of growth factors.

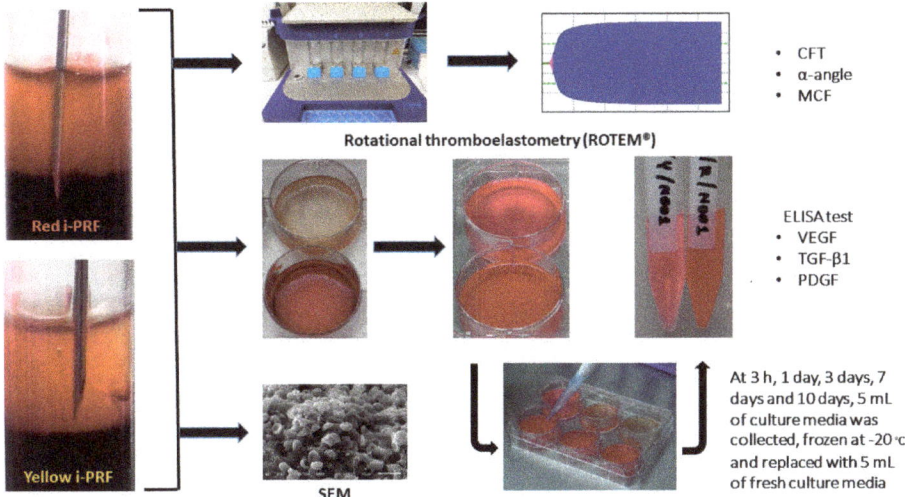

Figure 4. Study workflow of three experiments comparing the red and yellow i-PRF and their fractionation protocol used in this study.

4.2. Viscoelastic Property Analysis

The viscoelastic properties of the i-PRF samples were analyzed using rotational thromboelastometry (ROTEM®, Tem International GmbH, Munich, Germany) [18]. ROTEM® generates output from transducing changes reflecting from the viscoelastic strength of the blood sample while a constant rotational force is applied. Three parameters were analyzed and digitally recorded: Clot formation time (CFT), α-angle, and maximum clot firmness (MCF). The CFT represents the speed at which a solid clot forms, which is primarily influenced by the platelet function, fibrinogen, and coagulation factors. The CFT measures a duration from the clot initiation until it reaches the 20 mm amplitude. The α-angle is a measure to observe the dynamics of clot formation, which represents the acceleration of the fibrin network formation and the amount of cross-linking build up. The MCF is a measure for stability of the clot following the polymerization process.

4.3. Scanning Electron Microscopy (SEM)

The qualitative analysis of the morphological changes of the yellow and red i-PRF was performed using SEM (JEOL, Peabody, MA, USA). The SEM examination was used to observe blood elements present in the fibrin extracellular matrix. Cellular components such as leukocytes, platelets, and erythrocytes were examined and recorded. The density of the fibrin network was also analyzed. The i-PRF samples immediately after completing their clot formation were fixed in 2% glutaraldehyde in Dulbecco's phosphate buffered saline (DPBS) buffer with a pH of 7.4 overnight. Next, samples were dried in the desiccator, then sputtered with 20 nm gold. SEM photography was then performed at 5 to 10 kV using 100× to 2000× magnifications.

4.4. Enzyme-Linked Immunosorbent Assay (ELISA)

ELISA kits were used to quantify the released growth factors: VEGF, TGF-β1, and PDGF (R&D systems Minnesota, United States). Approximately 9 mL of collected i-PRF from each protocol was transferred into a 6-well plate. The plate was then placed in a 5% CO_2 incubator at 37 °C for 30 min to allow for complete clotting formation. Afterwards, 5 mL of Minimum Essential Medium Eagle-Alpha Modification (α-MEM) with nucleosides (Gibco, USA) was added to each sample. The samples were further incubated at 37 °C to allow for the release of growth factors during a 3 h to 14 days period of study. At 3 h, 24 h, 72 h (3 days), 168 h (7 days), and 336 h (14 days), 5 mL of culture media was collected, frozen at −20 °C, and replaced with 5 mL of additional fresh culture media. The concentrations of VEGF, TGF-β1, and PDGF were measured according to the manufacturer's instructions. Optical density was assessed using a microplate reader at 450 nm. The measurements were performed in triplicates.

4.5. Statistical Analysis

Statistical assessment was performed using SPSS version 17.0. All data were presented as the mean value ± standard deviation (SD). The data were tested for the distribution by using a Kolmogorov-Smirnov test and Levene's test to confirm the normality and equal variance assumptions of the data. The data, assuming normal distribution, were then analyzed using paired t-tests to compare the differences between two groups in each experiment. The level of significance used was $p < 0.05$.

5. Conclusions

This study suggests that fractionation protocols in i-PRF preparation can alter the final product's physical and biological properties. Clinicians should pay attention to the applications of i-PRF in combinational use with other grafting materials. For the handling and stability of grafting material application, the yellow and red may provide the final i-PRF product with a similar outcome. For the use of i-PRF in enhancing biological properties, the red i-PRF collected from the sample with the buffy coat, which released a higher level of growth factors, in particular, PDGF, should be used.

Author Contributions: Author P.T. designed the study, performed experiments, as well as collected and analyzed the data. Author R.S. designed the study and interpreted the data. Author S.B. helped with study designs, data interpretation, and manuscript writing. Author N.R. oversaw the entire project and supervised Author P.T. in all experiments. All authors contributed to the manuscript writing and approved the final draft.

Funding: The study was funded by Ph.D. Research Grant, Faculty of Dentistry Mahidol University and it was part of the research project supported by Mahidol University (043/2562).

Acknowledgments: We thank Chareerut Phruksaniyom for technical assistance and Center of Research, Faculty of Dentistry, Mahidol University for providing devices.

Conflicts of Interest: The authors declare no conflict of interest.

References

1. Choukroun, J.; Diss, A.; Simonpieri, A.; Girard, M.O.; Schoeffler, C.; Dohan, S.L.; Dohan, A.J.; Mouhyi, J.; Dohan, D.M. Platelet-rich fibrin (prf): A second-generation platelet concentrate. Part V: Histologic evaluations of prf effects on bone allograft maturation in sinus lift. *Oral Surg. Oral Med. Oral Pathol. Oral Radiol. Endod.* **2006**, *101*, 299–303. [CrossRef]
2. Choukroun, J.; Diss, A.; Simonpieri, A.; Girard, M.O.; Schoeffler, C.; Dohan, S.L.; Dohan, A.J.; Mouhyi, J.; Dohan, D.M. Platelet-rich fibrin (prf): A second-generation platelet concentrate. Part IV: Clinical effects on tissue healing. *Oral Surg. Oral Med. Oral Pathol. Oral Radiol. Endod.* **2006**, *101*, e56–e60. [CrossRef]
3. Dohan, D.M.; Choukroun, J.; Diss, A.; Dohan, S.L.; Dohan, A.J.; Mouhyi, J.; Gogly, B. Platelet-rich fibrin (prf): A second-generation platelet concentrate. Part III: Leucocyte activation: A new feature for platelet concentrates? *Oral Surg. Oral Med. Oral Pathol. Oral Radiol. Endod.* **2006**, *101*, e51–e55. [CrossRef]
4. Dohan, D.M.; Choukroun, J.; Diss, A.; Dohan, S.L.; Dohan, A.J.; Mouhyi, J.; Gogly, B. Platelet-rich fibrin (prf): A second-generation platelet concentrate. Part II: Platelet-related biologic features. *Oral Surg. Oral Med. Oral Pathol. Oral Radiol. Endod.* **2006**, *101*, e45–e50. [CrossRef]
5. Dohan, D.M.; Choukroun, J.; Diss, A.; Dohan, S.L.; Dohan, A.J.; Mouhyi, J.; Gogly, B. Platelet-rich fibrin (prf): A second-generation platelet concentrate. Part I: Technological concepts and evolution. *Oral Surg. Oral Med. Oral Pathol. Oral Radiol. Endod.* **2006**, *101*, e37–e44. [CrossRef]

6. Ghanaati, S.; Booms, P.; Orlowska, A.; Kubesch, A.; Lorenz, J.; Rutkowski, J.; Landes, C.; Sader, R.; Kirkpatrick, C.; Choukroun, J. Advanced platelet-rich fibrin: A new concept for cell-based tissue engineering by means of inflammatory cells. *J. Oral Implantol.* **2014**, *40*, 679–689. [CrossRef]
7. Del Corso, M.; Vervelle, A.; Simonpieri, A.; Jimbo, R.; Inchingolo, F.; Sammartino, G.; Dohan Ehrenfest, D.M. Current knowledge and perspectives for the use of platelet-rich plasma (prp) and platelet-rich fibrin (prf) in oral and maxillofacial surgery Part 1: Periodontal and dentoalveolar surgery. *Curr. Pharm. Biotechnol.* **2012**, *13*, 1207–1230. [CrossRef]
8. Kuo, T.F.; Lin, M.F.; Lin, Y.H.; Lin, Y.C.; Su, R.J.; Lin, H.W.; Chan, W.P. Implantation of platelet-rich fibrin and cartilage granules facilitates cartilage repair in the injured rabbit knee: Preliminary report. *Clinics* **2011**, *66*, 1835–1838. [CrossRef]
9. Miron, R.J.; Fujioka-Kobayashi, M.; Bishara, M.; Zhang, Y.; Hernandez, M.; Choukroun, J. Platelet-rich fibrin and soft tissue wound healing: A systematic review. *Tissue Eng. Part B Rev.* **2017**, *23*, 83–99. [CrossRef]
10. Miron, R.J.; Fujioka-Kobayashi, M.; Hernandez, M.; Kandalam, U.; Zhang, Y.; Ghanaati, S.; Choukroun, J. Injectable platelet rich fibrin (i-prf): Opportunities in regenerative dentistry? *Clin. Oral Investig.* **2017**, *21*, 2619–2627. [CrossRef]
11. Miron, R.J.; Zucchelli, G.; Pikos, M.A.; Salama, M.; Lee, S.; Guillemette, V.; Fujioka-Kobayashi, M.; Bishara, M.; Zhang, Y.; Wang, H.L.; et al. Use of platelet-rich fibrin in regenerative dentistry: A systematic review. *Clin. Oral Investig.* **2017**, *21*, 1913–1927. [CrossRef]
12. Torul, D.; Bereket, M.C.; Onger, M.E.; Altun, G. Comparison of the regenerative effects of platelet-rich fibrin and plasma rich in growth factors on injured peripheral nerve: An experimental study. *J. Oral Maxillofac. Surg.* **2018**, *76*, 1823.e1–1823.e12. [CrossRef]
13. He, L.; Lin, Y.; Hu, X.; Zhang, Y.; Wu, H. A comparative study of platelet-rich fibrin (prf) and platelet-rich plasma (prp) on the effect of proliferation and differentiation of rat osteoblasts in vitro. *Oral Surg. Oral Med. Oral Pathol. Oral Radiol. Endod.* **2009**, *108*, 707–713. [CrossRef]
14. Borie, E.; Olivi, D.G.; Orsi, I.A.; Garlet, K.; Weber, B.; Beltran, V.; Fuentes, R. Platelet-rich fibrin application in dentistry: A literature review. *Int. J. Clin. Exp. Med.* **2015**, *8*, 7922–7929.
15. Kumar, Y.R.; Mohanty, S.; Verma, M.; Kaur, R.R.; Bhatia, P.; Kumar, V.R.; Chaudhary, Z. Platelet-rich fibrin: The benefits. *Br. J. Oral Maxillofac. Surg.* **2016**, *54*, 57–61. [CrossRef]
16. Choukroun, J.; Ghanaati, S. Reduction of relative centrifugation force within injectable platelet-rich-fibrin (prf) concentrates advances patients' own inflammatory cells, platelets and growth factors: The first introduction to the low speed centrifugation concept. *Eur. J. Trauma Emerg. Surg.* **2018**, *44*, 87–95. [CrossRef]
17. Wend, S.; Kubesch, A.; Orlowska, A.; Al-Maawi, S.; Zender, N.; Dias, A.; Miron, R.J.; Sader, R.; Booms, P.; Kirkpatrick, C.J.; et al. Reduction of the relative centrifugal force influences cell number and growth factor release within injectable prf-based matrices. *J. Mater. Sci. Mater. Med.* **2017**, *28*, 188. [CrossRef]
18. Whiting, D.; DiNardo, J.A. Teg and rotem: Technology and clinical applications. *Am. J. Hematol.* **2014**, *89*, 228–232. [CrossRef]
19. De Almeida Barros Mourao, C.F.; Calasans-Maia, M.D.; de Mello Machado, R.C.; de Brito Resende, R.F.; Alves, G.G. The use of platelet-rich fibrin as a hemostatic material in oral soft tissues. *Oral Maxillofac. Surg.* **2018**, *22*, 329–333. [CrossRef]
20. Ghanaati, S.; Herrera-Vizcaino, C.; Al-Maawi, S.; Lorenz, J.; Miron, R.J.; Nelson, K.; Schwarz, F.; Choukroun, J.; Sader, R. Fifteen years of platelet rich fibrin (prf) in dentistry and oromaxillofacial surgery: How high is the level of scientific evidence? *J. Oral Implantol.* **2018**, *44*, 471–492. [CrossRef]
21. Fujioka-Kobayashi, M.; Miron, R.J.; Hernandez, M.; Kandalam, U.; Zhang, Y.; Choukroun, J. Optimized platelet-rich fibrin with the low-speed concept: Growth factor release, biocompatibility, and cellular response. *J. Periodontol.* **2017**, *88*, 112–121. [CrossRef]
22. Abd El Raouf, M.; Wang, X.; Miusi, S.; Chai, J.; Mohamed AbdEl-Aal, A.B.; Nefissa Helmy, M.M.; Ghanaati, S.; Choukroun, J.; Choukroun, E.; Zhang, Y.; et al. Injectable-platelet rich fibrin using the low speed centrifugation concept improves cartilage regeneration when compared to platelet-rich plasma. *Platelets* **2017**, *30*, 213–221. [CrossRef]
23. Wang, X.; Zhang, Y.; Choukroun, J.; Ghanaati, S.; Miron, R.J. Effects of an injectable platelet-rich fibrin on osteoblast behavior and bone tissue formation in comparison to platelet-rich plasma. *Platelets* **2018**, *29*, 48–55. [CrossRef]

24. Wang, X.; Zhang, Y.; Choukroun, J.; Ghanaati, S.; Miron, R.J. Behavior of gingival fibroblasts on titanium implant surfaces in combination with either injectable-prf or prp. *Int. J. Mol. Sci.* **2017**, *18*, 331. [CrossRef]
25. Sohn, D.S.; Huang, B.; Kim, J.; Park, W.E.; Park, C.C. Utilization of Autologous Concentrated Growth Factors (CGF) Enriched Bone Graft Matrix (Sticky Bone) and CGF-Enriched Fibrin Membrane in Implant Dentistry. *J. Implant Adv. Clin. Dent.* **2015**, *7*, 11–29.
26. Kelly, J.M.; Rizoli, S.; Veigas, P.; Hollands, S.; Min, A. Using rotational thromboelastometry clot firmness at 5 minutes (rotem((r)) extem a5) to predict massive transfusion and in-hospital mortality in trauma: A retrospective analysis of 1146 patients. *Anaesthesia* **2018**, *73*, 1103–1109. [CrossRef]
27. Franchini, M.; Mengoli, C.; Cruciani, M.; Marietta, M.; Marano, G.; Vaglio, S.; Pupella, S.; Veropalumbo, E.; Masiello, F.; Liumbruno, G.M. The use of viscoelastic haemostatic assays in non-cardiac surgical settings: A systematic review and meta-analysis. *Blood Transfusion Trasfusione del Sangue* **2018**, *16*, 235–243.
28. Longstaff, C. Measuring fibrinolysis: From research to routine diagnostic assays. *J. Thromb. Haemost.* **2018**, *16*, 652–662. [CrossRef]
29. Meledeo, M.A.; Peltier, G.C.; McIntosh, C.S.; Voelker, C.R.; Bynum, J.A.; Cap, A.P. Functional stability of the teg 6s hemostasis analyzer under stress. *J. Trauma Acute Care Surg.* **2018**, *84*, S83–S88. [CrossRef]
30. Remy, K.E.; Yazer, M.H.; Saini, A.; Mehanovic-Varmaz, A.; Rogers, S.R.; Cap, A.P.; Spinella, P.C. Effects of platelet-sparing leukocyte reduction and agitation methods on in vitro measures of hemostatic function in cold-stored whole blood. *J. Trauma Acute Care Surg.* **2018**, *84*, S104–S114. [CrossRef]
31. Baksaas-Aasen, K.; Van Dieren, S.; Balvers, K.; Juffermans, N.P.; Naess, P.A.; Rourke, C.; Eaglestone, S.; Ostrowski, S.R.; Stensballe, J.; Stanworth, S.; et al. Data-driven development of rotem and teg algorithms for the management of trauma hemorrhage: A prospective observational multicenter study. *Ann. Surg.* **2018**. [CrossRef] [PubMed]
32. Lang, T.; Bauters, A.; Braun, S.L.; Potzsch, B.; von Pape, K.W.; Kolde, H.J.; Lakner, M. Multi-centre investigation on reference ranges for rotem thromboelastometry. *Blood Coagul. Fibrinolysis* **2005**, *16*, 301–310. [CrossRef]
33. Madurantakam, P.; Yoganarasimha, S.; Hasan, F.K. Characterization of leukocyte-platelet rich fibrin, a novel biomaterial. *J. Vis. Exp.* **2015**. [CrossRef]
34. Schar, M.O.; Diaz-Romero, J.; Kohl, S.; Zumstein, M.A.; Nesic, D. Platelet-rich concentrates differentially release growth factors and induce cell migration in vitro. *Clin. Orthop. Relat. Res.* **2015**, *473*, 1635–1643. [CrossRef]
35. Martino, M.M.; Briquez, P.S.; Ranga, A.; Lutolf, M.P.; Hubbell, J.A. Heparin-binding domain of fibrin(ogen) binds growth factors and promotes tissue repair when incorporated within a synthetic matrix. *Proc. Natl. Acad. Sci. USA* **2013**, *110*, 4563–4568. [CrossRef]
36. Ansarizadeh, M.; Mashayekhan, S.; Saadatmand, M. Fabrication, modeling and optimization of lyophilized advanced platelet rich fibrin in combination with collagen-chitosan as a guided bone regeneration membrane. *Int. J. Biol. Macromol.* **2019**, *125*, 383–391. [CrossRef] [PubMed]
37. Kardos, D.; Hornyak, I.; Simon, M.; Hinsenkamp, A.; Marschall, B.; Vardai, R.; Kallay-Menyhard, A.; Pinke, B.; Meszaros, L.; Kuten, O.; et al. Biological and mechanical properties of platelet-rich fibrin membranes after thermal manipulation and preparation in a single-syringe closed system. *Int. J. Mol. Sci.* **2018**, *19*, 3433. [CrossRef]

© 2019 by the authors. Licensee MDPI, Basel, Switzerland. This article is an open access article distributed under the terms and conditions of the Creative Commons Attribution (CC BY) license (http://creativecommons.org/licenses/by/4.0/).

Article

Covalently-Linked Hyaluronan versus Acid Etched Titanium Dental Implants: A Crossover RCT in Humans

Saturnino Marco Lupi [1,*], Arianna Rodriguez y Baena [1], Clara Cassinelli [2], Giorgio Iviglia [2], Marco Tallarico [3], Marco Morra [2] and Ruggero Rodriguez y Baena [1]

1. Department of Clinico-Surgical, Diagnostic and Pediatric Sciences, Dental Clinic, University of Pavia, P.le Golgi 2, 27100 Pavia, Italy; arianna_rodriguez@hotmail.it (A.R.yB.); ruggero.rodriguez@unipv.it (R.R.yB.)
2. Nobil Bio Ricerche srl, V. Valcastellana 26, 14037 Portacomaro, Italy; ccassinelli@nobilbio.it (C.C.); giviglia@nobilbio.it (G.I.); mmorra@nobilbio.it (M.M.)
3. Private Practice, V. Vincenzo Ussani 86, 00151 Roma, Italy; me@studiomarcotallarico.it
* Correspondence: saturninomarco.lupi@unipv.it; Tel.: +39-382-516-255

Received: 30 December 2018; Accepted: 6 February 2019; Published: 11 February 2019

Abstract: Biochemical modification of titanium surfaces (BMTiS) entails immobilization of biomolecules to implant surfaces in order to induce specific host responses. This crossover randomized clinical trial assesses clinical success and marginal bone resorption of dental implants bearing a surface molecular layer of covalently-linked hyaluronan in comparison with control implants up to 36 months after loading. Patients requiring bilateral implant rehabilitation received hyaluronan covered implants in one side of the mouth and traditional implants in the other side. Two months after the first surgery, a second surgery was undergone to uncover the screw and to place a healing abutment. After two weeks, the operator proceeded with prosthetic procedures. Implants were evaluated by periapical radiographs and the crestal bone level was recorded at mesial and distal sites—at baseline and up to 36 months. One hundred and six implants were positioned, 52 HY-coated, and 48 controls were followed up. No differences were observed in terms of insertion and stability, wound healing, implant success, and crestal bone resorption at any time considered. All interventions had an optimal healing, and no adverse events were recorded. This trial shows, for the first time, a successful use in humans of biochemical-modified implants in routine clinical practice and in healthy patients and tissues with satisfactory outcomes.

Keywords: dental implants; surface modification; hyaluronan; clinical trial

1. Introduction

Ever since the pioneering studies of Professor Brånemark, osseointegration of titanium implant fixtures has been recognized as an interfacial event [1]. The clinical practice of implant dentistry has been based on the intimate apposition of newly formed bone tissue to the titanium surface. A great deal of literature has been devoted to the relationship between titanium surface properties and new bone formation [2–4]. The primary physical-chemical variables, titanium oxide surface chemistry, and implant surface topography dictate relevant surface parameters—such as surface charge and wettability—and have been deeply investigated in relation to cellular events leading to peri-implant bone regeneration [5–11]. The understanding of such relationships prompted the clinical evolution of titanium implants from the original turned to present day micro- and nano-rough surfaces [12–14].

Parallel to the growth and widespread acceptance of dental implantology, rising knowledge framed relevant interfacial biological events within a broader picture [15–18]. Cellular mechanisms leading to new bone formation and soft tissue healing, as well as inflammatory response leading to loss of supporting soft and hard tissue, are mediated by biological molecules and relevant signaling. The presentation of biomolecular cues, rather than the comparatively rough inorganic chemistry of titanium, seems a reasonable highway towards the evolution of better and innovative implant surfaces [19]. Accordingly, surface engineering of medical devices has long since been involved with the immobilization of a wide range of biomolecules to medical materials surfaces [20,21]. Biochemical modification of titanium surfaces (BMTiS) was defined by Puleo and Nancy [22] as the immobilization of proteins, enzymes, or peptides with the purpose of inducing specific cell and tissue responses using critical organic components of bone to affect tissue response. BMTiS are generally based on surface modification either by peptides, ECM proteins, or polysaccharides, all from animal and vegetal sources [23–34]. Despite a huge number of studies and promising in vitro and pre-clinical results [35], no practical application exists so far, and clinical performances of present-day dental implants are still dictated by the titanium/host tissue interface and mostly based on stimulation of cell behavior by surface topography.

This work presents the results of the first clinical trial involving bio-molecular modification of titanium implant surfaces. Implant fixtures used in the present trial bear a surface nano-layer (a few nm thick) of covalently-linked hyaluronic acid or hyaluronan [while its common acronym is HA, it is indicated as HY in the present paper to avoid any risk of confusion with the widely used inorganic HA (hydroxyapatite)-coatings]. This means that they present a complex macromolecular chemistry and not the relatively simple inorganic chemistry of titanium surfaces at the implant/tissue interface.

HY (Figure 1) is a linear polysaccharide consisting of the repeating disaccharide unit N-acetyl-D-glucosamine-D-glucuronate linked by β1-4 and β1-3 linkages. It is involved with a huge number of cellular processes [36–39].

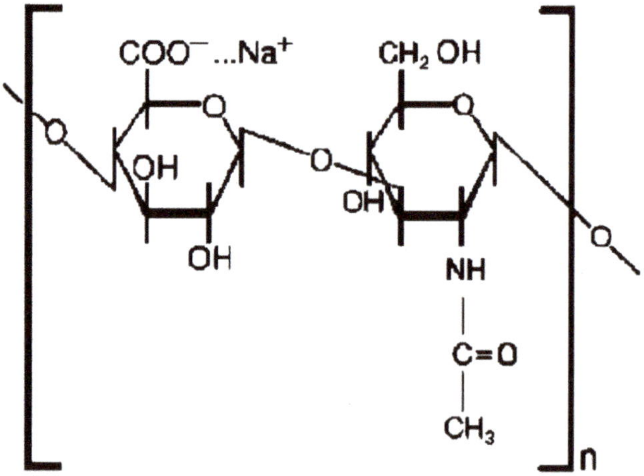

Figure 1. The repeating unit of Hyaluronan (HY).

Contrary to more complex glycosaminoglycans, HY is not sulfated, and it does not occur as part of a proteoglycan linked to a protein carrier. Significant interest in HY as a biomaterial and in biomaterials surface modification exists [40]. HY in medicine is mostly exploited because of its physical properties (hydration, viscosity, space filling), or by taking advantage of the hydration-promoted ability to reduce non-specific adhesion. Growing knowledge on HY as a key molecule in the regulation of many cellular processes involved with wound healing and tissue regeneration suggests that even more opportunities lie in the exploitation of its specific biological and bioactive properties [41–43].

Several literature reports indicate the potential interest of HY in BMTiS. Concerning bone regeneration, since the fifties it has been known that considerable HY is synthesized in the early stages of callus formation during the repair of fractured long bones [44]. Iwata and Urist [45] found that large amounts of HY were secreted when implants of decalcified bone underwent remineralization as bone. Bernard and coworkers presented studies aimed at developing "a foundation for the use of HY as a superior carrier for osteotropic substrates, even as HY acts to enhance osteogenesis due to its own molecular structure" [46]. Their in vitro studies using fetal calvarial cells and bone marrow osteogenic stem cells show that osteogenesis in vitro is significantly enhanced by HY 30–160 kDa, while high Mw HY (550–1300 kDa) shows weak inhibitory effects compared to the control. Zou and coworkers reported that 800 kDa HY added to bone marrow stromal cells cultured in vitro accelerates cell proliferation, increases alkaline phosphatase activity and osteocalcin gene expression, and that HY interacts with BMP-2 to generate direct and specific cellular effects [47]. Ito and coworkers showed that locally applied 900 kDa HY has a positive effect in bone ingrowth in Titanium fiber mesh implant in rats [48]. According to Cho, HY shows a positive effect in early bone consolidation in distraction osteogenesis [49]. HY based scaffolds aid in the regeneration of cartilage and bone defects in tissue engineering applications. The hypothesis of an active role played by local HY delivery upon scaffold degradation was suggested [50]. Zhao et al. [51] investigated the role of molecular weight and concentration of HY on the proliferation and osteogenic differentiation of rabbit bone marrow-derived stem cells in vitro. Factorial analysis indicated that molecular weight (MW) and concentration had an interactive effect on alkaline phosphatase mRNA expression ($p < 0.05$). HY of higher MW and higher concentration promoted bone formation. Regarding the in vivo studies on HY-coated implants, Aebli et al. [52] did not find bone growth increase in tests involving a sheep model. It should be noted that in the quoted study, the water-soluble HY was simply applied from solution to hydroxyapatite-coated implants without any intervening chemical bond to prevent rapid wash off [40]. An in vivo study on surface-engineered titanium implants bearing instead of a covalently-linked HY molecular surface layer in a four week rabbit model showed improvement of both bone to implant contact and bone ingrowth by hystomorphometry, while mechanical testing and evaluation of interfacial bone micro-hardness confirmed a faster bone maturation around HY coated implants [53]. Based on these and other encouraging pre-clinical results, the present study was conducted to investigate the clinical potential of HY covalently-linked implants and to set a starting point for future developments. The main goal was to confirm, in clinical practice and adopting objective clinical evaluation criteria, that "it is possible to do without the titanium surface chemistry". Once this point is set in routine clinics, pathways to actual exploitation of biomolecular signaling properties in compromised or challenging cases can be explored.

The paper presents first a detailed investigation of HY-coated titanium implants surfaces and relevant uncoated controls by SEM, energy dispersive X-ray spectroscopy (EDX), and X-ray photoelectron spectroscopy (XPS). Results of the clinical trial are then presented and discussed.

This split-mouth randomized clinical trial is aimed at assessing the clinical success and marginal bone resorption of dental implants bearing a surface molecular layer of covalently-linked hyaluronan in comparison with traditional sand-blasted and etched titanium implants up to 36 months after loading.

2. Results

2.1. Scanning Electron Microscopy/EDX Analysis

The surface topography of control and HY-coated implants was evaluated by SEM. Figures 2 and 3 show obtained results at $3000\times$ and $10,000\times$.

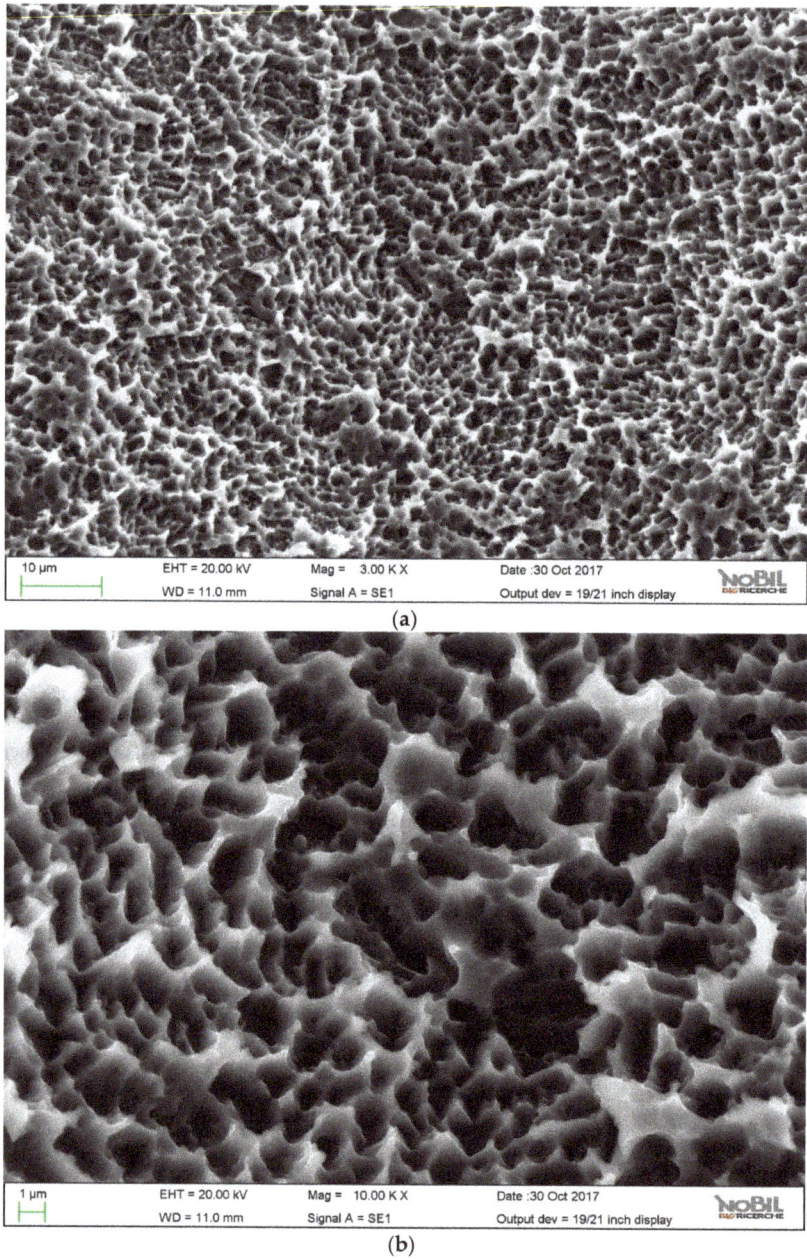

Figure 2. SEM images of the control implant surface, (**a**) 3000×; (**b**) 10,000×.

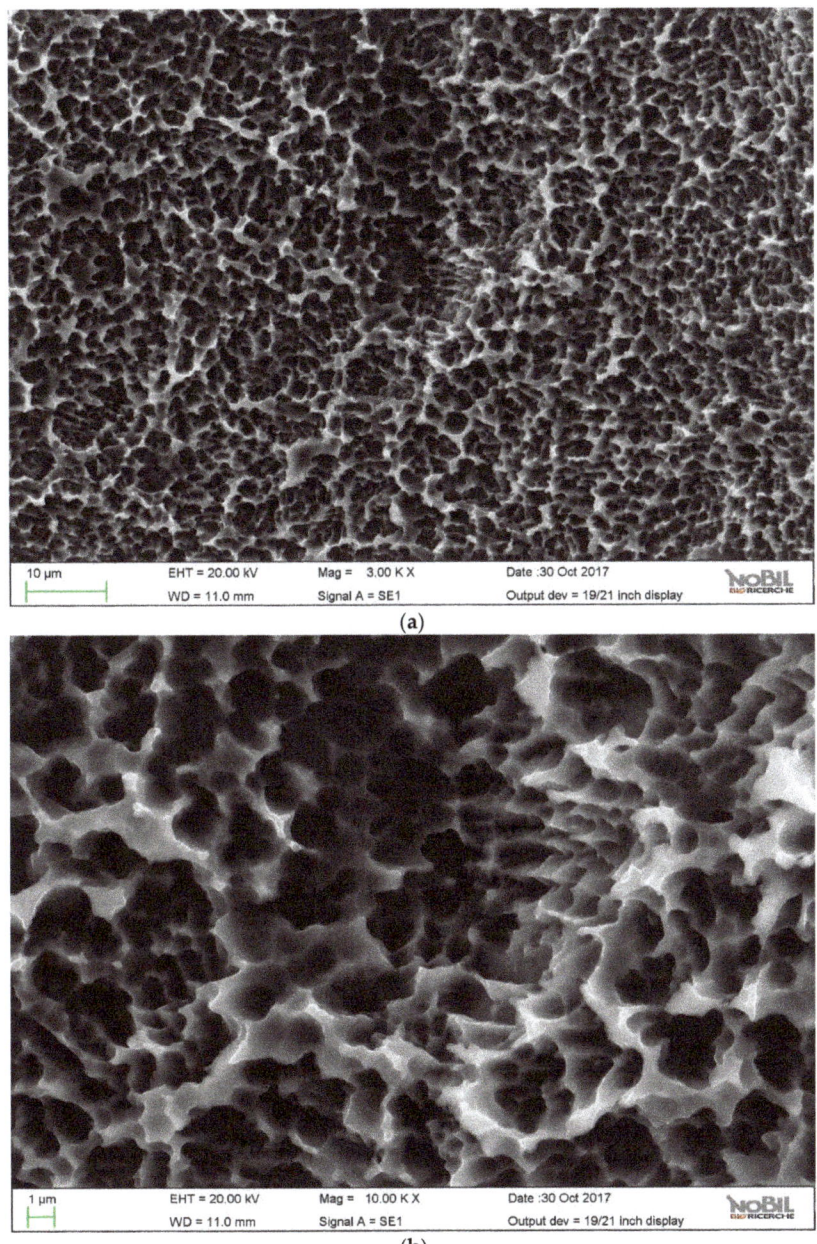

Figure 3. SEM images of the hyaluronan- (HY) coated implant surface, (**a**) 3000×; (**b**) 10,000×.

Both samples showed the typical microtopography of doubly acid etched surfaces—a microrough surface where the distance between peaks was of the order of the micrometer. As reported in the literature, a peak distance lower than the typical cell length can stimulate cell behavior, thus promoting accelerated osteogenesis [54,55]. No evidence of the HY coating was observed, even at 10,000×. Surface linking of HY involved just molecular layers, whose thicknesses were approximately a few nanometers at most [40]. At this level of magnification, it was not detected over the microrough landscape provided by the doubly acid etched titanium surface. Table 1 reports the roughness parameters according to ISO 4287 that were obtained by the stereo-SEM reconstruction of the surface topography obtained from 2000× images, as described in the Materials section. A 900 µm path length and an 80 µm cut-off wavelength were used. No significant difference between samples was observed.

Table 1. Roughness parameters obtained by stereo-SEM. Data are expressed in µm as average and standard deviation of three measurements.

Parameter	Control		HY-Coated		Description
	Mean	Std	Mean	Std	
Ra	0.97	0.19	0.93	0.24	Average roughness of profile
Rq	1.26	0.29	1.23	0.34	Root-Mean-Square roughness of profile
Rt	7.96	0.77	6.87	1.13	Maximum peak to valley height of roughness profile
Rz	5.67	1.23	5.77	1.15	Mean peak to valley height of roughness profile
Rmax	7.93	0.75	6.41	0.99	Maximum peak to valley height of roughness profile within a sampling length
Rp	4.33	0.63	3.93	0.73	Maximum peak height of roughness profile
Rv	3.63	0.14	2.93	0.45	Maximum valley height of roughness profile
Rc	3.69	0.64	3.62	1.03	Mean height of profile irregularities of roughness profile
Rsm	43.22	9.69	53.77	19.90	Mean spacing of profile irregularities of roughness profile

Figure 4 shows the energy dispersive X-ray spectroscopy (EDX) spectra obtained from the two samples. Both of them showed a very strong signal from titanium. In the case of the HY-coated implant, small but significant peaks due to carbon and oxygen were detected as well, suggesting the presence of an organic surface layer.

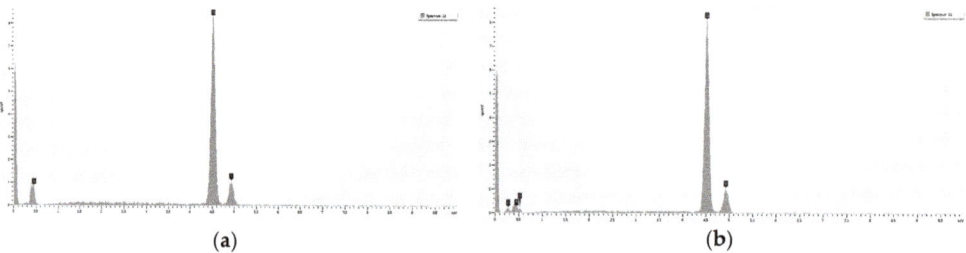

Figure 4. EDX spectra of control (**a**) and HY-coated (**b**) titanium implant.

2.2. X-Ray Photoelectron Spectroscopy

XPS wide-scan spectra of control and HY-coated dental implants used in the present clinical trial are shown in Figure 5.

The control implant showed the typical composition of titanium surfaces, yielding signals due to photoemitted electrons from core levels of the expected elements: titanium, oxygen (because of the native titanium oxide surface layer), and carbon (due to surface adsorption of ubiquitous carbon-containing compounds from the atmosphere and nitrogen). The overall surface composition, reported in Table 2, was in good quantitative agreement with data from the literature. It showed no unexpected elements or contamination, and its C/Ti ratio was indicative of excellent surface cleanliness [7,56].

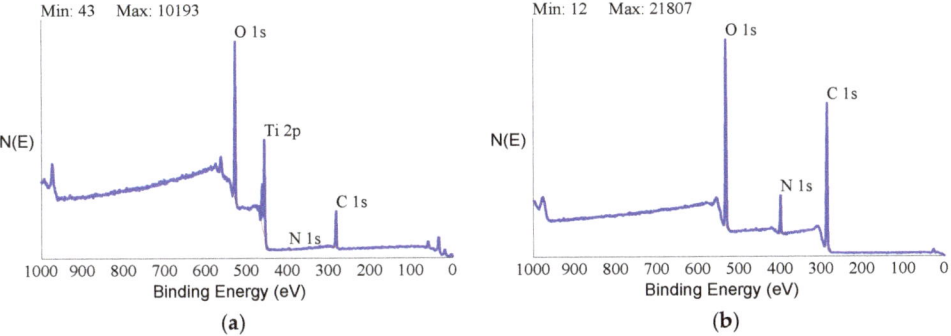

Figure 5. X-ray photoelectron spectroscopy (XPS) wide-scan spectra of control (**a**) and HY-coated (**b**) titanium implant.

Table 2. Surface composition (% at.) of Control and HY-coated implants.

Sample	O	Ti	N	C
Control	51.4	17.1	1.0	30.5
HY-coated	28.7	-	7.9	63.4

The XPS spectrum of the HY-coated implant, on the other hand, showed a completely different picture. First and foremost, there was no signal from titanium at all. The only elements detected in the nanometer-thick sampling depth probed by XPS were the basic elements of organic chemistry—oxygen, carbon, and nitrogen. The elemental ratio reported in Table 2 was consistent with typical polysaccharides stoichiometry, showing a high O/C ratio. Perusal of the literature confirmed that the detected surface stoichiometry was consistent with the reported composition of covalently-linked HY through aminated spacers [57]. Further clues were offered by the sensitivity to the carbon chemical environment provided by the high-resolution peak of C1s photoelectrons. Figure 6 shows C1s peaks of the tested control and HY-coated dental implants.

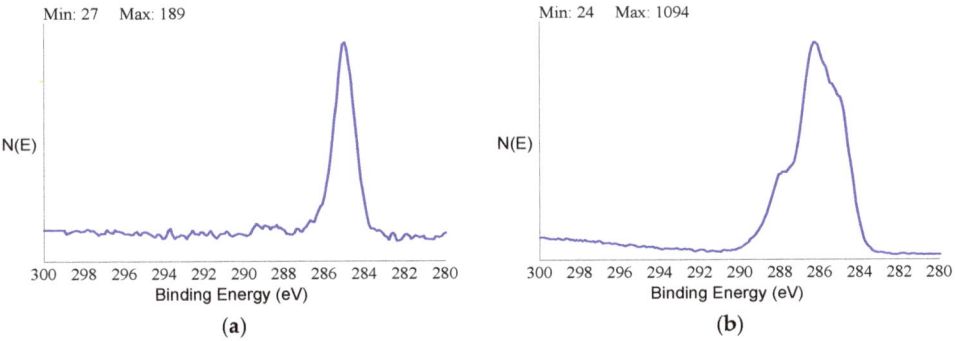

Figure 6. XPS high-resolution C1s peak of control (**a**) and HY-coated (**b**) titanium implant.

In the former case, the peak was relatively simple and symmetrical, confirming that it was mostly due to C-C and C-H functionalities from adventitious hydrocarbons. The C1s peak of the HY-coated implant was much broader and clearly contained different components. De-convolution of the C1s peak of the HY-coated implant following general XPS practice [58] is shown in Figure 7.

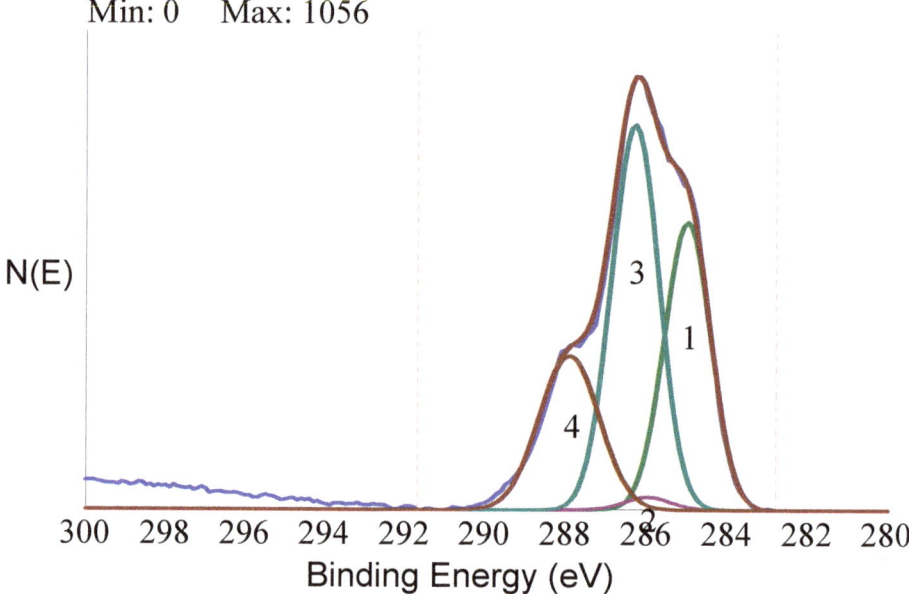

Figure 7. Curve fitting of high-resolution C1s peak of the HY-coated titanium implant by the four components expected from the molecular structure of hyaluronan.

The experimental curve was fitted according to literature indication [59] by four components located at 285.00 eV (C-C, C-H), 286.15 eV (C-N), 286.50 eV (C-O), and 288.10 eV (C=O, N-C=O, and O-C=O functionalities). Peak de-convolution showed that the carbon chemical environment at the HY-coated titanium implant surface was dominated by the carbon single bond oxygen functionality typical of polysaccharides, and that the experimental C1s peak contained all features expected from the molecular structure of HY. Taken together, these data convincingly showed that the outermost nanometers of the HY-coated implant surface presented to the external environment the molecular cues stemming from the HY repeating unit, which was a completely different chemical nature when compared to the titanium oxide surface of the conventional control implant.

2.3. Clinical Trial

From 8 April 2013 to 17 October 2014, 106 implants were positioned in 30 patients (demographic data in Table 3).

Table 3. Demographic data and implant location. Data in absolute values and (%).

Demographic Data	Total	HY	C
Males	21 (70)		
Females	9 (30)		
Patients mean age	59.8 ± 10.6		
Implant location			
Anterior maxilla	18 (17.0)	9 (16.4)	9 (17.6)
Posterior maxilla	41 (38.7)	24 (43.6)	17 (33.3)
Anterior mandibula	10 (9.4)	5 (9.0)	5 (9.8)
Posterior mandibula	37 (34.9)	17 (30.9)	20 (39.2)

During surgery, no differences were observed in regards to the insertion and stability between the two types of implants (test and control), nor were differences observed in the post-operative visits in regards to the indices of inflammation and wound healing. The healing was uneventful in all of the patients in the variability of the surgical situation, and no adverse events whatsoever were recorded. Only 102 implants in 29 subjects out of the 106 implanted were loaded and taken as baselines for X-ray values (mesial and distal). Two implants (one control and one HY coated) in two different patients failed and were removed. In both cases, the mucosa appeared edematous and bleeding, indicating signs of an infection. In one case, the implant was not replaced, and since no matching controls were available, the subject (Subject N°34) was excluded from the efficacy analysis. In the second subject, the failed implant was relocated in the same place after two months, reaching healing and tissue stabilization. Therefore, the subject was maintained in the statistics. One subject with two implants (one control and one HA) dropped out between the 12 months and the 18 months follow up visits due to a desire for pregnancy and therefore the unethical use of X-ray procedures. Results are summarized in Table 4.

Table 4. Bone resorption.

Time Points	HY		C		p (between-Group)	95% Confidence Interval	
	n	Mean ± sd	n	Mean ± sd		Min	Max
Mesial							
3	51	0.55 ± 0.46	47	0.51 ± 0.65	0.64	−0.24	0.55
6	24	0.72 ± 0.38	20	0.65 ± 0.39	0.58	−0.30	0.17
12	35	0.83 ± 0.61	35	0.66 ± 0.58	0.25	−0.45	0.12
18	16	0.65 ± 0.63	17	0.36 ± 0.43	0.14	−0.67	0.10
24	34	0.80 ± 0.87	32	0.50 ± 0.57	0.10	−0.66	0.06
36	21	0.55 ± 0.40	19	0.32 ± 0.36	0.06	−0.47	0.01
Distal							
3	51	0.71 ± 0.60	47	0.70 ± 0.62	0.97	−0.25	0.24
6	24	0.77 ± 0.60	20	0.93 ± 0.46	0.30	−0.15	0.49
12	35	0.83 ± 0.63	35	0.85 ± 0.60	0.90	−0.28	0.31
18	16	0.72 ± 0.53	17	0.47 ± 0.45	0.16	−0.59	0.11
24	34	0.76 ± 0.42	32	0.63 ± 0.62	0.33	−0.39	0.14
36	21	0.62 ± 0.53	19	0.55 ± 0.62	0.70	−0.44	0.30

Time points: months; n: number of implants; mean ± standard deviation (sd) expressed in mm.

The non-inferiority analysis did not show any significant differences between the HY- and the control- in terms of mesial and distal bone resorption at any time point.

3. Discussion

In the present clinical trial, a widely adopted microrough titanium surface and the same surface further modified by a covalently-linked nanolayer of hyaluronan were compared in terms of clinical success in routine clinical practice.

Analytical data confirmed the exquisitely superficial nature of the surface modification process that was adopted. The analytical signal captured in EDX analysis stemmed from a surface volume that was a few micrometers deep [60]. Thus, within the EDX sampling depth, the signal by the nanometers-thick HY surface layer was diluted within the micrometer-thick analytical layer, and the resulting spectrum contained convoluted contributions from the underlying Ti implant and the overlying HY molecules in roughly a 1000:1 ratio. As a consequence, the elements composing the surface layer were barely detectable. For this reason, a more surface-sensitive analytical technique, namely XPS, was conducted to fully appreciate the biomolecular modification of the HY-coated implants. Contrary to EDX analysis, the physics of photoelectrons escaped from solids endowed XPS with just a few nanometers of sampling depth, consequently providing chemical information on the outermost atomic layers of materials [58]. For this reason, XPS was extensively used in the surface analysis of dental implants [7,61–63].

Despite the widely different surface chemistry, objective optimal healing was observed for both groups, and no differences were detected in the clinical outcome for all tested parameters. Far from being uneventful, the described results suggested some important reflections. From a chemical-physical point of view, it would be difficult to imagine two more different surface structures—the titanium surface (in particular, the outermost few nanometers of oxidized titanium) presented a hard, impervious interface to the host tissue, whose chemical behavior was controlled by titanium oxide interfacial chemistry, as widely described in many scientific papers [7,8,61]. On the contrary, the HY-coated implant aqueous interface was diffuse, soft, and hydrated [64]. Its chemistry stemmed from the molecular details of the N-acetyl-D-glucosamine-D-glucuronate repeating unit. Shortly, one of them (the control arm) belonged to inorganic chemistry. The HY-coated surface belonged instead to organic biomacromolecular chemistry. Despite being two worlds apart, no significant difference was detected in the tested clinical variables, and both of them led to successful osseointegration and clinical success. The first significant reflection from the data of the present work is that the titanium surface chemistry is not necessary to achieve a clinically satisfactory load-bearing capacity by a titanium dental implant.

The previous point could be interpreted as a no-effect of surface chemistry on osseointegration. As long as the implant surface is inert and no disturbance is introduced in the bone healing mechanism—that is, as long as no toxic or irritating compounds are released in the evolving new bone matrix independent from specific details of surface chemistry—clinically effective osseointegration will occur. If this is the case, no advancement is expected by BMTiS over conventional surfaces, and the sole opportunity to direct tissue response is through surface topography.

The last sentence falls short when compared to existing scientific evidence gathered from animal data, which shows direct effects of interfacial chemistry on bone healing in terms of peri-implant bone volume, bone to implant contact, or gene expression by peri-implant bone cells [35,53,65,66]. In the present clinical trial, a patient's selection and surgery addressed comparatively routine clinical practice. The control arm, involving a state-of-the-art doubly acid etched microrough surface, obviously provided optimal healing and clinical success. Endpoint variables aimed at general clinical evidence expectedly confirmed that both the well-known titanium interfacial chemistry and the biologically relevant HY molecular cued direct cellular events involved with peri-implant tissue healing towards proper clinical response.

HY is a key molecule in many tissue regeneration processes; it is involved with most of the mammalian cells' healing mechanisms [67–69] in a concentration and size-dependent way. The permanent linking of HY to materials surfaces avoids quick wash-off of the water soluble HY and aims at providing these regenerative properties at the peri-implant interface, as confirmed by several in vivo evidences. It is not clear yet how surface-immobilized HY compares with HY in solution in terms of the effect of the hindered conformational freedom on the multiple ligand-receptor interactions required to trigger a biological response [40]. It would be of interest to check the stimulation of bone tissue regeneration at machined interfaces of hybrid implants, the effect on soft tissue healing in the transmucosal section of tissue level implants, or the control of inflammatory responses of periodontal patients [70] and relevant clinical implications. The positive evidence supplied by the present clinical trial in standard practice opens the path to comprehensive investigation of the merits of BMTiS in clinical implantology by further finely-targeted clinical trials.

A last observation involves the relationship of present results with in vitro investigation of dental implant surfaces. A widely adopted approach involves the "adhesion and growth" paradigm, meaning that prospective surface structures are screened in terms of effects on adhesion and growth of osteoblast cells—the faster and more extensive the surface colonization by cells, the better the properties. Against this view, HY-coated surfaces are notoriously anti-adhesive in vitro [64,71], meaning that they prevent cell adhesion of a number of cell lines, including osteoblasts. Several applications of HY-coated surfaces are based on tissue-anti-adhesive properties. In vitro tests of present clinically successful HY-coated implants would provide unsatisfactory results if judged according to the "adhesion and growth" paradigm. Yet, clinical evidence as supplied by the present trial indicates that they are fit for the intended use. Far from being inconsistent, this evidence simply indicates the complexity of the peri-implant environment as opposed to the comparatively simple biological environment of in vitro tests. In clinics, osteoblasts do not adhere and grow onto the implant surface. Rather, they come to a complex peri-implant milieu after the blood clot and relevant blood cells, inflammatory cells, and the sequela of cytokine and growth factors are released in the initial stage of inflammation and healing [72]. Rather than the direct effect of the interfacial HY (or of any surface-linked biomolecular layer) on osteoblasts, it is the effect of molecular signaling on the evolution of the peri-implant biochemical environment that directs clinical outcome and that holds the potential merits of BMTiS in clinics.

In summary, the clinical trial did not record any significant difference between the HY- and the C-group in terms of clinical success and marginal bone resorption. In the present study, for the first time in clinics (to the author's knowledge), dental implants with a biomolecular nanolayer on the surface showed a behavior similar to the commercial pure titanium one. This study presents the limits to having a follow up limited to 36 months. The duration of the follow up was chosen because in this span of time, the HY layer should be completely integrated within the newly formed interfacial tissue. Further studies should be conducted to evaluate the long-term success of HY-implants. Present data build up the ethical basis for the investigation into the merits of HY-coated implants in more challenging and compromised cases where signaling and regenerative properties encoded within the HY molecular structure could play a role that is presently not supplied by the comparatively rough titanium surface chemistry.

4. Materials and Methods

4.1. Titanium Implant

The fixture used in the clinical trial was a CE marked titanium grade 4, internal hexagon, doubly acid etched implant (Ornaghi Luigi & C, Brugherio, Italy). The implant fixture was further coated by covalently-linked HY by Nobil Bio Ricerche (Nobil Bio Ricerche srl, Portacomaro, Italy) through a proprietary process, as described in the Results section. Covalent-linking prevented the rapid wash-off of the water-soluble HY molecules, degradation by hyaluronidase, and release. The thickness of the coating, as discussed in the following sections, was a few nanometers and did not modify the nominal dimensions. The control was the same implant not coated with covalently-linked HY.

4.2. Surface Characterization

4.2.1. Scanning Electron Microscopy

The surface topography of the implants was evaluated by SEM. Analysis was conducted using an EVO MA 10 SEM (Carl Zeiss Microscopy GmbH, Jena, Germany). The electron acceleration voltage was maintained at 20 kV with the working distance between 10 and 12.5 mm. These parameters are reported in the images, along with level of magnification (MAG) and the kind of detector utilized (Signal A = SE1 or CZ BSD).

Images were acquired in both conventional mode (Signal A = SE1) and in backscattered electron mode (Signal A = CZ BSD), allowing improved contrast between different chemical elements.

Roughness was evaluated quantitatively by stereo-SEM (SSEM) using dedicated software to convert conventional SEM images into three-dimensional data (Mex 6.0, Alicona Imaging, Raaba/Graz, Austria). A stereo-pair was built by the acquisition of the same field of view at zero and after five degrees of eccentric tilting. Following the principles of stereoscopic vision, the stereo pair was transformed into a three-dimensional reconstruction of the surface by the quoted software, providing height profiles values used as input data for the calculation of roughness parameters according to ISO 4287. The stereo-pairs were built from 2000× images taken from three randomly selected areas for each implant.

4.2.2. X-ray Photoelectron Spectroscopy (XPS)

XPS analysis was performed using a Perkin Elmer PHI 5600 ESCA system (PerkinElmer Inc., Waltham, MA, USA). The instrument was equipped with a monochromatized Al anode operating at 10 kV and 200 W. The diameter of the analyzed spot was approximately 500 micrometers, and the analyzed depth was about 5 nanometers. The base pressure was maintained at 10–8 Pa. The angle between the electron analyzer and the sample surface was 45°. Analysis was performed by acquiring a wide-range survey spectra (0–1000 eV binding energy) and detailed high-resolution peaks of relevant elements. Quantification of elements was accomplished using the software and sensitivity factors supplied by the manufacturer. High-resolution C1s peaks were acquired using pass energy of 11.75 eV and a resolution of 0.100 eV/step.

4.3. Clinical Trial

The study "Blind Comparison of Covalently-Linked Hyaluronan versus Control-Dental Implants in a Randomized Crossover Clinical Investigation" is a post market clinical follow-up according to MEDDEV 2.12-2 rev 2 January 2012 conducted at the Dental Service, Department of Clinical, Surgical, Diagnostic, and Pediatric Sciences, University of Pavia, Pavia, Italy. The clinical trial was conducted according to the ISO 14155-11 and the Good Clinical Practice Guidelines (GCP). The study was carried out following the rules of the Declaration of Helsinki, as revised in 2013 and approved by the Ethical Committee of the University of Pavia on 13 December 2012. On 28 February 2013, the initiation of the study was communicated to the Italian Ministry of Health with the reference number 000017. All subjects gave their informed consent for inclusion before they participated in the study. The primary objective of this non-inferiority, crossover, fixed-size, and single-center trial was to assess the dental implant success (survival rate) of HA coating implants in comparison with control implants at one year. The secondary objective was to assess marginal bone resorption of HA coating implants in respect to traditional titanium implants measured at 3, 6, 12, 18, 24, and 36 months. To test the non-inferiority hypothesis of the HY-coated implant in respect to control implants, the null hypothesis was that the HY-coated implant resulted in a lower survival rate and a higher marginal bone loss in respect to the control implant. The P level of significance was set at 0.05. Patients with bilateral partial or full edentulism requiring implant rehabilitation were enrolled; inclusion and exclusion criteria are shown in Tables 5 and 6.

Table 5. Criteria for inclusion in the clinical trial.

Inclusion Criteria
Over 18 years of age;
Bilateral loss of one or more molars and bicuspids and/or bilateral loss of anterior teeth and the need for more than one implant in the same rehabilitation;
Edentulousness of both upper or lower jaw with need of at least 2 bilateral implants to stabilize a denture or to support a fixed prosthesis;
In general good health condition and with physical ability to tolerate surgical and prosthetic procedure (ASA 1 and 2);
Good plaque control and oral hygiene;
The subject who agree to return to the center for follow-up.

Table 6. Criteria for exclusion from the clinical trial.

Exclusion Criteria
Active infection or severe inflammation or suspected lesions in the areas intended for implant installation; Diabetes (regardless of control); Need for concomitant bone grafting and/or having less than 1 mm bone available at the buccal, lingual, and apical aspects of the implant. Less than 3 mm distance between implant and other dentition; Under treatment and/or within the past 12 months with radiotherapy to the head or chemotherapy; Suspected hypersensitivity and/or contraindication to any ingredients of the Investigational Device (ID)/Control ID; Subjects under any study medication treatment in the last 30 days; Pregnancy, breast feeding, oocyte donation, or oocyte implantation planned during the study; Subjects not able to follow study procedures, e.g., language problems, psychological disorders; Clinically relevant abnormal laboratory values suggesting an unknown disease and requiring further clinical evaluation (as assessed by the investigators); Female subjects of childbearing potential, not using and not willing to continue using a medically reliable method of contraception for the entire study duration; Any other untoward medical condition that could interfere with the participation of the subject in the trial.

In this study, concealment and randomization were implemented centrally by an operator (G.O.) unrelated to the surgical team, to the clinical examiner, and to the statistician. After manufacturing, implants were packaged in consecutively numbered boxes. Each box contained six implants, three with the HY surface (marked with A or B) and three with the control surface (marked with B if the HY implants were marked with A, or marked with A on the contrary). All of the implants were 4 mm in diameter and were 8, 10, and 12 mm long. The two surfaces were macroscopically indistinguishable. The assignment of the label (A or B) to the surfaces was randomly assigned in each box by a computer random number generator. Data were communicated to the statistician, omitting the surface characteristic. The list of randomization was concealed until after statistical analysis was concluded. Due to the split-mouth design of the study and the randomization process, the allocation ratio was 1:1. No block restrictions were applied.

4.3.1. Surgical Procedure

During surgery, implant(s) with one surface were placed in one side of the mouth and implant(s) with the other surface were placed in the other side of the mouth. When a patient required placement of the number of implants equal in both sides of the mouth, each couple of implants came out from a different box. When the number of implants required in one side of the mouth was greater than the other side, implants were chosen among those available in the opened boxes. If the odd implants were not available in the opened boxes, no experimental implants were positioned. This procedure was implemented after protocol approval because the surgical team signaled the problem of re-operating in patients with a different number of edentulous sites in the two sides of the mouth. Even though it was not necessary because randomization was implemented before surgery, the implant label choice for the sides of the mouth was randomized by a coin toss. The surgical protocol of installing the implants was well documented and consisted of a full thickness mucosal flap elevation, the preparation of the implant site, the installation of the implant with the prosthetic platform (that is, the edge of the collar) at the level of the bone crest, the positioning of the surgical cover screw, and the closing of the flap with sutures [73–80]. The surgery was done under coverage of antibiotics (2 g of amoxicillin and clavulanic acid per os one hour before surgery, followed by administration of 1 g after 6 and 18 h). The post-operative instructions indicated rinses with chlorhexidine 0.12% × three times a day from the day after surgery and for 30 days and control of the pain with anti-inflammatory drugs as needed. The removal of the sutures was carried out from the tenth to the fourteenth day.

The second stage of surgery was conducted after two months, both in the mandibular and in the maxillary arch. After about two weeks, the operator proceeded with the prosthesis installation. During follow-up visits, normal hygienic procedures were carried out when needed [81].

4.3.2. Investigation Hypothesis or Pass/Fall Criteria

The primary objective of the present study was to assess the success of HY coated implants in comparison with control implants after one year. The implant is considered successful (primary endpoints) if: (A) the implant is still present without any sign of mobility and (B) there is no evidence of radiolucency by means of periapical X-rays and (C) there is no clinical sign of peri-implantitis. The secondary objective of the present study was to demonstrate a non-inferiority of the HY-coated implants compared to the control implants considering the marginal bone resorption parameter. To do this, the null hypothesis is that control implants demonstrate an inferior crestal bone resorption in respect to HY coated implants.

Sample Size Calculation

The sample size was estimated based on the crestal bone level at 12 months post implant.

Since the objective of this study was the non-inferiority of the investigational device (ID) compared to the Control ID in a crossover design from data published by Mumcu [82], inferior limits for the Control ID were estimated.

Practically, from pooling data reported in Table 1 of Mumcu and Coll., an inferior limit of 0.7 mm was estimated and considered clinically appropriate. This implies that if the lower bond of the 95% confidence interval for the estimated difference in crestal bone level between implants is above −0.7 mm, then the ID is considered non-inferior to the Control ID.

Sample size was calculated under the following assumptions:

- Lower limits in difference between implants: −0.70 mm (Δ)
- Hypothesis testing:
- H0: μControl ID- μID ≤ Δ
- H1: μControl ID- μID > Δ (ID implant is non-inferior with respect to the mean response)
- α level of probability: 0.025 for one-sided test
- 1-β (power) level: 90% for a conservative approach
- Standard deviation for mean difference: 0.66, which corresponded to 80% of the pooled mean calculated from Mumcu (0.83 mm). In this estimate, it was considered that the ID presented on the crestal bone level a greater variability in respect to the Control ID and that reported in the publication.

By applying the formula reported in Julius [83] for non-inferiority trials in crossover design:

$$n = \frac{2\sigma^2 \left(Z_{(1-\beta)} + Z_{(1-\alpha)}\right)^2}{\left((\mu_{Control\ ID} - \mu_{ID}) - d\right)^2}$$

18 evaluable subjects are needed to demonstrate the non-inferiority of the ID in respect to the Control ID. Considering that a couple of devices are implanted into each subject, we need to have at least 18 implant pairs in order to reach the trial primary objective.

5. Conclusions

The clinical trial "Blind Comparison of Covalently-Linked Hyaluronan versus Control-Dental Implants in a Randomized Crossover Clinical Investigation" evaluated clinical success up to 36 months with HY-coated dental implants compared to the control—uncoated microrough Ti grade 4 implants. Results showed a lack of differences between the two arms of the study. Both of them provided optimal healing.

Substitution of the clinically accepted and consolidated titanium surface chemistry with the molecular structure of HY and ensuing interfacial interactions results in a satisfactory clinical outcome.

Author Contributions: Conceptualization: S.M.L., A.R.yB., C.C., G.I., M.T., M.M. and R.R.yB.; methodology: S.M.L., C.C., M.M. and R.R.yB.; validation: A.R.yB., G.I., M.T. and R.R.yB.; formal analysis, S.M.L., C.C., M.M. and R.R.yB.; investigation, A.R.yB., G.I., M.T. and R.R.yB.; resources, A.R.yB.; data curation, S.M.L., M.M. and R.R.yB.; writing—original draft preparation, S.M.L. and M.M.; writing—review and editing, S.M.L., A.R.yB., C.C., G.I., M.T., M.M. and R.R.yB.; supervision, R.R.yB.; project administration, M.M.; funding acquisition, C.C., M.M. and R.R.yB.

Funding: The study was supported by Nobil Bio Ricerche srl, the profit company that executes the surface treatment process.

Acknowledgments: The authors thank: BSDpharma srl, under the responsibility of Claudio F. Omini, for the role in CRO duties and for monitoring of the study; Gabriele Omini for randomization list and labels and for monitoring of the study; Guido Fedele for statistical support; Giuseppe Zuccari for Quality Assurance; AA Gandolfi for monitoring of the study.

Conflicts of Interest: Marco Morra and Clara Cassinelli own shares of the funding company Nobil Bio Ricerche srl. Giorgio Iviglia is an employee of Nobil Bio Ricerche srl. Saturnino Marco Lupi, Arianna Rodriguez y Baena and Ruggero Rodriguez y Baena declare no conflict of interest.

Abbreviations

BMTiS	Biochemical modification of titanium surfaces
ECM	Extra Cellular Matrix
EDX	Energy Dispersive X-ray Analysis
GCP	Good Clinical Practice
HY	hyaluronic acid or hyaluronan
ID	Investigational Device
MAG	level of magnification
RCT	Randomized Clinical Trial
SEM	Scanning Electron Microscopy
XPS	X-ray Photoelectron Spectroscopy

References

1. Branemark, P.I. Osseointegration and its experimental background. *J. Prosthet. Dent.* **1983**, *50*, 399–410. [CrossRef]
2. Brunette, D.M.; Tengvall, P.; Textor, M.; Thomsen, P. *Titanium in Medicine: Material Science, Surface Science, Engineering, Biological Responses and Medical Applications*; Springer: Berlin, Germany, 2001.
3. Davies, J.E. *The Bone-Biomaterial Interface*; University of Toronto Press: Toronto, ON, Canada, 1991.
4. Ellingsen, J.E.; Lyngstadaas, S.P. *Bio-Implant Interface: Improving Biomaterials and Tissue Reactions*; CRC Press: London, UK, 2003.
5. Kasemo, B.; Lausmaa, J. Biomaterial and implant surfaces: A surface science approach. *Int. J. Oral Maxillofac. Implants* **1988**, *3*, 247–259. [PubMed]
6. Kasemo, B.; Lausmaa, J. Biomaterial and implant surfaces: On the role of cleanliness, contamination, and preparation procedures. *J. Biomed. Mater. Res.* **1988**, *22*, 145–158. [CrossRef] [PubMed]
7. Kasemo, B.; Lausmaa, J. Material-tissue interfaces: The role of surface properties and processes. *Environ. Health Perspect.* **1994**, *102*, 41–45. [PubMed]
8. Larsson Wexell, C.; Thomsen, P.; Aronsson, B.O.; Tengvall, P.; Rodahl, M.; Lausmaa, J.; Kasemo, B.; Ericson, L.E. Bone response to surface-modified titanium implants: Studies on the early tissue response to implants with different surface characteristics. *Int. J. Biomater.* **2013**, *2013*, 412482. [CrossRef] [PubMed]
9. Thomsen, P.; Larsson, C.; Ericson, L.E.; Sennerby, L.; Lausmaa, J.; Kasemo, B. Structure of the interface between rabbit cortical bone and implants of gold, zirconium and titanium. *J. Mater. Sci. Mater. Med.* **1997**, *8*, 653–665. [CrossRef] [PubMed]
10. Rupp, F.; Gittens, R.A.; Scheideler, L.; Marmur, A.; Boyan, B.D.; Schwartz, Z.; Geis-Gerstorfer, J. A review on the wettability of dental implant surfaces i: Theoretical and experimental aspects. *Acta Biomater.* **2014**, *10*, 2894–2906. [CrossRef] [PubMed]
11. Rupp, F.; Scheideler, L.; Olshanska, N.; de Wild, M.; Wieland, M.; Geis-Gerstorfer, J. Enhancing surface free energy and hydrophilicity through chemical modification of microstructured titanium implant surfaces. *J. Biomed. Mater. Res. A* **2006**, *76*, 323–334. [CrossRef]

12. Wennerberg, A.; Albrektsson, T. On implant surfaces: A review of current knowledge and opinions. *Int. J. Oral Maxillofac. Implants* **2010**, *25*, 63–74.
13. Rupp, F.; Liang, L.; Geis-Gerstorfer, J.; Scheideler, L.; Huttig, F. Surface characteristics of dental implants: A review. *Dent. Mater.* **2018**, *34*, 40–57. [CrossRef]
14. Rodriguez y Baena, R.; Rizzo, S.; Manzo, L.; Lupi, S.M. Nanofeatured titanium surfaces for dental implantology: Biological effects, biocompatibility, and safety. *J. Nanomater.* **2017**. [CrossRef]
15. Nishimura, I. Genetic networks in osseointegration. *J. Dent. Res.* **2013**, *92*, 109S–118S. [CrossRef] [PubMed]
16. Smeets, R.; Stadlinger, B.; Schwarz, F.; Beck-Broichsitter, B.; Jung, O.; Precht, C.; Kloss, F.; Grobe, A.; Heiland, M.; Ebker, T. Impact of dental implant surface modifications on osseointegration. *Biomed. Res. Int.* **2016**, *2016*, 6285620. [CrossRef] [PubMed]
17. Brett, E.; Flacco, J.; Blackshear, C.; Longaker, M.T.; Wan, D.C. Biomimetics of bone implants: The regenerative road. *Biores. Open Access* **2017**, *6*, 1–6. [CrossRef] [PubMed]
18. Insua, A.; Monje, A.; Wang, H.L.; Miron, R.J. Basis of bone metabolism around dental implants during osseointegration and peri-implant bone loss. *J. Biomed. Mater. Res. A* **2017**, *105*, 2075–2089. [CrossRef] [PubMed]
19. Bryers, J.D.; Giachelli, C.M.; Ratner, B.D. Engineering biomaterials to integrate and heal: The biocompatibility paradigm shifts. *Biotechnol. Bioeng.* **2012**, *109*, 1898–1911. [CrossRef]
20. Morra, M. Biochemical modification of titanium surfaces: Peptides and ecm proteins. *Eur. Cell. Mater.* **2006**, *12*, 1–15. [CrossRef]
21. Morra, M. Biomolecular modification of implant surfaces. *Expert Rev. Med. Devices* **2007**, *4*, 361–372. [CrossRef]
22. Puleo, D.A.; Nanci, A. Understanding and controlling the bone-implant interface. *Biomaterials* **1999**, *20*, 2311–2321. [CrossRef]
23. Morra, M.; Cassinelli, C.; Cascardo, G.; Cahalan, P.; Cahalan, L.; Fini, M.; Giardino, R. Surface engineering of titanium by collagen immobilization. Surface characterization and in vitro and in vivo studies. *Biomaterials* **2003**, *24*, 4639–4654. [CrossRef]
24. Morra, M.; Cassinelli, C.; Cascardo, G.; Mazzucco, L.; Borzini, P.; Fini, M.; Giavaresi, G.; Giardino, R. Collagen i-coated titanium surfaces: Mesenchymal cell adhesion and in vivo evaluation in trabecular bone implants. *J. Biomed. Mater. Res. A* **2006**, *78*, 449–458. [CrossRef] [PubMed]
25. Morra, M.; Cassinelli, C.; Cascardo, G.; Nagel, M.D.; Della Volpe, C.; Siboni, S.; Maniglio, D.; Brugnara, M.; Ceccone, G.; Schols, H.A.; et al. Effects on interfacial properties and cell adhesion of surface modification by pectic hairy regions. *Biomacromolecules* **2004**, *5*, 2094–2104. [CrossRef] [PubMed]
26. Kim, H.W.; Li, L.H.; Lee, E.J.; Lee, S.H.; Kim, H.E. Fibrillar assembly and stability of collagen coating on titanium for improved osteoblast responses. *J. Biomed. Mater. Res. A* **2005**, *75*, 629–638. [CrossRef] [PubMed]
27. Rammelt, S.; Heck, C.; Bernhardt, R.; Bierbaum, S.; Scharnweber, D.; Goebbels, J.; Ziegler, J.; Biewener, A.; Zwipp, H. In vivo effects of coating loaded and unloaded ti implants with collagen, chondroitin sulfate, and hydroxyapatite in the sheep tibia. *J. Orthop. Res.* **2007**, *25*, 1052–1061. [CrossRef] [PubMed]
28. Rammelt, S.; Illert, T.; Bierbaum, S.; Scharnweber, D.; Zwipp, H.; Schneiders, W. Coating of titanium implants with collagen, rgd peptide and chondroitin sulfate. *Biomaterials* **2006**, *27*, 5561–5571. [CrossRef] [PubMed]
29. Stadlinger, B.; Pilling, E.; Huhle, M.; Mai, R.; Bierbaum, S.; Bernhardt, R.; Scharnweber, D.; Kuhlisch, E.; Hempel, U.; Eckelt, U. Influence of extracellular matrix coatings on implant stability and osseointegration: An animal study. *J. Biomed. Mater. Res. B Appl. Biomater.* **2007**, *83*, 222–231. [CrossRef] [PubMed]
30. Stadlinger, B.; Pilling, E.; Mai, R.; Bierbaum, S.; Berhardt, R.; Scharnweber, D.; Eckelt, U. Effect of biological implant surface coatings on bone formation, applying collagen, proteoglycans, glycosaminoglycans and growth factors. *J. Mater. Sci. Mater. Med.* **2008**, *19*, 1043–1049. [CrossRef] [PubMed]
31. Stadlinger, B.; Pilling, E.; Huhle, M.; Mai, R.; Bierbaum, S.; Scharnweber, D.; Kuhlisch, E.; Loukota, R.; Eckelt, U. Evaluation of osseointegration of dental implants coated with collagen, chondroitin sulphate and bmp-4: An animal study. *Int. J. Oral Maxillofac. Surg.* **2008**, *37*, 54–59. [CrossRef]
32. Kokkonen, H.; Verhoef, R.; Kauppinen, K.; Muhonen, V.; Jorgensen, B.; Damager, I.; Schols, H.A.; Morra, M.; Ulvskov, P.; Tuukkanen, J. Affecting osteoblastic responses with in vivo engineered potato pectin fragments. *J. Biomed. Mater. Res. A* **2012**, *100*, 111–119. [CrossRef]

33. Gurzawska, K.; Svava, R.; Jorgensen, N.R.; Gotfredsen, K. Nanocoating of titanium implant surfaces with organic molecules. Polysaccharides including glycosaminoglycans. *J. Biomed. Nanotechnol.* **2012**, *8*, 1012–1024. [CrossRef]
34. Gurzawska, K.; Svava, R.; Syberg, S.; Yihua, Y.; Haugshoj, K.B.; Damager, I.; Ulvskov, P.; Christensen, L.H.; Gotfredsen, K.; Jorgensen, N.R. Effect of nanocoating with rhamnogalacturonan-i on surface properties and osteoblasts response. *J. Biomed. Mater. Res. A* **2012**, *100*, 654–664. [CrossRef] [PubMed]
35. Kellesarian, S.; Malignaggi, V.; Kellesarian, T.; Bashir Ahmed, H.; Javed, F. Does incorporating collagen and chondroitin sulfate matrix in implant surfaces enhance osseointegration? A systematic review and meta-analysis. *IJOMS* **2018**, *47*, 241–251. [CrossRef] [PubMed]
36. Evered, D.E.; Whelan, J.E. *The Biology of Hyaluronan*; Wiley: Chichester, UK, 1989.
37. Laurent, T.C. *The Chemistry, Biology and Medical Applications of Hyaluronan and Its Derivative*; Portland Press: London, UK, 1998.
38. Abatangelo, G.; Weigel, P.H. New Frontiers in Medical Sciences: Redefining Hyaluronan. In Proceedings of the Symposium on New Frontiers in Medical Sciences, Padua, Italy, 17–19 June 1999; Elsevier: Oxford, UK, 2000.
39. Kennedy, J.F.; Balazs, E.A.F.; Phillips, G.O.E.; Williams, P.A.E. *Hyaluronan 2000: An International Meeting Celebrating the 80th Birthday of Endre a Balazs*; Woodhead: Cambridge, CA, USA, 2002.
40. Morra, M. Engineering of biomaterials surfaces by hyaluronan. *Biomacromolecules* **2005**, *6*, 1205–1223. [CrossRef] [PubMed]
41. Valachova, K.; Volpi, N.; Stern, R.; Soltes, L. Hyaluronan in medical practice. *Curr. Med. Chem.* **2016**, *23*, 3607–3617. [CrossRef] [PubMed]
42. Salwowska, N.M.; Bebenek, K.A.; Zadlo, D.A.; Wcislo-Dziadecka, D.L. Physiochemical properties and application of hyaluronic acid: A systematic review. *J. Cosmet. Dermatol.* **2016**, *15*, 520–526. [CrossRef] [PubMed]
43. Casale, M.; Moffa, A.; Vella, P.; Sabatino, L.; Capuano, F.; Salvinelli, B.; Lopez, M.A.; Carinci, F.; Salvinelli, F. Hyaluronic acid: Perspectives in dentistry. A systematic review. *Int. J. Immunopathol. Pharmacol.* **2016**, *29*, 572–582. [CrossRef] [PubMed]
44. Maurer, P.H.; Hudack, S.S. The isolation of hyaluronic acid from callus tissue of early healing. *Arch. Biochem. Biophys.* **1952**, *38*, 49–53. [CrossRef]
45. Iwata, H.; Urist, M.R. Hyaluronic acid production and removal during bone morphogenesis in implants of bone matrix in rats. *Clin. Orthop. Relat. Res.* **1973**, 236–245.
46. Bernard, G.W.; Pilloni, A.; Kang, M.; Sison, J.; Hunt, D.; Jovanovic, S. Osteogenesis in vitro and in vivo with hyaluronan and bone morphogenetic protein-2. In *New Frontiers in Medical Sciences: Redefining Hyaluronan*; Abatangelo, G., Weigel, P.H., Eds.; Elsevier: New York, NY, USA, 2000; pp. 215–231.
47. Zou, X.; Li, H.; Chen, L.; Baatrup, A.; Bunger, C.; Lind, M. Stimulation of porcine bone marrow stromal cells by hyaluronan, dexamethasone and rhbmp-2. *Biomaterials* **2004**, *25*, 5375–5385. [CrossRef]
48. Itoh, S.; Matubara, M.; Kawauchi, T.; Nakamura, H.; Yukitake, S.; Ichinose, S.; Shinomiya, K. Enhancement of bone ingrowth in a titanium fiber mesh implant by rhbmp-2 and hyaluronic acid. *J. Mater. Sci. Mater. Med.* **2001**, *12*, 575–581. [CrossRef]
49. Cho, B.C.; Park, J.W.; Baik, B.S.; Kwon, I.C.; Kim, I.S. The role of hyaluronic acid, chitosan, and calcium sulfate and their combined effect on early bony consolidation in distraction osteogenesis of a canine model. *J. Craniofac. Surg.* **2002**, *13*, 783–793. [CrossRef] [PubMed]
50. Solchaga, L.A.; Temenoff, J.S.; Gao, J.; Mikos, A.G.; Caplan, A.I.; Goldberg, V.M. Repair of osteochondral defects with hyaluronan- and polyester-based scaffolds. *Osteoarthr. Cartilage* **2005**, *13*, 297–309. [CrossRef]
51. Zhao, N.; Wang, X.; Qin, L.; Guo, Z.; Li, D. Effect of molecular weight and concentration of hyaluronan on cell proliferation and osteogenic differentiation in vitro. *Biochem. Biophys. Res. Commun.* **2015**, *465*, 569–574. [CrossRef] [PubMed]
52. Aebli, N.; Stich, H.; Schawalder, P.; Theis, J.C.; Krebs, J. Effects of bone morphogenetic protein-2 and hyaluronic acid on the osseointegration of hydroxyapatite-coated implants: An experimental study in sheep. *J. Biomed. Mater. Res. A* **2005**, *73*, 295–302. [CrossRef] [PubMed]
53. Morra, M.; Cassinelli, C.; Cascardo, G.; Fini, M.; Giavaresi, G.; Giardino, R. Covalently-linked hyaluronan promotes bone formation around ti implants in a rabbit model. *J. Orthop. Res.* **2009**, *27*, 657–663. [CrossRef]

54. Gomi, K.; Davies, J.E. Guided bone tissue elaboration by osteogenic cells in vitro. *J. Biomed. Mater. Res.* **1993**, *27*, 429–431. [CrossRef] [PubMed]
55. Lossdorfer, S.; Schwartz, Z.; Wang, L.; Lohmann, C.H.; Turner, J.D.; Wieland, M.; Cochran, D.L.; Boyan, B.D. Microrough implant surface topographies increase osteogenesis by reducing osteoclast formation and activity. *J. Biomed. Mater. Res. A* **2004**, *70*, 361–369. [CrossRef]
56. Morra, M.; Cassinelli, C.; Bruzzone, G.; Carpi, A.; Di Santi, G.; Giardino, R.; Fini, M. Surface chemistry effects of topographic modification of titanium dental implant surfaces: 1. Surface analysis. *Int. J. Oral Maxillofac. Implants* **2003**, *18*, 40–45.
57. Morra, M.; Cassinelli, C. Simple model for the xps analysis of polysaccharides coated surface. *Surf. Interface Anal.* **1998**, *26*, 742–747. [CrossRef]
58. Briggs, D.; Seah, M.P. Practical surface analysis. In *Auger and X-ray Photoelectron Spectroscopy*; Wiley: Chichester, UK, 1996.
59. Shard, A.; Davies, M.; Tendler, S.; Bennedetti, L.; Purbrick, M.; Paul, A.; Beamson, G. X-ray photoelectron spectroscopy and time-of-flight sims investigations of hyaluronic acid derivatives. *Langmuir* **1997**, *13*, 2808–2814. [CrossRef]
60. Goldstein, J. *Scanning Electron Microscopy and X-ray Microanalysis*; Plenum Press: New York, NY, USA, 2003.
61. Lausmaa, J. Surface spectroscopic characterization of titanium implant materials. *J. Electron. Spectros. Relat. Phenomena* **1996**, *81*, 343–361. [CrossRef]
62. Sawase, T.; Hai, K.; Yoshida, K.; Baba, K.; Hatada, R.; Atsuta, M. Spectroscopic studies of three osseointegrated implants. *J. Dent.* **1998**, *26*, 119–124. [CrossRef]
63. Kang, B.S.; Sul, Y.T.; Oh, S.J.; Lee, H.J.; Albrektsson, T. Xps, aes and sem analysis of recent dental implants. *Acta Biomater.* **2009**, *5*, 2222–2229. [CrossRef] [PubMed]
64. Cassinelli, C.; Morra, M.; Pavesio, A.; Renier, D. Evaluation of interfacial properties of hyaluronan coated poly(methylmethacrylate) intraocular lenses. *J. Biomater. Sci. Polym. Ed.* **2000**, *11*, 961–977. [CrossRef]
65. Sverzut, A.T.; Crippa, G.E.; Morra, M.; de Oliveira, P.T.; Beloti, M.M.; Rosa, A.L. Effects of type i collagen coating on titanium osseointegration: Histomorphometric, cellular and molecular analyses. *Biomed. Mater.* **2012**, *7*, 035007. [CrossRef] [PubMed]
66. Sartori, M.; Giavaresi, G.; Parrilli, A.; Ferrari, A.; Aldini, N.N.; Morra, M.; Cassinelli, C.; Bollati, D.; Fini, M. Collagen type i coating stimulates bone regeneration and osseointegration of titanium implants in the osteopenic rat. *Int. Orthop.* **2015**, *39*, 2041–2052. [CrossRef]
67. Litwiniuk, M.; Krejner, A.; Speyrer, M.S.; Gauto, A.R.; Grzela, T. Hyaluronic acid in inflammation and tissue regeneration. *Wounds* **2016**, *28*, 78–88.
68. Maytin, E.V. Hyaluronan: More than just a wrinkle filler. *Glycobiology* **2016**, *26*, 553–559. [CrossRef]
69. Zhao, N.; Wang, X.; Qin, L.; Zhai, M.; Yuan, J.; Chen, J.; Li, D. Effect of hyaluronic acid in bone formation and its applications in dentistry. *J. Biomed. Mater. Res. A* **2016**, *104*, 1560–1569. [CrossRef]
70. Gurgel, B.C.V.; Montenegro, S.C.L.; Dantas, P.M.C.; Pascoal, A.L.B.; Lima, K.C.; Calderon, P.D.S. Frequency of peri-implant diseases and associated factors. *Clin. Oral Implants Res.* **2017**, *28*, 1211–1217. [CrossRef]
71. Morra, M.; Cassineli, C. Non-fouling properties of polysaccharide-coated surfaces. *J. Biomater. Sci. Polym. Ed.* **1999**, *10*, 1107–1124. [CrossRef] [PubMed]
72. Davies, J.E. *Bone Engineering*; University of Toronto Press: Toronto, ON, Canada, 2000.
73. Rizzo, S.; Zampetti, P.; Rodriguez, Y.B.R.; Svanosio, D.; Lupi, S.M. Retrospective analysis of 521 endosseous implants placed under antibiotic prophylaxis and review of literature. *Minerva. Stomatol.* **2010**, *59*, 75–88. [PubMed]
74. Lupi, S.M.; Rodriguez, Y.B.A.; Cervino, G.; Todaro, C.; Rizzo, S. Long-term effects of acute myeloid leukemia treatment on the oral system in a pediatric patient. *Open Dent. J.* **2018**, *12*, 230–237. [CrossRef] [PubMed]
75. Rodriguez y Baena, R.; Pastorino, R.; Gherlone, E.F.; Perillo, L.; Lupi, S.M.; Lucchese, A. Histomorphometric evaluation of two different bone substitutes in sinus augmentation procedures: A randomized controlled trial in humans. *Int. J. Oral Maxillofac. Implants* **2017**, *32*, 188–194. [CrossRef] [PubMed]
76. Lupi, S.M.; Cislaghi, M.; Rizzo, S.; Rodriguez, Y.B.R. Rehabilitation with implant-retained removable dentures and its effects on perioral aesthetics: A prospective cohort study. *Clin. Cosmet. Investig. Dent.* **2016**, *8*, 105–110. [PubMed]

77. RR, Y.B.; Lupi, S.M.; Pastorino, R.; Maiorana, C.; Lucchese, A.; Rizzo, S. Radiographic evaluation of regenerated bone following poly(lactic-co-glycolic) acid/hydroxyapatite and deproteinized bovine bone graft in sinus lifting. *J. Craniofac. Surg.* **2013**, *24*, 845–848.
78. Rodriguez, Y.B.R.; D'Aquino, R.; Graziano, A.; Trovato, L.; Aloise, A.C.; Ceccarelli, G.; Cusella, G.; Pelegrine, A.A.; Lupi, S.M. Autologous periosteum-derived micrografts and plga/ha enhance the bone formation in sinus lift augmentation. *Front. Cell Dev. Biol.* **2017**, *5*, 87. [CrossRef] [PubMed]
79. Ceccarelli, G.; Presta, R.; Lupi, S.M.; Giarratana, N.; Bloise, N.; Benedetti, L.; Cusella De Angelis, M.G.; Rodriguez, Y.B.R. Evaluation of poly(lactic-co-glycolic) acid alone or in combination with hydroxyapatite on human-periosteal cells bone differentiation and in sinus lift treatment. *Molecules* **2017**, *22*. [CrossRef] [PubMed]
80. Lupi, S.M.; Rodriguez, Y.B.A.; Todaro, C.; Ceccarelli, G.; Rodriguez, Y.B.R. Maxillary sinus lift using autologous periosteal micrografts: A new regenerative approach and a case report of a 3-year follow-up. *Case Rep. Dent.* **2018**, *2018*, 3023096. [CrossRef]
81. Lupi, S.M.; Granati, M.; Butera, A.; Collesano, V.; Rodriguez, Y.B.R. Air-abrasive debridement with glycine powder versus manual debridement and chlorhexidine administration for the maintenance of peri-implant health status: A six-month randomized clinical trial. *Int. J. Dent. Hyg.* **2017**, *15*, 287–294. [CrossRef]
82. Mumcu, E.; Bilhan, H.; Cekici, A. Marginal bone loss around implants supporting fixed restorations. *J. Oral Implantol.* **2011**, *37*, 549–558. [CrossRef] [PubMed]
83. Julious, S.A. Sample sizes for clinical trials with normal data. *Stat. Med.* **2004**, *23*, 1921–1986. [CrossRef] [PubMed]

 © 2019 by the authors. Licensee MDPI, Basel, Switzerland. This article is an open access article distributed under the terms and conditions of the Creative Commons Attribution (CC BY) license (http://creativecommons.org/licenses/by/4.0/).

Article

Potential for Drug Repositioning of Midazolam for Dentin Regeneration

Takeo Karakida [1], Kazuo Onuma [2], Mari M. Saito [1], Ryuji Yamamoto [1], Toshie Chiba [3], Risako Chiba [1], Yukihiko Hidaka [4], Keiko Fujii-Abe [4], Hiroshi Kawahara [4] and Yasuo Yamakoshi [1,*]

[1] Department of Biochemistry and Molecular Biology, School of Dental Medicine, Tsurumi University, 2-1-3 Tsurumi, Tsurumi-ku, Yokohama 230-8501, Japan; karakida-t@tsurumi-u.ac.jp (T.K.); saito-mari@tsurumi-u.ac.jp (M.M.S.); yamamoto-rj@tsurumi-u.ac.jp (R.Y.); chiba-r@tsurumi-u.ac.jp (R.C.)

[2] National Institute of Advanced Industrial Science & Technology, Central 6, 1-1-1 Higashi, Tsukuba, Ibaraki 305-8566, Japan; k.onuma@aist.go.jp

[3] Research Center of Electron Microscopy, School of Dental Medicine, Tsurumi University, 2-1-3 Tsurumi, Tsurumi-ku, Yokohama 230-8501, Japan; chiba-t@tsurumi-u.ac.jp

[4] Department of Dental Anesthesiology, School of Dental Medicine, Tsurumi University, 2-1-3 Tsurumi, Tsurumi-ku, Yokohama 230-8501, Japan; 2911002@stu.tsurumi-u.ac.jp (Y.H.); fujii-keiko@tsurumi-u.ac.jp (K.F.-A.); kawahara-h@tsurumi-u.ac.jp (H.K.)

* Correspondence: yamakoshi-y@tsurumi-u.ac.jp; Tel.: +81-45-580-8479; Fax: +81-45-573-9599

Received: 13 December 2018; Accepted: 31 January 2019; Published: 4 February 2019

Abstract: Drug repositioning promises the advantages of reducing costs and expediting approval schedules. An induction of the anesthetic and sedative drug; midazolam (MDZ), regulates inhibitory neurotransmitters in the vertebrate nervous system. In this study we show the potential for drug repositioning of MDZ for dentin regeneration. A porcine dental pulp-derived cell line (PPU-7) that we established was cultured in MDZ-only, the combination of MDZ with bone morphogenetic protein 2, and the combination of MDZ with transforming growth factor-beta 1. The differentiation of PPU-7 into odontoblasts was investigated at the cell biological and genetic level. Mineralized nodules formed in PPU-7 were characterized at the protein and crystal engineering levels. The MDZ-only treatment enhanced the alkaline phosphatase activity and mRNA levels of odontoblast differentiation marker genes, and precipitated nodule formation containing a dentin-specific protein (dentin phosphoprotein). The nodules consisted of randomly oriented hydroxyapatite nanorods and nanoparticles. The morphology, orientation, and chemical composition of the hydroxyapatite crystals were similar to those of hydroxyapatite that had transformed from amorphous calcium phosphate nanoparticles, as well as the hydroxyapatite in human molar dentin. Our investigation showed that a combination of MDZ and PPU-7 cells possesses high potential of drug repositioning for dentin regeneration.

Keywords: drug repositioning; cell; dentin; hydroxyapatite; nanorod; nanoparticle

1. Introduction

Tissue engineering is a multidisciplinary science. Applications of tissue engineering are founded on three components; a cell source, a scaffold, and bioactive molecules. In the field of dental tissue engineering, various soft and hard dental tissues have been regenerated in vitro using stem cells [1]. Dental pulp stem cells (DPSCs) have been isolated with various techniques and used for studies related to the cell differentiation potential and scaffolding for tissue regeneration [2,3]. DPSCs can differentiate into multiple cell types, including odontogenic and osteogenic cells, osteocytes, chondrocytes, vascular cells, neurons, and hepatocytes. In addition, DPSCs are used to generate induced pluripotent stem (iPS) cells.

In the dental research field, dental pulp-derived cells that express alkaline phosphatase (ALP) and odontoblastic marker genes, and that form precipitated nodules when cultured in a medium containing β-glycerophosphate and ascorbic acid, are generally assumed to be differentiated odontoblast-like cells. A number of odontoblastic cell lines have been established from rat [4,5], mouse [6], cattle [7], pig [8], and human [9,10] dental pulp tissues. Odontoblastic differentiation is induced by many factors, which are associated with ectoderm–mesenchyme molecular interactions. Various factors participate in the regulation of dental pulp cell differentiation, such as bone morphogenetic protein (BMP) [11,12], fibroblast growth factors (FGF) [13], and transforming growth factor beta (TGF-β) [14,15].

Drug repositioning is when new therapeutic applications are identified for existing drugs. In addition to the studies on dental pulp cells expressing cytokines, such as BMP, FGF, and TGF-β, existing drugs for treating Alzheimer's disease have been reported to promote dentin regeneration [16]. In Japan, Alzheimer's drugs such as donepezil hydrochloride (Aricept) have been widely used as acetylcholinesterase inhibitors that prevents acetylcholine, an excitatory neurotransmitter in the vertebrate nervous system. This study focuses on repositioning midazolam (MDZ), which controls gamma-aminobutyric acid (GABA), the principal inhibitory neurotransmitter in the mammalian central nervous system. Midazolam (MDZ) is a chemically synthesized imidazobenzodiazepine derivative that possesses pharmacological effects as a hypnotic, sedative, and anesthetic, has anti-anxiety and anticonvulsant properties, and acts as muscle relaxant [17]. In the dental field, MDZ has been used mainly as a sedative prior to dental anesthesia. The intravenous MDZ formulation has been recommended as a first-line drug for treating convulsive status epilepticus [18]. Intravenous MDZ has not been approved to treat status epilepticus in most countries, but it has been used off-label for patients in Japan. In tumor and cancer cells, MDZ induces cellular apoptosis by regulating the caspase pathways, endoplasmic reticulum stress, autophagy, and the cell cycle [19–22]. Few studies have focused on the effect of MDZ on dental pulp cells, although an in vitro study using human mesenchymal stem cells (hMSCs) has shown that MDZ inhibits ALP activity and calcium deposition in hMSCs, suggesting a suppressive effect on osteogenic differentiation [23]. In this study, using dental pulp cells, we examine the potential for drug repositioning of MDZ for dentin regeneration at the cell biological, genetic, protein, and crystal engineering levels.

2. Results

2.1. Combined Effect of MDZ with BMP2 or TGF-β1 on Differentiation of the PPU-7 Cell Line

MDZ is a short-acting benzodiazepine derivative with an imidazole structure and a molecular weight of 325.77 g/mol (Figure 1a). To investigate the effect of MDZ on dentin regeneration, we used a porcine animal model to establish dental pulp-derived cell lines (PPU) and ultimately selected the PPU-7 cell line for the present study (see Figures S1–S3). Because alkaline phosphatase (ALP) is considered the initial marker for the differentiation of mesenchymal cells into hard tissue-forming cells such as osteoblasts or odontoblasts [24,25], we investigated the effects of MDZ on ALP activity in the PPU-7 cell line by using a quantitative colorimetric method with a p-nitrophenylphosphate as the substrate and examined the combination of MDZ (0-10 µM) with a recombinant human BMP2 (rhBMP2) (500 ng/mL) or a recombinant human TGF-β1 (rhTGF-β1) (1 ng/mL) (Figure 1b). When the ALP activity level of the control (i.e., 0 µM MDZ, 0 ng/mL rhBMP2, 0 ng/mL rhTGF-β1) was set at 1.0, the addition of MDZ-only significantly enhanced ALP activity in PPU-7 cells in a concentration-dependent manner, especially the ALP activity at 10 µM MDZ (i.e., 10 µM MDZ, 0 ng/mL rhBMP2, 0 ng/mL rhTGF-β1), which was 1.75-fold higher than the control. BMP2-only (i.e., 0 µM MDZ, 500 ng/mL rhBMP2, 0 ng/mL rhTGF-β1) and TGF-β1-only (i.e., 0 µM MDZ, 0 ng/mL rhBMP2, 1 ng/mL rhTGF-β1) also slightly or significantly enhanced the ALP activity. The combination of rhBMP2 or rhTGF-β1 with MDZ (MDZ and BMP2 or MDZ and TGF-β1) (5 or 10 µM) slightly increased ALP activity (1.14–1.27-fold for MDZ and BMP2, and 1.28–1.29-fold for MDZ and TGF-β1) compared to the control. ALP staining for the mineral-induced PPU-7 cells displayed blue colored staining images (Figure 1c).

The cells were densely distributed on the plate of MDZ-only, whereas the combinations of MDZ and BMP2 as well as MDZ and TGF-β1 displayed a low density.

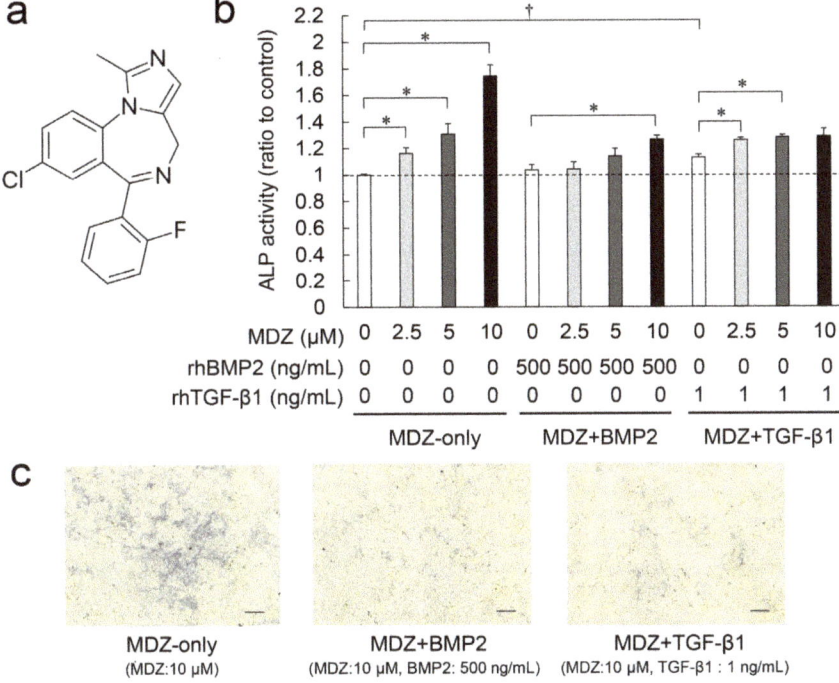

Figure 1. Combined effect of midazolam (MDZ) with bone morphogenetic protein 2 (BMP2) or transforming growth factor beta 1 (TGF-β1) on alkaline phosphatase (ALP) activity in the porcine dental pulp-derived (PPU-7) cell line. (**a**) Structure of MDZ. (**b**) ALP-inducing activity of midazolam without BMP2 or TGF-β1 (MDZ-only), midazolam with BMP2 (MDZ and BMP2) and midazolam with TGF-β1 (MDZ and TGF-β1) in PPU-7 cells. ALP activities are indicated as increasing or decreasing ratios relative to the level of the control (i.e., 0 μM MDZ, 0 ng/mL BMP2, 0 ng/mL TGF-β1), which was set at 1 (dotted line). Values are the means ± standard error of six culture wells. Significant differences are indicated by an asterisk (* $p < 0.05$, Steel's test) or a dagger († $p < 0.05$, Mann–Whitney U-test). (**c**) ALP staining for PPU-7 cells cultured with MDZ-only, MDZ and BMP2, MDZ and TGF-β1 (Scale bar: 200 μm).

2.2. Effect of MDZ on Temporal Changes in Gene Expression of PPU-7 Cell Line

Since the MDZ-only treatment was more effective in enhancing ALP activity and inducing mineralization in PPU-7 cells than the combination treatment of MDZ with BMP2 or TGF-β1, we investigated the effect of MDZ-only on gene expression in the PPU-7 cell line. The gene expression of a panel of odontoblastic, osteoblastic and chondrocytic markers in PPU-7 cells at one and seven days after MDZ treatment was analyzed by quantitative polymerase chain reaction (qPCR) (Figure 2). For odontoblastic markers (Figure 2a), we quantified the mRNA expression levels of matrix metalloprotease 2 (*MMP2*) and two products from the full-length *DSPP* transcript: a segment containing the dentin glycoprotein and dentin phosphoprotein (DGP and DPP) coding region (*DSPP-v1*) and a smaller segment specific for the dentin sialoprotein (DSP)-only transcript (*DSPP-v2*). At day one, the expression levels of *MMP2*, *DSPP-v1*, and *DSPP-v2* in cells cultured with MDZ were not significantly different from those in cells cultured without MDZ. In contrast, at day seven, the three mRNA expression levels in cells cultured with MDZ were significantly higher (3.8-fold for *MMP2*, 1.67-fold for *DSPP-v1*, and 2.2-fold for *DSPP-v2*) than those in cells cultured without MDZ. We also amplified osteocalcin (OC) and runt-related transcription factor 2 (*RUNX2*) as osteoblastic markers (Figure 2b) and type II collagen (Col II) and aggrecan (*ACAN*) as chondrocytic markers (Figure 2c). There were no significant differences in the four mRNA expression levels between cells cultured with and without MDZ at either day one or day seven. We interpret these findings as evidence that MDZ is required for the differentiation of PPU-7 cells into odontoblasts.

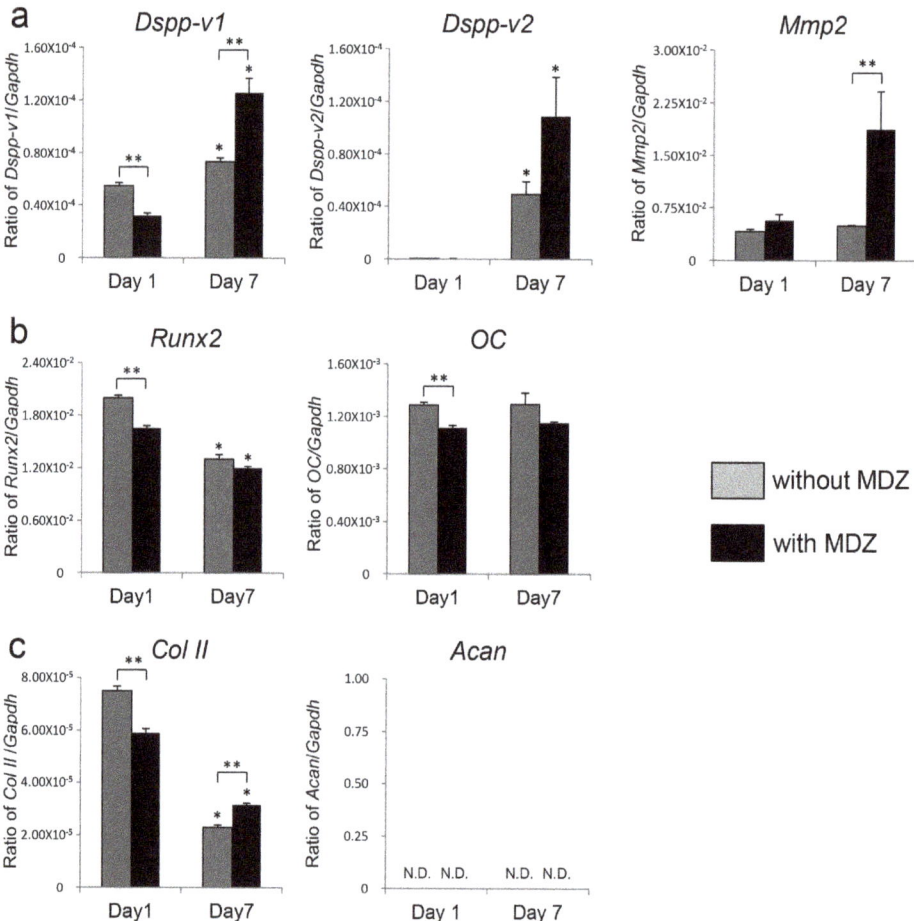

Figure 2. Effect of MDZ on temporal changes in gene expression of the PPU-7 cell line. The mRNA expression by quantitative polymerase chain reaction (qPCR) analysis of (**a**) odontoblastic differentiation markers, i.e., *DSPP*-variant 1 (*DSPP-v1*), *DSPP*-variant 2 (*DSPP-v2*) and matrix metalloprotease 2 (*MMP2*); (**b**) osteoblastic differentiation markers, i.e., osteocalcin (OC) and runt-related transcription factor 2 (*RUNX2*); and (**c**) chondrogenic differentiation markers, i.e., type II collagen (Col II) and aggrecan (*ACAN*). Each ratio was normalized to glyceraldehyde-3-phosphate dehydrogenase (*GAPDH*) as a reference gene, and the relative quantification data of *DSPP-v1*, *DSPP-v2*, *MMP2*, OC, *RUNX2*, Col II and *ACAN* in PPU-7 cell line were generated on the basis of a mathematical model for relative quantification in a qPCR system. Values are the means ± standard error of six culture wells. The asterisk (*) on the bar graph indicates a significant difference ($p < 0.05$, Steel's test) between day one and day seven. The double asterisk (**) on the bar graph indicates a significant difference ($p < 0.05$, Mann–Whitney U-test) between cells cultured with and without MDZ.

2.3. Effect of MDZ on Mineralization Induction of the PPU-7 Cell Line

To obtain additional information about the effect of MDZ on mineralization inducibility, we cultured PPU-7 cells in a mineralization-inducing culture medium (Figure 3). The nodule formation and mineralization capacities of the cells were assessed with both Alizarin Red S and von Kossa staining (Figure 3a). At seven days following mineralization induction, in contrast to that of the cells not subjected to mineralization induction, the plate of the cells cultured in mineralization-inducing culture medium clearly displayed precipitated nodules by both staining methods, regardless of the addition of MDZ.

We also quantitatively analyzed the calcium content in PPU-7 cells (Figure 3b). At five days following mineralization induction, relative to the control cells without MDZ (i.e., no MDZ), the cells administered with MDZ-only displayed a dramatically increased amount of calcium deposition (approximately 2.0-fold).

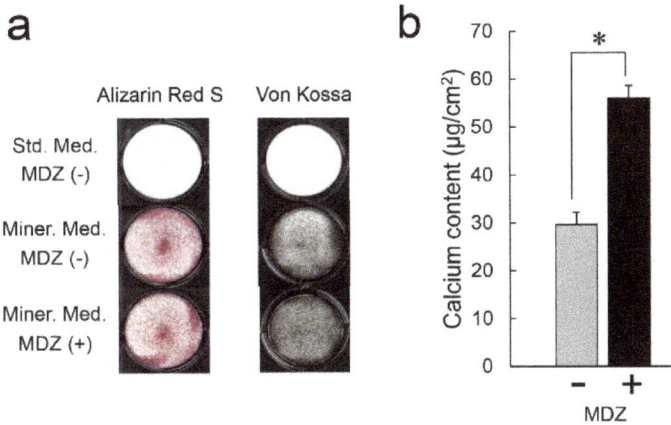

Figure 3. Effect of MDZ on nodule formation in the PPU-7 cell line. Nodule cultures were stained with (**a**) Alizarin Red S (left) and von Kossa (right) staining on day seven. In contrast to PPU-7 cells not subjected to mineralization induction, PPU-7 cells cultured in mineralization-inducing culture media clearly exhibited nodule formation regardless of the addition of MDZ. (**b**) Calcium contents in PPU-7 cells were determined on day five after the mineralization induction. Values are the means ± standard error of six culture wells. The asterisk (*) on the bar graph indicates a significant difference ($p < 0.05$, Mann–Whitney U-test) between the cells incubated with and without MDZ. Std. Med.: Standard culture medium, Miner. Med.: Mineralization-inducing culture medium.

2.4. Detection of DPP in Precipitated Nodules from PPU-7 Cells

We attempted to detect a dentin-specific protein, dentin phosphoprotein (DPP), in precipitated nodules induced by MDZ treatment in the PPU-7 cell line at the protein level. The precipitated nodules in PPU-7 cells were subjected to a series of three extractions: With a tris-guanidine buffer (G1), with hydrochloric acid–formic acid solution (HF), and with tris-guanidine buffer again (G2). These extractions yielded three fractions, the G1, HF, and G2 extracts, which were analyzed by SDS-polyacrylamide gel electrophoresis (SDS-PAGE) stained with SimplyBlue SafeStain (Figure 4a) and Stains-all stain (Figure 4b). In the G2 fraction obtained from PPU-7 cells incubated in mineralization-inducing culture medium, Stains-all-positive DPP doublet bands were observed at approximately 100 kDa by SDS-PAGE (Figure 4b, lanes 8 and 9). The addition of MDZ increased the intensity of the DPP doublets bands (Figure 4b, lane 8), whereas no DPP bands were detected in the extracts from PPU-7 cells cultured with the standard medium (Figure 4b, lane 7).

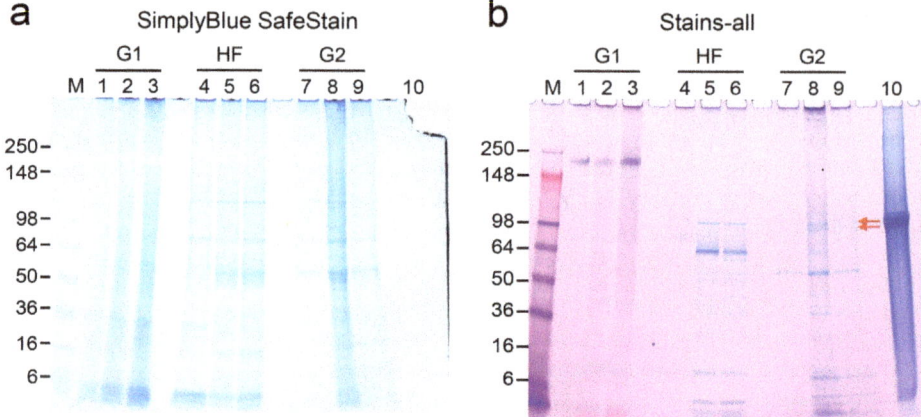

Figure 4. Sequential extraction of proteins from precipitated nodules induced by PPU-7 cells. Precipitated nodules formed in PPU-7 cells cultured in mineralization-inducing culture medium with (MDZ(+)) or without (MDZ(−)) MDZ were sequentially extracted by tris-guanidine (G1 extraction), formic acid–HCl (HF extraction) and tris-guanidine again (G2 extraction). The PPU-7 cells cultured with only standard medium were also extracted as a control. Analysis of the G1, HF, and G2 extracts by 5–20% gradient SDS-PAGE were stained with (**a**) SimplyBlue SafeStain and (**b**) Stains-all stain. Dentin phosphoprotein (DPP) doublet bands in lanes 5, 6, 8, 9, and 10 detected by Stains-all staining are indicated with a red arrow. Lanes 1, 4 and 7: Control; lanes 2, 5 and 8: MDZ(+); lanes 3, 6 and 9: MDZ(-); lane 10: DPP purified from porcine dentin. M: Molecular weight marker (SeeBlue Plus2 Pre-Stained standard).

2.5. X-ray Diffraction (XRD) Patterns of Precipitated Nodules

Based on all the above results, we further characterized the precipitated nodules from PPU-7 cells at the crystal engineering level. Figure 5a shows microbeam XRD patterns for samples grown with (magenta curve) and without (blue curve) MDZ in the culture solution. The pattern for a cell sheet (green curve) is shown only for reference. The cell sheet exhibited a broad intense peak at approximately $2\theta = 10°$, which tailed to higher 2θ values. Peaks attributed to the precipitates appeared at $2\theta = 26.0, 28.7,$ and $32.0°$ (black circles); they were independent of the presence or absence of MDZ. The entire XRD pattern of the precipitates from the solution with MDZ was the same as that of the precipitates from the solution without MDZ, signifying that MDZ did not affect the kind of phase precipitated. The peak widths in the precipitate patterns were broad, indicating low crystallinity and/or small crystallites. Figure 5b–e shows ideal XRD patterns for dicalcium phosphate dihydrate ($CaHPO_4·2H_2O$; DCPD) (b), octacalcium phosphate ($Ca_8(HPO_4)_2(PO_4)_4·5H_2O$; OCP) (c), β-tricalcium phosphate (β-$Ca_3(PO_4)_2$; β-TCP) (d), and hydroxyapatite ($Ca_{10}(PO_4)_6(OH)_2$; HAP) (e), which were derived using the 2θ versus diffraction-intensity relationships in the corresponding Joint Committee on Powder Diffraction Standards (JCPDS) cards. The asterisks indicate the particular peaks used to identify each calcium phosphate phase. The XRD patterns of the precipitates did not show these particular peaks; however, the profile from $2\theta = 25$ to $35°$ resembled those for OCP and HAP.

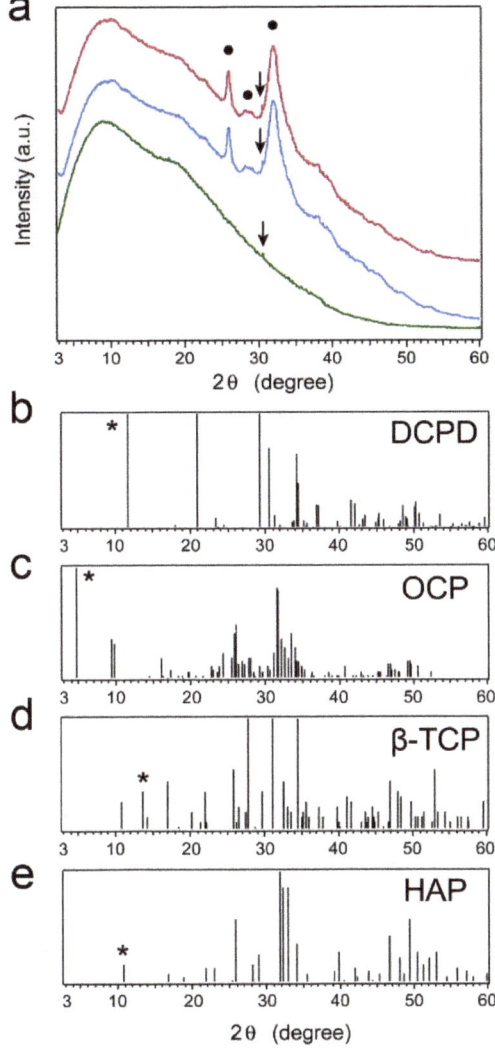

Figure 5. (**a**) Microbeam XRD patterns for samples grown with (magenta curve) and without (blue curve) MDZ in culture solution and for a cell sheet (green curve) for reference. Peaks attributed to the precipitates are indicated by black circles; spike peaks attributed to damage to the imaging plate are indicated by arrows. Ideal XRD patterns for (**b**) dicalcium phosphate dihydrate (DCPD), (**c**) octacalcium phosphate (OCP), (**d**) β-tricalcium phosphate (β-TCP), and (**e**) hydroxyapatite (HAP) derived using 2θ versus diffraction–intensity relationships in the corresponding JCPDS cards. The peaks used to identify each calcium phosphate phase are indicated by asterisks.

2.6. Scanning Electron Microscopy (SEM) Observations of Precipitated Nodules

A macroscopic SEM image of a cross-sectional sample (Figure 6a) showed that the morphology of the precipitates was not uniform. Both irregularly (white arrow) and regularly shaped precipitates (i.e., μm-scale precipitates with a ball-like appearance, yellow arrows) were observed. A magnified image of a region with these ball-like precipitates (Figure 6b) showed that they consisted of aggregated nanoparticles. The size of each nanoparticle was approximately 50 nm or less. The isotropic morphology of the elementary particles made it difficult to ascertain the phase of the calcium phosphate, so the precipitates were observed using TEM.

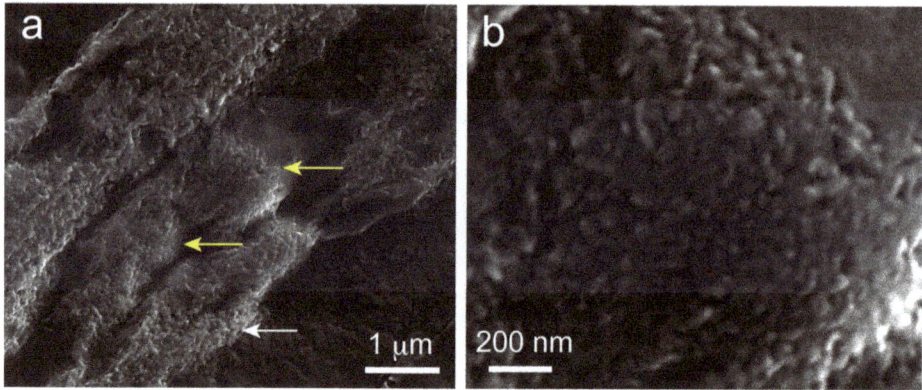

Figure 6. SEM images of a cross-sectional sample. (**a**) Low-magnification image showing irregularly shaped precipitates (white arrow) and ball-like ones (yellow arrows). (**b**) High-magnification images of ball-like precipitate showing nanoparticles sized approximately 50 nm.

2.7. Transmission Electron Microscopy (TEM) Observation and Selected-Area Electron Diffraction (SAED) Measurement of Precipitated Nodules in a Wide Area

A macroscopic TEM image of a microtome-cut section (Figure 7a) showed an irregularly shaped powder-like precipitate. A higher magnification image (Figure 7b) of the precipitate revealed that it consisted of nanorods (light blue arrows) and bulky materials (black arrows). These nanorods were not observed in the SEM image, probably because of their limited width (less than 10 nm).

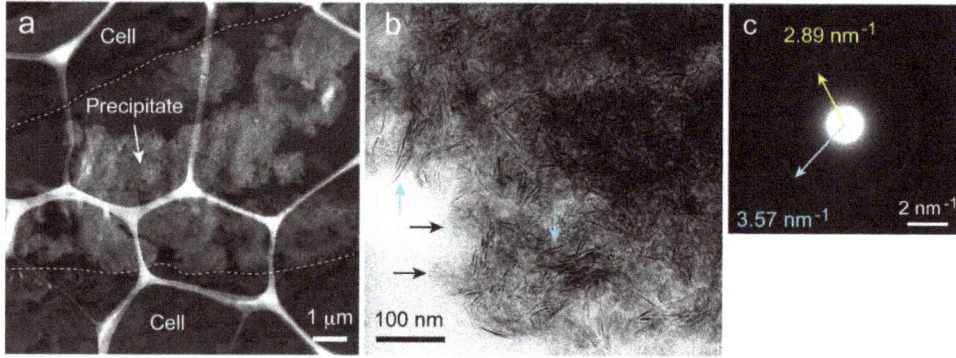

Figure 7. Wide-area TEM images of precipitates. (**a**) Low-magnification image of a microtome-cut sample. The dotted line shows the boundary between the cell sheet and precipitate region. (**b**) High-magnification image of precipitates. Many nanorods (light blue arrows) and bulky materials (black arrows) were observed. (**c**) Selected-Area Electron Diffraction (SAED) pattern measured at 800 nm ϕ. Two Debye rings (low and high intensities) were observed.

The Selected-Area Electron Diffraction (SAED) pattern of the precipitates measured at 800 nm ϕ showed two Debye rings (Figure 7c), indicating that the crystals in the precipitates were randomly oriented. The rings were 2.89 nm^{-1} (low-intensity ring) and 3.57 nm^{-1} (high-intensity ring) from the center. These characteristic distances were converted to interplanar distances, d, of 0.346 and 0.280 nm, respectively. For HAP, the diffraction plane corresponding to $d = 0.346$ nm (with a margin of error of <1.5%) was {002}. For OCP, there were several corresponding planes, such as {121} and {5$\bar{2}$0}. β-TCP also had several corresponding planes, including {1010} and {01$\bar{1}$0}. DCPD, however, had no diffraction plane corresponding to a d of 0.346 nm.

The d of 0.280 nm corresponded to the {211}, {121}, and {112} diffraction planes for HAP, {511}, {7$\bar{1}$0}, etc. for OCP, and {002} for the DCPD. β-TCP had no diffraction plane corresponding to a d of 0.280 nm.

The relative intensity of the diffraction plane in each calcium phosphate crystal provided important information for identifying the calcium phosphate phase when the orientations of the crystals were random. The relative intensities of the {002}, {211}, {121}, and {112} diffraction planes of HAP were 49, 100, 43, and 70%, respectively. The diffraction intensities of each plane of the OCP corresponding to d = 0.346 or 0.280 nm were all less than 35%. Similarly, low diffraction intensities were also observed for the plane corresponding to d = 0.346 nm for β-TCP and the plane corresponding to d = 0.280 nm for DCPD. This result strongly suggests that the precipitates were HAP.

Since high-resolution TEM (HR-TEM) observation of the precipitates was difficult due to the overlapping crystals, a lattice image of each crystal in the precipitates was observed using a crushed sample.

2.8. High-Resolution TEM (HR-TEM) Observation and Fast Fourier Transform (FFT) Image Analysis of Precipitated Nodules in a Narrow Area

Figure 8a shows a magnified image of nanorods, a major component of the precipitates. Randomly oriented nanorods (white arrows) with a width less than 10 nm were identified, and lattice images were observed for each rod.

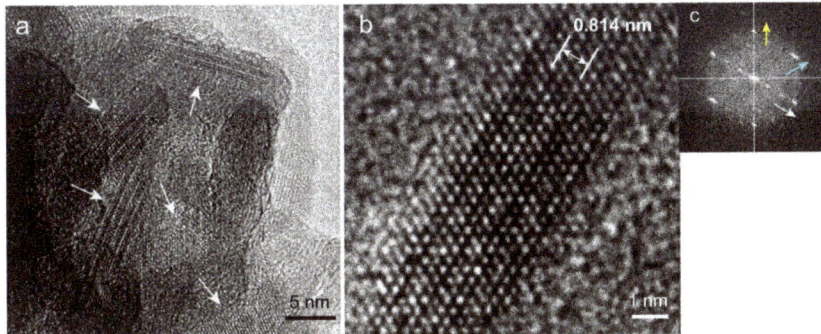

Figure 8. Narrow-area TEM images of nanorods. (**a**) Aggregation of nanorods. The orientation of each rod was random, and lattice fringes were observed in the rods. (**b**) HR-TEM image of a rod. The lattice image of the rod is visible. (**c**) FFT image of a rod. Three directions of the crystal planes are indicated by white, blue, and yellow arrows. The interplanar distance corresponding to the direction of the white arrow (0.814 nm) is superimposed in (**b**).

HR-TEM observation of a nanorod (Figure 8b) revealed a lattice pattern corresponding to the atomic arrangement. An FFT image of a nanorod (Figure 8c) revealed three periodic directions corresponding to d = 0.814, 0.277, and 0.276 nm (white, blue, and yellow arrows, respectively). The d of 0.814 nm corresponded to the {26} plane for HAP and the {012}, {10$\bar{2}$}, and {1$\bar{1}$2} planes for β-TCP with a margin of error of <1.5%. In contrast, OCP and DCPD had no planes corresponding to d = 0.814 nm. The d values of 0.277 and 0.276 nm corresponded with many diffraction planes for HAP, β-TCP, OCP, and DCPD. For HAP, the corresponding planes were {112}, {2$\bar{1}$2}, and {$\bar{2}$12}; for β-TCP, {128}, {3$\bar{2}$8}, etc.; for OCP, {601}, {$\bar{2}$22}, etc., and for DCPD, the d of 0.277 nm corresponded to {002}.

The angle of intersection between the directions of the white and blue arrow was 59.0°, and between the directions of the white and yellow arrows was 120.3°. Although the d-value analysis indicated the possibility that the nanorod was HAP or β-TCP, comparison of the measured intersection angles with theoretical ones indicated that the nanorod was HAP (Table S3) The measured d values and intersection angles were consistent with the theoretical values with a margin of error of <1.5%. The directions of the white, blue, and yellow arrows corresponded to [100], [112], and [$\bar{2}$12] for HAP. The image in Figure 8b therefore shows the HAP lattice viewed from the [02$\bar{1}$] zone axis.

We performed similar observations and analyses for ten different nanorods in several areas and found that all of them were HAP.

Figure 9a shows a high-magnification TEM image corresponding to the bulky materials observed in Figure 7b. They consisted of nanoparticles (white arrows) with a diameter less than approximately 10 nm. This observation means that the cell-induced precipitates consisted of a mixture of nanorods and nanoparticles. The 50 nm particles observed by SEM (Figure 6) therefore were secondary particles consisting of smaller nanoparticles.

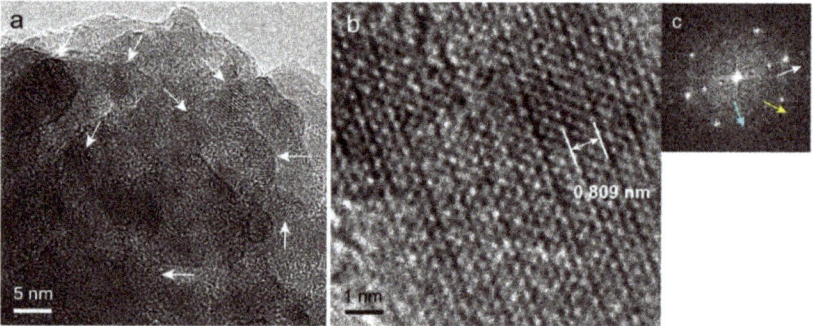

Figure 9. Narrow-area TEM images of bulky precipitates. (**a**) The precipitates consisted of nanoparticles sized less than approximately10 nm. (**b**) HR-TEM image of a nanoparticle. (**c**) FFT image of a nanoparticle. Three directions of crystal planes are indicated by white, blue, and yellow arrows. The interplanar distance corresponding to the direction of the white arrow (0.809 nm) is superimposed in (**b**).

A HR-TEM image of a nanoparticle (Figure 9b) revealed a lattice arrangement, and a FFT image of nanoparticles (Figure 9c) revealed at least three periodic directions. The d values corresponding to each direction (indicated by white, blue, and yellow arrows) were 0.809, 0.385, and 0.285 nm, respectively. The d of 0.809 nm corresponded to [26] for HAP, {012}, {10$\bar{2}$}, and {1$\bar{1}$2} for β-TCP with a margin of error of <1.5%. OCP and DCPD had no planes corresponding to d = 0.809. The d of 0.385 nm corresponded to {1$\bar{2}$1}, {111}, and {$\bar{2}$11} for HAP with a margin of error of <1.5%. OCP corresponded to several planes, such as {1$\bar{2}$1} and {2$\bar{2}$1}. β-TCP and DCPD had no corresponding planes. The d of 0.281 nm corresponded to many planes for both HAP and OCP: {3$\bar{2}$1}, {21$\bar{1}$}, etc. for HAP, and {130}, {420}, etc. for OCP. DCPD corresponded to {002}, whereas β-TCP had no corresponding plane. The d-value analysis indicated that only HAP had planes corresponding to three directions. The intersection angle between the directions of the white (d = 0.809 nm) and blue (d = 0.385 nm) arrows was 90.7°, and between the directions of the white and yellow (d = 0.281 nm) arrows was 46.9°. These angles corresponded to those between [100] and [1$\bar{2}$1] and between [100] and [3$\bar{2}$1] for HAP, respectively. The measured d values and intersection angles between each plane were consistent with the theoretical values (with a margin of error <1.5%; see Table S3). The image in Figure 9b therefore corresponded to the HAP lattice viewed from the [012] zone axis. We performed similar observations and analyses for ten different nanoparticles in several areas and found that all of them were HAP.

2.9. Element Content in Precipitated Nodules Measured Using Scanning TEM Energy-Dispersive X-ray Spectroscopy (STEM-EDS)

Figure 10 shows a low-magnification (approximately 12 µm square) TEM image (a) and two-dimensional elemental mappings of Ca (b), P (c), and N (d) in the area in (a). Ca and P were observed in only the precipitates, and N was mostly observed in the cell region. A small amount of N was detected in the precipitate region; it was probably due to cell contamination. The energy-dispersive X-ray spectroscopy (EDS) spectrum (Figure 10e, magenta curve) corresponded to the region in Figure 10a where the average Ca/P atomic % ratio was 1.41 ± 0.01. This Ca/P ratio increased slightly with a decrease in the area size and reached equilibrium when the area size was less than a few µm. The blue and green curves show the spectra measured for 1.5 µm and 200 nm squares, the corresponding Ca/P ratios of which were 1.44 ± 0.01 and 1.45 ± 0.01. The detected elements (except C, O, N, and Cu) and their contents for the 200 nm square are shown in Table S4.

We performed EDS measurements on five different areas in the 1.5 µm square and determined that the average Ca/P atomic % ratio was 1.45 with a deviation of ± 0.02.

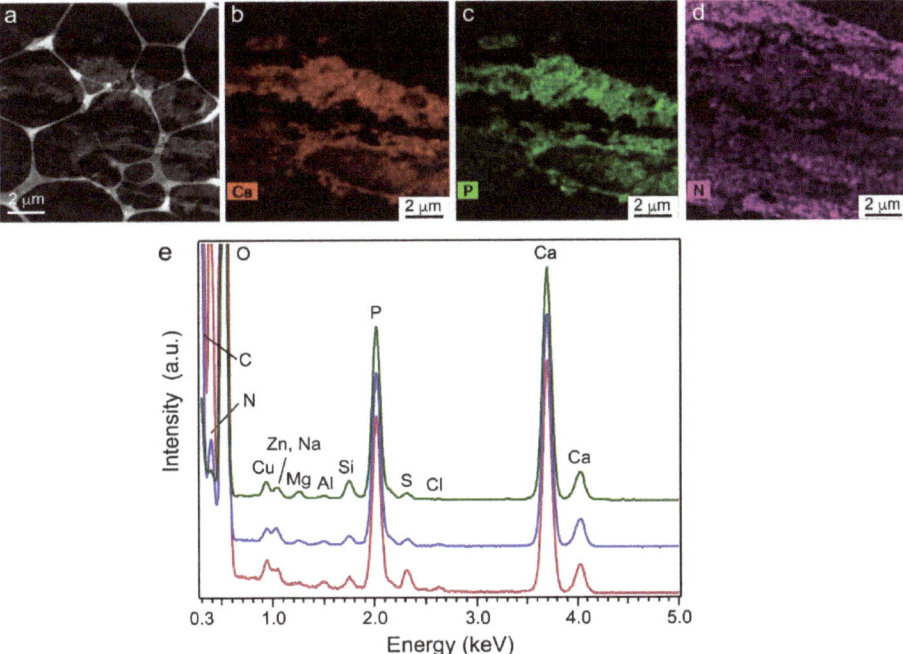

Figure 10. Scanning TEM energy-dispersive X-ray spectroscopy (STEM-EDS) analysis results for precipitates. (**a**) Low-magnification TEM image for energy-dispersive X-ray spectroscopy (EDS) analysis. Two-dimensional elemental mappings of (**b**) Ca, (**c**) P, and (**d**) N. (**e**) STEM-EDS spectra for three measured areas: approximately 12 µm, 1.5 µm, and 200 nm (magenta, blue, and green curves, respectively). Cu is attributed to the TEM grid, and Al is equipment specific. Each spectrum was normalized by the intensity of Ca at 3.7 keV.

3. Discussions

Cytokines such as BMP and TGF-β play important roles in the regulation of odontoblast differentiation [14,15,27,28]. When we established the PPU-7 cell line, we demonstrated that both BMP2 and TGF-β1 significantly affected cell proliferation, cell population doubling time, mRNA expression of dentin sialophosphoprotein (DSPP), which is one of the differentiation marker for odontoblasts, and ALP activity in the PPU-7 cell line. These findings are in agreement with the reports described above. In addition, considering the inherent ALP activity in the PPU-7 cell line, our data suggest that some level of ALP activity is required for the induction of the differentiation of PPU-7 cells into hard tissue-forming cells by BMP2 and/or TGF-β1.

In vitro study using human mesenchymal stem cells (hMSCs) shows that MDZ inhibits ALP activity and calcium deposition in hMSCs, suggesting a suppressive effect on osteogenic differentiation [23]. However, little was known about the effect of MDZ on the properties of dental pulp tissues. We therefore hypothesized that MDZ affects odontoblastic differentiation rather than the osteogenic differentiation of dental pulp stem cells. We demonstrated that the addition of MDZ-only to PPU-7 cells dramatically enhanced ALP activity and mineralization induction. This finding suggests that MDZ possesses the ability to induce the differentiation of PPU-7 cells into hard tissue-forming cells. Moreover, the results led us to investigate the combination of MDZ and cytokines, namely, BMP2 and TGF-β1.

In general, many drugs express their drug efficacy through a receptor. MDZ controls the GABA, which is the main inhibitory neurotransmitter. The GABAA receptor is a multi-subunit chloride channel that mediates the fastest inhibitory synaptic transmission in the central nervous system. The benzodiazepine class, which includes MDZ, enhances the effect of the neurotransmitter GABA at the GABAA receptor [17]. The GABAA receptor has at least 19 different subunits and shows different susceptibility to drugs depending on the combination of units [29,30]. The GABAA receptor beta 1 subunit (GABRB1) is strongly expressed in the human coronal pulp and permanent predentin/odontoblasts [31,32]. Although we demonstrated that MDZ-only treatment enhanced ALP activity and mineralization induction, combinations of MDZ with BMP2 or TGF-β1 tended to decrease both of them. Our findings suggest that the pharmacological interaction of the combination of MDZ with BMP2 or TGF-β1 reduces the susceptibility and/or reactivity of GABRB1.

Dental pulp stem cells possess a pluripotency for various types of specialized cells, such as neuron, cardiomyocytes, chondrocytes, odontoblasts, osteoblasts, and liver cells [33,34]. The starting population of these cells and their differentiated progeny are evaluated by analyzing the expression of specific gene markers. DSPP is an important marker of odontoblastic differentiation. We previously generated two amplification products from the full-length *DSPP* (*DSPP-v1*) and the DSP-only (*DSPP-v2*) transcripts and found that both the *DSPP-v1* and *DSPP-v2* products were predominantly observed in odontoblasts, whereas only trace expression of the *DSPP-v1* transcript was detected in dental pulp [35]. Our previous study showed that in addition to these two DSPP splice variants, matrix metalloproteases such as *MMP2* and *MMP20* can become odontoblast differentiation markers [36]. We demonstrated that the mRNA levels of *DSPP-v1*, *DSPP-v2* and *MMP2* were dramatically enhanced by MDZ at day seven. In contrast, mRNA expression levels of the osteoblastic differentiation markers OC and *RUNX2*, and the chondrogenic differentiation markers Col II and *ACAN* decreased over time regardless of the presence or absence of MDZ. These findings suggest that MDZ promotes the differentiation of PPU-7 cells into odontoblasts.

Alizarin Red S staining has been widely used to evaluate calcium deposits in cell culture [37], whereas von Kossa staining has generally been used to detect both phosphates and carbonates in calcium deposits [38]. We demonstrated that the formation of precipitated nodules in a mineralization-inducing culture medium in the presence or absence of MDZ was evident on the plate by both staining methods and that the MDZ-only treatment showed the highest ability to induce calcification. These findings suggest that the precipitated nodules possess calcium carbonate-based or calcium phosphate-based components and that the use of MDZ-only is the most effective in inducing mineralization in the PPU-7 cell line.

DPP is a highly phosphorylated protein in dentin with an isoelectric point near 1.1. Due to the biased amino acid composition and extensive acidity (35–40% phosphoserine and 40–50% aspartic acid) [39], DPP stains very strongly with Stains-all but does not stain with Coomassie Brilliant Blue. Genetic studies have shown that for porcine DPP, the length variations in the DPP code are closely correlated with the length variations observed at the protein level [40]. Therefore, DPP samples separately isolated from 22 pigs appear as single or doublet bands migrating between 96 kDa and 100 kDa on SDS-PAGE [40]. Our previous study showed that full-length *DSPP* containing DPP is an odontoblast-predominant transcript and that little DPP is present in porcine dental pulp, suggesting that DPP is a dentin-specific protein [35]. In this study, we demonstrated that DPP doublet bands were present in the precipitated nodules in PPU-7 cells cultured with a mineralization-inducing culture medium and that the addition of MDZ increased the intensity of the DPP doublet bands. Our data suggest that the precipitated nodules possessed dentin-like characteristics.

The HR-TEM observations and FFT analyses indicated that the nanorods and nanoparticles were HAP. To support this conclusion, SAED simulation of a HAP crystal viewed from a particular zone axis was performed using the ReciPro application. Figure S4 shows the SAED patterns of a HAP crystal viewed from $\{02\bar{1}\}$ (Figure S4a) and $\{012\}$ (Figure S4b). The pattern corresponding to the $\{02\bar{1}\}$ direction (dotted hexagon) was essentially the same as the FFT image in Figure 8c. The diffraction spots corresponding to $\{012\}$ and $\{\bar{1}12\}$ were not clearly visible in the FFT image, probably due to the width of the nanorods in those directions not having sufficient periodicity in the image. The simulated SAED corresponding to the $\{012\}$ direction was consistent with the FFT image in Figure 9c, indicating that the nanorods and nanoparticles were indeed HAP. The two Debye rings observed in the wide-area SAED pattern (Figure 7c) revealed the most likely d values in randomly oriented HAP, and the XRD pattern of the precipitates showed peaks corresponding to those of HAP. These findings indicated that the precipitates consisted of a single phase of low-crystalline HAP. The STEM-EDS results showed that the HAP in the precipitates was calcium deficient.

In our previous study, an enamel-like HAP nanorod array was constructed on an amorphous calcium phosphate (ACP) substrate consisting of nanoparticles (<80 nm diameter) that had been compression molded [41]. When the substrate was immersed in a calcium phosphate solution without cells under pseudo-physiological conditions, ACP nanoparticles close to the substrate surface quickly transformed into HAP ones and grew perpendicularly as rods due to geometric selection. During this reaction, ACP nanoparticles located in the middle to deep regions of the substrate were transformed into low-crystalline HAP. We found that the macroscopic and microscopic morphologies of cell-induced HAP closely resembled those of HAP transformed from ACP nanoparticles. Figure S5 in Supplementary shows the TEM images, SAED patterns, and STEM-EDS spectra for both types of HAP crystal. The macroscopic morphology of the grown material was fibrous for both types (Figure S5a,c) and consisted of a mixture of nanorods and nanoparticles (Figure S5b,d). Regarding the morphology in elemental crystals and their aggregates, the two types of HAP crystal were similar. The SAED patterns for the grown materials measured at 800 nm ϕ were also similar (Figure S5e). The average Ca/P atomic % ratio of cell-induced HAP (Ca/P = 1.45) was slightly less than that of HAP transformed from ACP nanoparticles (Ca/P = 1.50) (Figure S5f) and that of human dentin [42]. One reason for this difference is the substitution of several cations, such as Zn and Mg, at Ca-ion sites in the cell-induced HAP, which occurred due to the complex composition of the initial growing fluid. Based on the content of each element shown in Table S2, the Ca/(Zn + Na + Mg + Si) atomic % ratio was approximately 7.48, meaning that the x component in $Ca_{10-x}(Zn, Na, Mg, Si)_x(PO_4)_6(OH)_2$ was approximately 1.18. Therefore, the calculated Ca/P molar ratio of the cell-induced HAP was 1.47, which was close to the measured ratio for the precipitates.

The morphological, structural, and chemical similarities between the cell-induced HAP and the ACP-transformed HAP strongly suggest that the cells, which had high ALP activity, produced locally concentrated phosphate ions, which nucleated ACP nanoparticles in the initial reaction stage. These particles subsequently transformed into low-crystalline HAP. Biological analysis after HAP precipitation showed that the initial pulpal cells differentiated into odontoblast cells and that the present system simulated the dentin matrix formation process, suggesting that the HAP in tooth dentin is formed by phase transformation from ACP nanoparticles. Indeed, a TEM image of human molar dentin (Figure S6) showed that it consisted of nanorods and nanoparticles, the same as for the images in Figure S5a,c.

4. Materials and Methods

These studies have received approval from the Institutional Ethics Review Committee of the Tsurumi University School of Dental Medicine (Project identification code #1318, 1 December 2015). Additionally, all animal experiments were approved by the Institutional Animal Care Committee and the Recombination DNA Experiment and Biosafety Committee of the Tsurumi University School of Dental Medicine. We have confirmed that all experiments were performed in accordance with relevant guidelines and regulations.

4.1. Alkaline Phosphatase (ALP) Activity Assay of the PPU-7 Cell Line

The PPU-7 cell line was established from porcine dental pulp cells by our group (see Figures S1–S3). We measured ALP activity in each well as described previously [43]. The cells were plated on a 96-well plate at a density of 3.16×10^4 cells/cm^2 and were cultured in the standard medium for 24 h. The medium was changed to growth medium supplemented with 0, 2.5, 5, or 10 µM of MDZ with 500 ng/mL of rhBMP2 (R&D Systems, Minneapolis, MN, USA) or with 1 ng/mL of rhTGF-β1 (Cell Signaling Technology, Danvers, MA, USA). After 72 additional hours of incubation, the cells were washed once with PBS, and ALP activity was assayed using 10 mM p-nitrophenylphosphate as the substrate in 100 mM 2-amino-2-methyl-1,3-propanediol-HCl buffer (pH 10.0) containing 5 mM MgCl$_2$ and incubated for 10 min at 37 °C. Adding 0.2 M NaOH quenched the reaction, and the absorbance at 405 nm was read on a plate reader.

4.2. Alkaline Phosphatase Staining

The cells were spread on a 24-well plate at density of 3.16×10^4 cells/cm^2. After incubation for 2 days, the cells were rinsed twice with PBS, fixed with 10% formaldehyde for 30 min, stained with 0.1 mg/mL of naphthol AS-MX phosphate (Sigma-Aldrich, St. Louis, MO, USA), 0.5% N,N-dimethylformamide, Fast Blue BB salt (Sigma-Aldrich), and 2 mM MgCl$_2$ in 0.1 M Tris–HCl buffer (pH 8.5) for 30 min at room temperature, and then washed with dH$_2$O and photographed.

4.3. Gene Expression of the PPU-7 Cell Line

The cells were extracted with an RNA extraction reagent (Isogen, Nippon Gene Co., Ltd., Tokyo, Japan). The purified total RNA (2 µg) was reverse transcribed; the reaction mixture consisted of SYBR Green PCR master mix (Roche), supplemented with 0.5 µM forward and reverse primers and 2 µL of cDNA as template. The specific primer sets were designed using Primer-BLAST software (URL: http://www.ncbi.nlm.nih.gov/tools/primer-blast). The specific primer sets and running conditions are shown in Table S1. *GAPDH* was used as the reference gene. Each ratio was normalized the relative quantification data of *DSPP-v1*, *DSPP-v2*, *MMP2*, OC, *RUNX2*, Col II and *ACAN* in comparison to a reference gene (*GAPDH*), generated on the basis of a mathematical model for relative quantification in qPCR system. All values were represented as means ± standard error. Statistical significance (*) was determined using a Steel's test, whereas (**) was determined using a Mann–Whitney U-test. In all cases, $p < 0.05$ was regarded as statistically significant.

4.4. Detection of Precipitated Nodules in PPU-7 Cells

PPU-7 cells were grown on a 12-well plate at an initial density of 3.16×10^4 cells/cm^2. After incubation for 24 h, the medium was changed to a mineralization-inducing medium containing 10 mM β-glycerophosphate and 50 µM ascorbic acid. The cells were cultured for up to 7 days. Mineralization was visualized by Alizarin Red S and von Kossa staining. After fixation with 4% paraformaldehyde neutral buffer solution for 30 min, the cells were stained with 1% Alizarin Red S (Sigma-Aldrich) solution for 10 min, then washed with distilled water and photographed. Alternatively, silver staining was performed using a von Kossa staining kit (Polysciences, Inc., Warrington, PA, USA) according to the manufacturer's instructions.

PPU-7 cells were grown on a 24-well plate at an initial density of 3.16×10^4 cells/cm^2. After incubation for 24 h, the medium was changed to a mineralization-inducing medium. The cells were cultured for up to 5 days. Each well on the plates was rinsed with PBS, and the calcium was dissolved in 0.5 mL of 0.5 N HCl by gentle rocking for 1 h. The calcium concentration in the eluate was spectrophotometrically determined at 570 nm by following the color development with a calcium assay kit (Calcium C-test Wako, Wako Pure Chemical Industries, Ltd.). All values were normalized against the cultivation area.

4.5. Statistical Analysis

For ALP assays and qPCR and calcium analyses, all values are presented as the mean ± standard error. Statistical significance (*) was determined using the Mann–Whitney U-test for qPCR analysis and a nonparametric Steel's test for the ALP assay and calcium analysis. $p < 0.05$ was regarded as statistically significant for Steel's test, and $p < 0.01$ or $p < 0.05$ was regarded as statistically significant for the Mann–Whitney U-test.

4.6. Characterization of Proteins in the Precipitated Nodules from PPU-7 Cells

The PPU-7 cell line was grown in standard medium or in mineralization-inducing culture medium on the 6 cm culture dish at an initial density of 3.16×10^4 cells/cm^2 for 14 days, and the cell sheet was scraped off with a cell scraper. The cell sample was suspended with 50 mM Tris–HCl/4M guanidine buffer (pH 7.4) containing Protease Inhibitor Cocktail Set III [1 mM 4-(2-aminoethyl) benzenesulfonyl fluoride hydrochloride (AEBSF), 0.8 mM aprotinin, 50 mM bestatin, 15 mM E-64, 20 mM leupeptin, and 10 mM pepstatin (Calbiochem EMD Chemicals Inc., Gibbstown, NJ, USA)] and 1 mM 1,10-phenanthroline (Sigma-Aldrich), and was homogenized using a Polytron (Capitol Scientific, Inc., Austin, TX, USA) homogenizer for 30 s at half speed. Insoluble material was pelleted by centrifugation (15,900g) and the supernatant (G1 extract) was stored at −80 °C. The guanidine-insoluble material was extracted against 0.17 N HCl and 0.95% formic acid containing 10 mM benzamidine (Sigma-Aldrich), 1 mM PMSF, and 1 mM 1,10-phenanthroline using the homogenizer. Following centrifugation, the acid-soluble supernatant (HF extract) was stored at −80 °C. The pellet was extracted with 50 mM Tris–HCl/4M guanidine buffer (pH 7.4) containing Protease Inhibitor Cocktail Set III and centrifuged, and the supernatant (G2 extract) was stored at −80 °C. The G1, HF, and G2 extracts were desalted with an Amicon Ultra centrifugal filter (0.5 mL, MW = 3000 cut off) (Merck Millipore, Darmstadt, Germany). Following the quantitative analysis of total protein in each extract using a Pierce 660 nm Protein Assay Reagent (Thermo Fisher Scientific, Waltham, MA, USA), the amount of total protein was normalized by dividing the amount of total protein in control (i.e., PPU-7 cell line was cultured in the standard medium), and characterized by sodium dodecyl sulfate–polyacrylamide gel electrophoresis (SDS-PAGE).

SDS-PAGE was performed using a 15% e-PAGEL mini gel (ATTO Corporation, Tokyo, Japan). The gel was stained with SimplyBlue SafeStain (Invitrogen) or Stains-all stain (Sigma-Aldrich). The apparent molecular weights of the protein bands were estimated by comparison with the SeeBlue Plus2 Pre-Stained standard (Life Technologies/Invitrogen).

4.7. Degree of Supersaturation with Respect to Each Calcium Phosphate Phase

The calcium phosphate phases that form under physiological conditions are DCPD, OCP, β-TCP, and HAP. We calculated the degree of supersaturation, σ, of the initial culture fluid with respect to each phase using:

$$\sigma = (IP/K_{sp})^{1/n} - 1 \quad (n = 2, 16, 5, \text{ and } 18 \text{ for DCPD, OCP, β-TCP, and HAP, respectively}) \quad (1)$$

where IP is the ionic activity of the solution and K_{sp} is the solubility product constant of the material at 37 °C [44]. The concentration of each ion in the fluid and the degree of supersaturation are given in Table S2.

4.8. XRD

Microbeam XRD (RAPID, Rigaku) with monochromated Cu-Kα radiation was used to characterize the precipitates at 50 kV and 30 mA. The incident beam was focused to a diameter of approximately 100 μm and irradiated perpendicular to the surface of the cell sheet. The diffraction was recorded on an imaging plate, and the digital data on the plate were converted to the intensity versus 2θ relationship using DISPLAY software (Rigaku). The diffraction peak positions were referenced to those of different calcium phosphate phases in the JCPDS cards (DCPD: Card 11–293, OCP: Card 26–1056, β-TCP: Card 9–169, HAP: Card 9–432).

4.9. SEM

Field-emission scanning electron microscopy (FE-SEM, JSM-7000F, JEOL, Ltd.) with an acceleration voltage of 5 kV was used to observe cross-sectional samples. The samples were coated with an osmium film (approximately 3 nm thick) before observation.

4.10. TEM Observation and Elemental Analysis Using STEM-EDS

Two types of samples were prepared for TEM observations. Ultramicrotome-cut samples were prepared by embedding cell sheets with precipitates in epoxy resin and placing them in a dark room for 48 h at room temperature. After the resin completely filled the gaps (precipitate regions) between cell sheets, the resin was solidified using ultraviolet irradiation. The solidified block was then sliced using a diamond knife into samples with a thickness of 70–100 nm. The samples were placed on a TEM grid with a Cu mesh for observation. Samples crushed in an agate mortar were dispersed in 99.5% ethanol solution, which was ultrasonically stirred for 5 min at 40 kHz. An aliquot of the stirred solution was placed on a TEM grid with a Cu mesh and allowed to dry naturally in air. An analytical TEM (Tecnai Osiris, FEI Co.) with an acceleration voltage of 200 kV was used to observe the samples. SAED measurement with a typical area of 800 nm or 200 nm ϕ was simultaneously performed to characterize the material.

Two-dimensional elemental mapping and quantitative analysis of the content of each element were performed using a Super-X EDS system in the TEM. The probe diameter used for analysis was approximately 0.5 nm, and the amplitude was approximately 0.55 nA. The beam residence time at each measurement position was 10 μs, and the analysis was completed within 5 min. The relative content of each element was calculated from the energy-dispersive spectra by using peak deconvolution.

5. Conclusions

Drug repositioning is the process of discovering, validating, and marketing previously approved drugs for new indications. Due to the promise of reduced costs and expedited approval schedules, the research field of drug repositioning has been attracting attention [45]. To find candidates for drug repositioning, a standard database consisting of both approved and failed drugs has been developed as a web application [46]. Without using the database, the present study demonstrated that MDZ enhances the differentiation of PPU-7 cells to odontoblast and promotes the formation of dentin-like hydroxyapatite. Further studies are required to elucidate the pharmacokinetic and pharmacological efficacy of MDZ in animal experiments. In the dental field, these findings support the repositioning of MDZ to promote dentin regeneration for endodontic treatments, such as pulp capping. Moreover, these findings support advancing research from pig to human experimental models using human DPSCs to discover MDZ's potential, not only for future dental treatments, but also for organ regenerative medicine.

Supplementary Materials: The following are available online at http://www.mdpi.com/1422-0067/20/3/670/s1.

Author Contributions: Conceptualization, T.K., K.O. and Y.Y.; Formal Analysis, T.K., K.O., M.M.S., R.Y., T.C., R.C. and Y.H.; Investigation, T.K., K.O. and Y.Y.; Data Curation, T.K., K.O., R.Y., T.C., R.C., K.F.-A., H.K. and Y.Y.; Writing-Original Draft Preparation, T.K., K.O. and Y.Y.; Writing-Review & Editing, K.O. and Y.Y.

Funding: This study was supported by a JSPS KAKENHI Grant-in-Aid for Scientific Research (C26462982, 16K04954, 17K11975 and 18K09630) and the MEXT-Supported Program for the Strategic Research Foundation at Private Universities (S1511018).

Acknowledgments: We thank Ichiro Saito from the Department of Pathology at the School of Dental Medicine, Tsurumi University, and Yoshinobu Asada from the Department of Pediatric Dentistry at the School of Dental Medicine, Tsurumi University for their support.

Conflicts of Interest: The authors declare no conflict of interest.

Abbreviations

mRNA	messenger ribonucleic acid
SDS	sodium dodecyl sulfate
kDa	kilodalton
Tris	2-amino-2-hydroxymethyl-propane-1,3-diol
PMSF	phenylmethylsulfonyl fluoride
PBS	phosphate buffered saline

References

1. Zafar, M.S.; Khurshid, Z.; Almas, K. Oral tissue engineering progress and challenges. *Tissue Eng. Regen. Med.* **2015**, *12*, 387–397. [CrossRef]
2. Hu, L.; Gao, Z.; Xu, J.; Zhu, Z.; Fan, Z.; Zhang, C.; Wang, J.; Wang, S. Decellularized Swine Dental Pulp as a Bioscaffold for Pulp Regeneration. *BioMed Res. Int.* **2017**, *2017*, 9342714. [CrossRef] [PubMed]
3. Yasui, T.; Mabuchi, Y.; Morikawa, S.; Onizawa, K.; Akazawa, C.; Nakagawa, T.; Okano, H.; Matsuzaki, Y. Isolation of dental pulp stem cells with high osteogenic potential. *Inflamm. Regen.* **2017**, *37*, 8. [CrossRef] [PubMed]
4. Hayashi, Y.; Imai, M.; Goto, Y.; Murakami, N. Pathological mineralization in a serially passaged cell line from rat pulp. *J. Oral Pathol. Med.* **1993**, *22*, 175–179. [CrossRef] [PubMed]
5. Nagata, T.; Yokota, M.; Ohishi, K.; Nishikawa, S.; Shinohara, H.; Wakano, Y.; Ishida, H. 1α,25-dihydroxyvitamin D_3 stimulation of osteopontin expression in rat clonal dental pulp cells. *Arch. Oral Biol.* **1994**, *39*, 775–782. [CrossRef]
6. MacDougall, M.; Selden, J.K.; Nydegger, J.R.; Carnes, D.L. Immortalized mouse odontoblast cell line MO6-G3 application for in vitro biocompatibility testing. *Am. J. Dent.* **1998**, *11*, S11–S16. [PubMed]
7. Thonemann, B.; Schmalz, G. Bovine dental papilla-derived cells immortalized with HPV 18 E6/E7. *Eur. J. Oral Sci.* **2000**, *108*, 432–441. [CrossRef] [PubMed]
8. Iwata, T.; Yamakoshi, Y.; Simmer, J.P.; Ishikawa, I.; Hu, J.C. Establishment of porcine pulp-derived cell lines and expression of recombinant dentin sialoprotein and recombinant dentin matrix protein-1. *Eur. J. Oral Sci.* **2007**, *115*, 48–56. [CrossRef]
9. Kamata, N.; Fujimoto, R.; Tomonari, M.; Taki, M.; Nagayama, M.; Yasumoto, S. Immortalization of human dental papilla, dental pulp, periodontal ligament cells and gingival fibroblasts by telomerase reverse transcriptase. *J. Oral Pathol. Med.* **2004**, *33*, 417–423. [CrossRef]
10. Galler, K.M.; Schweikl, H.; Thonemann, B.; D'Souza, R.N.; Schmalz, G. Human pulp-derived cells immortalized with Simian Virus 40 T-antigen. *Eur. J. Oral Sci.* **2006**, *114*, 138–146. [CrossRef]
11. Yang, X.; van der Kraan, P.M.; van den Dolder, J.; Walboomers, X.F.; Bian, Z.; Fan, M.; Jansen, J.A. STRO-1 selected rat dental pulp stem cells transfected with adenoviral-mediated human bone morphogenetic protein 2 gene show enhanced odontogenic differentiation. *Tissue Eng.* **2007**, *13*, 2803–2812. [CrossRef] [PubMed]
12. Ohazama, A.; Tucker, A.; Sharpe, P.T. Organized tooth-specific cellular differentiation stimulated by BMP4. *J. Dent. Res.* **2005**, *84*, 603–606. [CrossRef] [PubMed]

13. Shimabukuro, Y.; Ueda, M.; Ozasa, M.; Anzai, J.; Takedachi, M.; Yanagita, M.; Ito, M.; Hashikawa, T.; Yamada, S.; Murakami, S. Fibroblast growth factor-2 regulates the cell function of human dental pulp cells. *J. Endod.* **2009**, *35*, 1529–1535. [CrossRef] [PubMed]
14. Tjaderhane, L.; Koivumaki, S.; Paakkonen, V.; Ilvesaro, J.; Soini, Y.; Salo, T.; Metsikko, K.; Tuukkanen, J. Polarity of mature human odontoblasts. *J. Dent. Res.* **2013**, *92*, 1011–1016. [CrossRef]
15. Li, Y.; Lu, X.; Sun, X.; Bai, S.; Li, S.; Shi, J. Odontoblast-like cell differentiation and dentin formation induced with TGF-β1. *Arch. Oral Biol.* **2011**, *56*, 1221–1229. [CrossRef]
16. Neves, V.C.; Babb, R.; Chandrasekaran, D.; Sharpe, P.T. Promotion of natural tooth repair by small molecule GSK3 antagonists. *Sci. Rep.* **2017**, *7*, 39654. [CrossRef]
17. Olkkola, K.T.; Ahonen, J. Midazolam and other benzodiazepines. In *Modern Anesthetics. Handbook of Experimental Pharmacology*; Schuttler, J., Schwilden, H., Eds.; Springer: Berlin, Germany, 2008; Volume 182, pp. 335–360.
18. Smith, R.; Brown, J. Midazolam for status epilepticus. *Aust. Prescr.* **2017**, *40*, 23–25. [CrossRef]
19. So, E.C.; Chen, Y.C.; Wang, S.C.; Wu, C.C.; Huang, M.C.; Lai, M.S.; Pan, B.S.; Kang, F.C.; Huang, B.M. Midazolam regulated caspase pathway, endoplasmic reticulum stress, autophagy, and cell cycle to induce apoptosis in MA-10 mouse Leydig tumor cells. *OncoTargets Ther.* **2016**, *9*, 2519–2533.
20. Ohno, S.; Kobayashi, K.; Uchida, S.; Amano, O.; Sakagami, H.; Nagasaka, H. Cytotoxicity and type of cell death induced by midazolam in human oral normal and tumor cells. *Anticancer Res.* **2012**, *32*, 4737–4747. [PubMed]
21. Stevens, M.F.; Werdehausen, R.; Gaza, N.; Hermanns, H.; Kremer, D.; Bauer, I.; Kury, P.; Hollmann, M.W.; Braun, S. Midazolam activates the intrinsic pathway of apoptosis independent of benzodiazepine and death receptor signaling. *Reg. Anesth. Pain Med.* **2011**, *36*, 343–349. [CrossRef]
22. Mishra, S.K.; Kang, J.H.; Lee, C.W.; Oh, S.H.; Ryu, J.S.; Bae, Y.S.; Kim, H.M. Midazolam induces cellular apoptosis in human cancer cells and inhibits tumor growth in xenograft mice. *Mol. Cells* **2013**, *36*, 219–226. [CrossRef] [PubMed]
23. Zhang, T.; Shao, H.; Xu, K.Q.; Kuang, L.T.; Chen, R.F.; Xiu, H.H. Midazolam suppresses osteogenic differentiation of human bone marrow-derived mesenchymal stem cells. *Eur. Rev. Med. Pharmacol. Sci.* **2014**, *18*, 1411–1418. [PubMed]
24. Sloan, A.J.; Rutherford, R.B.; Smith, A.J. Stimulation of the rat dentine-pulp complex by bone morphogenetic protein-7 in vitro. *Arch. Oral Biol.* **2000**, *45*, 173–177. [CrossRef]
25. Chen, S.; Gu, T.T.; Sreenath, T.; Kulkarni, A.B.; Karsenty, G.; MacDougall, M. Spatial expression of Cbfa1/Runx2 isoforms in teeth and characterization of binding sites in the dspp gene. *Connect. Tissue Res.* **2002**, *43*, 338–344. [CrossRef] [PubMed]
26. 1000 Genomes Project Consortium. A map of human genome variation from population-scale sequencing. *Nature* **2010**, *467*, 1061–1073. [CrossRef] [PubMed]
27. Qin, W.; Yang, F.; Deng, R.; Li, D.; Song, Z.; Tian, Y.; Wang, R.; Ling, J.; Lin, Z. Smad 1/5 is involved in bone morphogenetic protein-2-induced odontoblastic differentiation in human dental pulp cells. *J. Endod.* **2012**, *38*, 66–71. [CrossRef] [PubMed]
28. Li, J.; Huang, X.; Xu, X.; Mayo, J.; Bringas, P., Jr.; Jiang, R.; Wang, S.; Chai, Y. SMAD4-mediated WNT signaling controls the fate of cranial neural crest cells during tooth morphogenesis. *Development* **2011**, *138*, 1977–1989. [CrossRef]
29. Cope, D.W.; Hughes, S.W.; Crunelli, V. GABA$_A$ receptor-mediated tonic inhibition in thalamic neurons. *J. Neurosci.* **2005**, *25*, 11553–11563. [CrossRef]
30. Bormann, J. The 'ABC' of GABA receptors. *Trends Pharmacol. Sci.* **2000**, *21*, 16–19. [CrossRef]
31. Kim, J.H.; Jeon, M.; Song, J.S.; Lee, J.H.; Choi, B.J.; Jung, H.S.; Moon, S.J.; DenBesten, P.K.; Kim, S.O. Distinctive genetic activity pattern of the human dental pulp between deciduous and permanent teeth. *PLoS ONE* **2014**, *9*, e102893. [CrossRef]
32. Kim, S.H.; Kim, S.; Shin, Y.; Lee, H.S.; Jeon, M.; Kim, S.O.; Cho, S.W.; Ruparel, N.B.; Song, J.S. Comparative Gene Expression Analysis of the Coronal Pulp and Apical Pulp Complex in Human Immature Teeth. *J. Endod.* **2016**, *42*, 752–759. [CrossRef] [PubMed]
33. Potdar, P.D.; Jethmalani, Y.D. Human dental pulp stem cells: Applications in future regenerative medicine. *World J. Stem Cells* **2015**, *7*, 839–851. [CrossRef] [PubMed]

34. Shi, S.; Gronthos, S. Perivascular niche of postnatal mesenchymal stem cells in human bone marrow and dental pulp. *J. Bone Miner. Res.* **2003**, *18*, 696–704. [CrossRef] [PubMed]
35. Yamamoto, R.; Oida, S.; Yamakoshi, Y. Dentin Sialophosphoprotein-derived Proteins in the Dental Pulp. *J. Dent. Res.* **2015**, *94*, 1120–1127. [CrossRef] [PubMed]
36. Niwa, T.; Yamakoshi, Y.; Yamazaki, H.; Karakida, T.; Chiba, R.; Hu, J.C.; Nagano, T.; Yamamoto, R.; Simmer, J.P.; Margolis, H.C.; et al. The dynamics of TGF-β in dental pulp, odontoblasts and dentin. *Sci. Rep.* **2018**, *8*, 4450. [CrossRef]
37. Puchtler, H.; Meloan, S.N.; Terry, M.S. On the history and mechanism of alizarin and alizarin red S stains for calcium. *J. Histochem. Cytochem.* **1969**, *17*, 110–124. [CrossRef]
38. Puchtler, H.; Meloan, S.N. Demonstration of phosphates in calcium deposits: a modification of von Kossa's reaction. *Histochemistry* **1978**, *56*, 177–185. [CrossRef]
39. Saito, T.; Toyooka, H.; Ito, S.; Crenshaw, M.A. In vitro study of remineralization of dentin: effects of ions on mineral induction by decalcified dentin matrix. *Caries Res.* **2003**, *37*, 445–449. [CrossRef]
40. Yamakoshi, Y.; Lu, Y.; Hu, J.C.; Kim, J.W.; Iwata, T.; Kobayashi, K.; Nagano, T.; Yamakoshi, F.; Hu, Y.; Fukae, M.; et al. Porcine dentin sialophosphoprotein: length polymorphisms, glycosylation, phosphorylation, and stability. *J. Biol. Chem.* **2008**, *283*, 14835–14844. [CrossRef]
41. Onuma, K.; Iijima, M. Artificial enamel induced by phase transformation of amorphous nanoparticles. *Sci. Rep.* **2017**, *7*, 2711. [CrossRef]
42. Teruel Jde, D.; Alcolea, A.; Hernandez, A.; Ruiz, A.J. Comparison of chemical composition of enamel and dentine in human, bovine, porcine and ovine teeth. *Arch. Oral Biol.* **2015**, *60*, 768–775. [CrossRef] [PubMed]
43. Nagano, T.; Oida, S.; Suzuki, S.; Iwata, T.; Yamakoshi, Y.; Ogata, Y.; Gomi, K.; Arai, T.; Fukae, M. Porcine Enamel Protein Fractions Contain Transforming Growth Factor-β1. *J. Periodontol.* **2006**, *77*, 1688–1694. [CrossRef] [PubMed]
44. Wang, L.; Nancollas, G.H. Calcium orthophosphates: crystallization and dissolution. *Chem. Rev.* **2008**, *108*, 4628–4669. [CrossRef] [PubMed]
45. Rodriguez-Esteban, R. A Drug-Centric View of Drug Development: How Drugs Spread from Disease to Disease. *PLoS Comput. Biol.* **2016**, *12*, e1004852. [CrossRef] [PubMed]
46. Brown, A.S.; Patel, C.J. A standard database for drug repositioning. *Sci. Data* **2017**, *4*, 170029. [CrossRef] [PubMed]

© 2019 by the authors. Licensee MDPI, Basel, Switzerland. This article is an open access article distributed under the terms and conditions of the Creative Commons Attribution (CC BY) license (http://creativecommons.org/licenses/by/4.0/).

Article

Graphene-Induced Osteogenic Differentiation Is Mediated by the Integrin/FAK Axis

Han Xie [1], Tong Cao [1], Alfredo Franco-Obregón [2,3] and Vinicius Rosa [1,4,5,*]

1. Faculty of Dentistry, National University of Singapore, 9 Lower Kent Ridge Road, Singapore 119085, Singapore; xiehan512@foxmail.com (H.X.); dencaot@nus.edu.sg (T.C.)
2. Department of Surgery, Yong Loo Lin School of Medicine, National University of Singapore, NUHS Tower Block, Level 8, 1E Kent Ridge Road, Singapore 119228, Singapore; suraf@nus.edu.sg
3. BioIonic Currents Electromagnetic Pulsing Systems Laboratory, BICEPS, National University of Singapore, MD6, 14 medical Drive, #14-01, Singapore 117599, Singapore
4. Department of Materials Science and Engineering, National University of Singapore, Blk EA, #03-09 9 Engineering Drive 1, Singapore 117575, Singapore
5. Centre for Advanced 2D Materials and Graphene Research Centre, National University of Singapore, 6 Science Drive 2, Singapore 117546, Singapore
* Correspondence: denvr@nus.edu.sg

Received: 14 December 2018; Accepted: 23 January 2019; Published: 29 January 2019

Abstract: Graphene is capable of promoting osteogenesis without chemical induction. Nevertheless, the underlying mechanism(s) remain largely unknown. The objectives here were: (i) to assess whether graphene scaffolds are capable of supporting osteogenesis in vivo and; (ii) to ascertain the participation of the integrin/FAK mechanotransduction axis during the osteogenic differentiation induced by graphene. MSC-impregnated graphene scaffolds (n = 6) were implanted into immunocompromised mice (28 days). Alternatively, MSCs were seeded onto PDMS substrates (modulus of elasticity = 130, 830 and 1300 kPa) coated with a single monomolecular layer of graphene and cultured in basal medium (10 days). The ensuing expressions of FAK-p397, integrin, ROCK1, F-actin, Smad p1/5, RUNX2, OCN and OPN were evaluated by Western blot (n = 3). As controls, MSCs were plated onto uncoated PDMS in the presence of mechanotransduction inhibitors (echistatin, Y27632 and DMH1). MSC-impregnated graphene scaffolds exhibited positive immunoexpression of bone-related markers (RUNX2 and OPN) without the assistance of osteogenic inducers. In vitro, regardless of the stiffness of the underlying PDMS substrate, MSCs seeded onto graphene-coated PDMS substrates demonstrated higher expressions of all tested osteogenic and integrin/FAK proteins tested compared to MSCs seeded onto PDMS alone. Hence, graphene promotes osteogenesis via the activation of the mechanosensitive integrin/FAK axis.

Keywords: tissue engineering; differentiation; focal adhesion kinase; bone; nanomaterials; osteoblast; mechanical

1. Introduction

Graphene consists of a single sheet of carbon atoms arranged into a tightly packed two-dimensional (2D) honeycomb lattice. Graphene sheets have proven quite amenable to molecular chemical vapor deposition functionalization as well as present high specific surface area and ultrahigh mechanical strength. Graphene can also be produced by (CVD) that is a scalable method to produce high quality graphene substrates in two and three-dimensions [1].

The demonstrated ability of CVD-grown graphene to induce osteogenic differentiation in vitro has rendered this method as a promising platform for cell-based bone tissue engineering and regeneration [2–9]. Notably, stem cells cultured on graphene films with basal growth medium demonstrated elevated expression of osteogenic proteins including Runt-related transcription factor 2 (RUNX2), osteocalcin (OCN), alkaline phosphatase (ALP), and osteopontin (OPN) [2,4,8]. Similarly, three-dimensional graphene substrates (scaffolds and hydrogels) increased the expression of bone-related genes and proteins such as RUNX2, collagen I (COL-1), and OCN [4,10,11]. Graphene film has also been shown to augment the mineralization potential of different somatic and stem cell classes [2,3,7]. Remarkably, MG-63 cells cultured on graphene coated titanium exhibited higher amounts of calcium nodules after 7 days compared to the uncoated controls (12 vs. 8 Ca/ng of DNA) [7]. The osteogenic differentiation achieved with CVD-grown graphene in the absence of osteogenic inductors is highly advantageous for bone tissue engineering as these chemicals and biomolecules present the following shortcomings: (a) dexamethasone is relatively non-specific promoting both osteogenesis as well as adipogenesis in the long-term; (b) bone morphogenetic proteins (BMPs) are expensive and plagued by concerns over their safety and; (c) the systematic administration of glucocorticoids can paradoxically result in bone loss [12–14].

Although graphene has been shown to induce osteogenic differentiation in vitro the intracellular events involved in this phenomenon remain largely unresolved. The physical characteristic of graphene (elastic properties and superficial topographical features such as wrinkles and ripples) may facilitate cell anchorage and thereby, promote stem cell differentiation [2,8]. Previous studies have shown that mesenchymal stem cells (MSCs) undergoing osteogenic differentiation on graphene in the absence of osteogenic inductors, exhibiting an upregulation in the basal expression of collagen I, RUNX2, OCN and OPN as well as other osteogenic genes [2,4,6,10,15]. Graphene activates physiologically-relevant mechanotransduction pathways, stimulating the expression of transcripts for bone morphogenetic protein 2 (BMP-2) and myosin heavy chain [4,6] and supported by findings that myosin heavy-chain mRNA and protein are upregulated by mechanical loading [16]. NIH-3T3 fibroblasts cultured on graphene coated substrates (glass, silicone or polydimethylsiloxane) exhibited enhanced adhesion and spreading and higher expression of F-actin and vinculin. Moreover, fibroblasts grown on graphene exhibited more elaborated stress fiber cytoarchitecture than those fibroblasts grown on uncoated substrates [17]. Cells grown on graphene films or scaffolds exhibited greater activation of focal adhesion kinase (FAK) [17,18], a key component of focal adhesion-mediated mechanotransduction regulating proliferation, differentiation and apoptosis, and other key processes in cellular physiology [19]. Finally, FAK is an essential component of the signal transduction machinery controlling osteogenesis, whereby its suppression disrupts bone formation [20].

Stem cell lineage commitment is modulated by the elastic characteristics of their microenvironments [21]. For instance, MSCs cultured in a rigid hydrogel (G′ = 3600 Pa) exhibited higher mineralization potential than those cultured in hydrogels with elastic modulus of G′ ≤ 1570 Pa [22]. Similarly, stem cells from human exfoliated deciduous teeth (SHED) exhibit greater potential for osteogenic differentiation with increasing substrate rigidity [23]. Substrate matrix stiffness influences osteogenic differentiation of MSC via mechanosensing pathways conveyed through α-integrins, Rho-associated protein kinase (ROCK) and focal adhesion kinase (FAK) [24,25]. Indeed, an increase in the substrate stiffness enhances osteoblastic differentiation through the α2-integrin-ROCK-FAK-ERK1/2 axis [26]. As graphene has a very high elastic modulus (>1.0 TPa [27]), high in-plane stiffness [3] and remarkable flexibility for out-of-plane deformation [8], it is reasonable to assume that the osteogenic differentiation promoted by this material is similarly triggered and mediated by the activation of the mechanotransduction integrin-FAK axis. Unveiling the intracellular cascade(s) evoked by graphene and governing osteogenic differentiation is thus critical for the effective clinical translation of this carbonaceous and versatile material.

The objectives of this study were: (i) to assess the ability of a graphene scaffold to promote osteogenic differentiation of human MSCs in vivo and; (ii) to test the hypothesis that the osteogenic differentiation of MSCs on graphene is mediated via the activation of the integrin-FAK axis.

2. Results

2.1. Graphene Scaffold Induced Osteogenic Differentiation In Vivo

To evaluate whether graphene was capable of inducing osteogenesis in vivo we seeded MSCs onto a graphene scaffold (red arrows in H&E) before transplantation into the subcutaneous space of SCID mice for four weeks (Figure 1). The histological sections obtained from the scaffolds revealed the formation of bony-like structures resembling primary bone (white arrows in H&E). Immunohistochemical characterization showed that the tissues presented positive expression of the bone-related markers RUNX2 and OPN (green arrows) and for specific human mitochondria antibody (blue arrows).

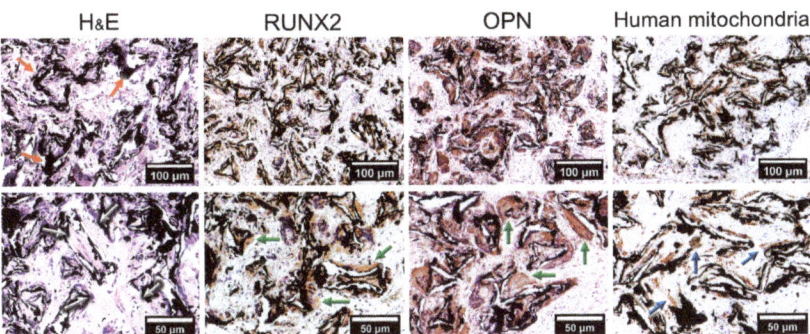

Figure 1. Graphene scaffold induce osteogenic differentiation in vivo. MSCs were seeded onto graphene scaffolds and transplanted into subcutaneous pockets of SCID mice. After four weeks, there was the formation of bony-like structures on the surface of the graphene scaffold (H&E, white arrows) that presented positive expression for the markers of osteogenic differentiation Runt-related transcription factor 2 (RUNX2) and osteopontin (OPN) (green arrows). The positive staining for specific for human mitochondria confirmed that the tissues formed were populated by human cells (blue arrows).

2.2. Activation of Integrin-FAK Axis during Graphene-Mediated Osteogenic Differentiation

To determine whether graphene promotes osteogenic differentiation via the actions of mechanosensitive pathways, we analyzed the expression of key proteins involved in the integrin-FAK axis in response to substrates of diverse mechanical properties. MSCs were seeded onto PDMS substrates with tuned elastic moduli (PDMS) or graphene-coated PDMS (Gp) in the presence or absence of inhibitors. The expression of the proteins was evaluated after 10 days in vitro.

The effective concentrations of inhibitors were ascertained by their ability to suppress proliferation. MSCs were administered different concentrations of echistatin (a Disintegrin), Y-27632 (p160ROCK inhibitor) or DMH1 (Activin receptor-like kinase 2 inhibitor) and their growth monitored for seven days. A decrease in cell proliferation of approximately 30% compared to the control was set as the threshold for effective inhibition. Echistatin significantly inhibited proliferation as early as 1 day ($p < 0.05$) at all concentrations used. However, at 10 nM proliferation was reduced by approximately 30% after seven days (Figure 2A). Effective proliferation inhibition was obtained at a concentration of 50 µM for both Y27632 and DMH1 (Figure 2B,C).

Next, we evaluated whether the integrin-FAK axis was activated during graphene-induced osteogenic differentiation. MSCs were cultured on PDMS substrates of varying stiffness that had been coated with a single monomolecular layer of graphene (Gp), or not. After 10 days, MSCs grown on Gp presented higher expression levels of FAK-p397, as well as all downstream proteins recruited in this axis compared to those seeded on PDMS alone. Highest expressions were observed on graphene-coated substrates (Gp) regardless of the stiffness of the underlying PDMS substrate. The expression of all mechanotransductory-related proteins was decreased by the presence of Echistatin (10 nM), strongly implicating the integrin-FAK axis in the osteogenic differentiation triggered by graphene (Figure 3A,B). The quantification of relative expressions showed that cells grown on Gp exhibited higher protein expression than cells cultured on PDMS alone of similar modulus of elasticity (Figure 3B).

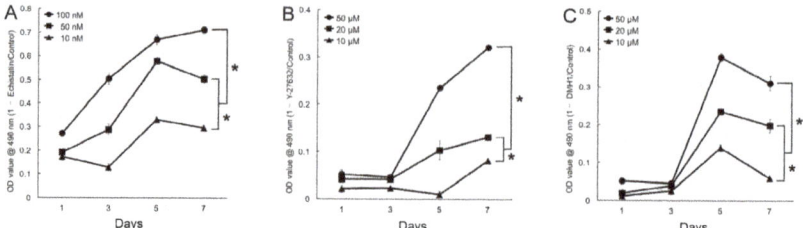

Figure 2. Effects of mechanotransduction inhibitors on cell proliferation. All inhibitors concentrations decreased cell proliferation at all time points compared to controls. After seven days, the proliferation decreased by approximately 30%, when cells were treated with 10 nM of echistatin (**A**) and 50 µM of Y27632 and DMH1 (**B**,**C**) comparing to the untreated control. (* denotes statistical difference between the groups, $p < 0.05$. For the sake of clarity, only the statistical significances at day seven are depicted).

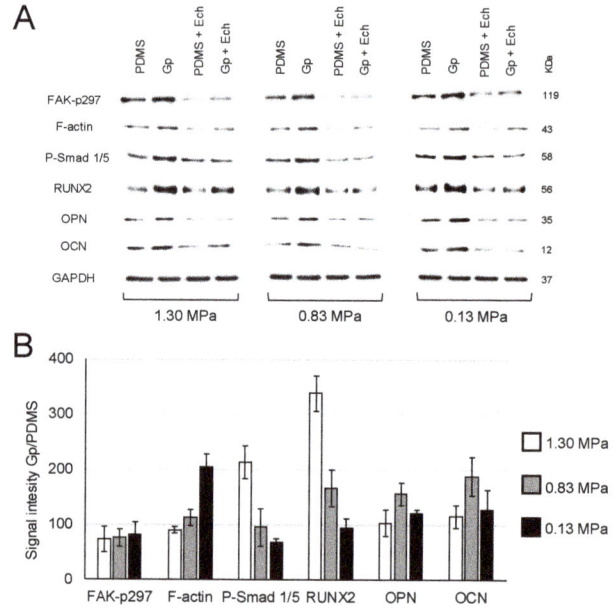

Figure 3. (**A**) Absolute and (**B**) relative expression levels of indicated proteins derived from MSCs grown on PDMS of different stiffnesses (determined by ratio of Sylgard 184 and 527) and graphene-coated PDMS (Gp). Regardless of the stiffness of the underlying substrates, MSC on Gp presented higher expression of physical stimuli-related proteins (FAK-p397, Smad p1/5 and F-actin) and bone-related markers (RUNX2, osteopontin (OPN) and osteocalcin (OCN)) compared to cells cultured on PDMS alone. OPN and OCN expression increased on Gp relative to PDMS (Gp/PDMS) for all stiffnesses tested. (**B**) relative quantification of all groups in the absence of inhibitors. Signal intensity is in arbitrary units. The presence of 10 nM echistatin attenuated the expression of all proteins examined. GAPDH represents housekeeping gene.

Y27632 (50 µM) was used to confirm a downstream role of ROCK1 in the osteogenic differentiation induced by graphene. As previously, regardless of the stiffness of the underlying polymer, MSCs on graphene-coated PDMS exhibited higher expression levels of ROCK1 in conjunction with its downstream affiliated transforming growth factor β modulating protein, Smad 1/5, and bone-related proteins (RUNX2, OPN and OCN), whose expressions were attenuated by the administration of Y27632 (Figure 4A,B).

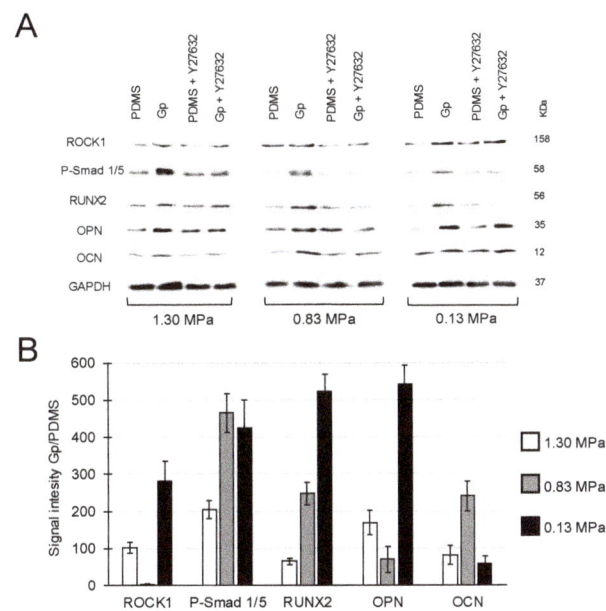

Figure 4. (**A**) Regardless of the stiffness of the underlying substrate (PDMS), Gp upregulated the expression levels of ROCK1, Smad p1/5 and F-actin and bone-related proteins. With the exception of ROCK1/0.83 MPa, Gp increased the expression of all proteins by >50%. (**B**) Relative quantification of all groups in the absence of inhibitors. Signal intensity is in arbitrary units.

Finally, we checked the expression levels of the selected proteins before and after inhibiting Smad p1/5 in response to treatment with DMH1 (50 µM). The expressions of Smad p1/5 and of the downstream bone-related proteins (RUNX2, OPN and OCN) were higher on Gp compared to all PDMS conditions tested. The presence of DMH1 suppressed the expression of all proteins confirming that the osteogenic differentiation on graphene is regulated by the activation of Smad p1/5 (Figure 5A). The quantification of protein expression showed that cells on Gp exhibited increased compared to PDMS for all modulus of elasticities studied (Figure 5B).

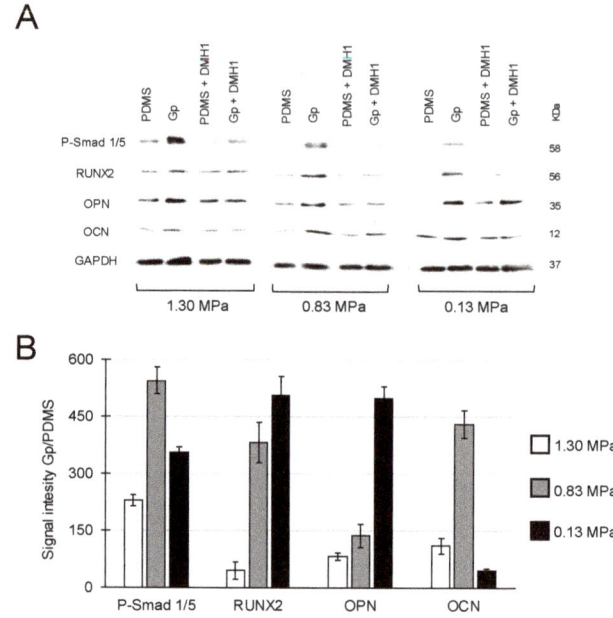

Figure 5. (**A**) MSCs grown on Gp exhibited greater increases of Smad p1/5 and of the classical markers for osteogenic differentiation, RUNX2, OPN and OCN. The expressions of p-SMAD increased by >180% on Gp compared to PDMS alone. The treatment 50 µM of DMHI decreased the expression of all protein studies. (**B**) Relative quantification of all groups in the absence of inhibitors. Signal intensity is in arbitrary units.

3. Discussion

CVD-grown graphene has emerged as a promising platform with which to enhance osteogenic outcome [5,6,8,9,23]. The osteogenic capabilities of CVD-grown graphene have been attributed to its molecular structure, whereby the potential for π–π stacking between the aromatic rings of given osteogenic inducers (e.g., dexamethasone, β-glycerophosphate) and the graphene has been hypothesized. This would act to effectively absorb and immobilize said biomolecules to its surface for more effective cell delivery and enhanced differentiation [3,24]. Seemingly at odds with the previous studies, several other publications have attested to the osteogenic potential of graphene films, scaffolds and hydrogels even in the absence of chemical inductors [2,4,6,8,12]. A resolution to these seemingly disparate discoveries may come with the recent finding that unstimulated paracrine cells plated onto graphene enhance their release of signaling molecules and proteins that are capable of promoting osteogenic differentiation of MCSc cultured onto inert glass substrate [2]. Hence, it is feasible that graphene both promotes the paracrine release of pro-osteogenic biomolecules in the immediate microenvironment as well as enhances their delivery to support osteogenic differentiation [2,4,8,11]. However, the potential of CVD-grown graphene promote osteogenesis in vivo and the cellular mechanisms evoked to allow differentiation remain largely unknown. In this report, the graphene scaffold induced the differentiation of human MSC in vivo (Figure 1). After four weeks from implantation, the histological analysis with H&E revealed the presence of bony-like tissues within the scaffolds resembling primary bone. Notably, the imunnohistochemical analysis of the tissues formed showed positive expression of the transcription factor RUNX2 and OPN. The first is essential for osteoblastic differentiation and skeletal morphogenesis, while OPN is an important bone matrix protein synthesized by osteoblastic cells [25,26]. Positive immunoreactivity to an antibody specific for human mitochondria indicated that the tissues formed was of human origin and confirmed that the cells remained viable for 28 days as well as that they underwent osteogenic differentiation in vivo.

Evidence exists for integrin-mediated mechanical signaling regulating stem cell paracrine factor production and release [10]. We investigated the role of the integrin-FAK-axis in the graphene-induced osteogenic differentiation. The FAK activity is related to alterations in the actin and microtubule filaments. As cells experience changes in forces through integrin contacts that link the extracellular matrix with the cytoskeleton, the FAK is activated modulating corrective cell responses to environmental stimuli [18,27]. Both integrins and FAK proteins are sensitive to physical stimuli and elastic properties of the substrates [28]. Accordingly, the high modulus of elasticity (up to 2.4 TPa) and high flexibility to out-of-plane deformation of graphene [8,22] could modulate the expression of these proteins and trigger a mechano-stimulated osteogenic differentiation. To test this hypothesis, MSC were cultured on PDMS substrates with modulus of elasticity ranging from 130 to 1.3 MPa with or without a coating of graphene. After 10 days, the expressions of the proteins involved in the integrin-FAK axis were evaluated by Western blot.

The expression of FAK-p397 was higher in all graphene-coated PDMS substrates (Gp) compared to PDMS alone (Figure 3). Similar enhancement in FAK expression was observed in fibroblasts cultured on stiff fibronectin gel (66 kPa) compared with a softer one (1050 Pa) [29]. Besides regulating cell adhesion, FAK-p397 is necessary for osteogenic differentiation, as its inhibition decreases RUNX2 expression by ~50% and mineralized matrix deposition by 30% [30].

Additionally, we observed that MSCs on Gp presented higher F-actin expression compared MSCs on uncoated PDMS (Figure 3). F-actin controls the cytoskeleton tension, promotes the phosphorylation of Smad and upregulates the downstream of osteogenic regulatory proteins [31]. The higher expression of RUNX2, OCN and OPN observed in this study may be related to greater substrate-mediated contractile forces reflected by increased F-actin stress fibers that enhances osteoblastic differentiation [18]. In addition, regardless of the stiffness of the underlying PDMS and presence of graphene, cells treated with the integrin inhibitor (echistatin) presented low expression of all proteins analyzed. This inhibitor binds to integrin $\alpha_v\beta_3$ receptor in a nondissociable manner [32] decreasing the expression of FAK, actin cytoskeleton and RhoA, hence, compromising the integrin-FAK signaling [33]. A similar trend ca be observed in Figure 3, confirming that the integrin-FAK complex is triggered by graphene, inducing osteogenic differentiation.

Contractile and mechanical forces associated by stiffer substrates activate RhoA/Rho-associated protein kinase (ROCK) that is essential for the formation of stress fibers and for the activation of Smad signaling imperative for osteogenesis [34,35]. MSCs on Gp exhibited higher expression of ROCK1, Smad p1/5 and both OCN and OPN compared to MSCs seeded onto PDMS alone (Figure 4). The latter are well-recognized signature proteins for the presence and function of mature osteoblasts [2,4]. Conversely, treatment with Y27632 decreased the expressions of ROCK1, Smad p1/5 and bone-related markers. Moreover, periodontal ligament cells treated with Y27632 also showed significant reductions in the expression of osteogenic genes (e.g., fibronectin 1, collagen type I and III, and biglycan) [36]. Together, these findings confirm that the graphene-induced osteogenic differentiation of MSC is mediated by the expression of ROCK1.

Finally, we checked the role of Smad (Figure 5), which is a sensitive regulatory protein that mediates the osteogenic differentiation in MSCs [37]. A single layer graphene increased the expression of Smad p1/5 independently of the stiffness of the underlying PDMS substrate. It is known that the phosphorylated Smad protein complex (e.g., Smad p1/5/8) can translocate into the cell nucleus and directly induce an upstream activator of RUNX2 [37,38], while the suppression of phosphorylation and nuclear translocation of Smad 1/5/8 inhibits the expression of osteogenic genes in MSCs [39]. Smad functions with RUNX2 through direct binding in a transcriptional activator complex during the osteogenic induction [40]. Accordingly, MSCs grown on Gp revealed the increased expression of Smad p1/5 and RUNX2 culminating in higher expressions of OCN and OPN proteins. The expressions decreased by the presence of DMH1 inhibitor confirming that the graphene-induced osteogenic differentiation is regulated by the upstream of Smad proteins.

Despite our provocative findings showing osteogenic induction by CVD-grown graphene in vivo and elucidating an underlying mechanotransduction-related signaling cascade, this work has limitations. For instance, PDMS with tunable properties has been shown to induce "inside-out" changes in cells to influence migration, stiffness and differentiation via adjusting the physical characteristics of the microenvironment [18,21]. Moreover, our synthetic material and the two-dimensional set-up of the experiments do not mimic the complex organic three-dimensional environments in which cells are imbedded in vivo. In addition, the contribution of the integrin-FAX axis in the graphene-induced osteogenic differentiation was elucidated in vitro, rather than in the animal. In addition, the osteogenic potential was evaluated in an ectopic model. Future studies shall evaluate the potential of graphene scaffold in situ and benchmark its potential to induce osteogenesis to other materials used for bone tissue engineering research. On other hand, these caveats provide us future avenues for investigation. Future studies will investigate the contribution of other integrin-relevant links (e.g., serine/threonine kinase and Rac pathways) and pathways involved in the differentiation of stem cells into osteoblasts (e.g., TGF-β/SMAD, Wnt and BMP-2). Despite of these limitations, our work provides the evidence that graphene induces the osteogenic differentiation by the upregulation of proteins sensitive to physical stimuli involved in the integrin-FAK axis. Thus, the molecular properties of graphene are capable of activating the integrin-FAK transmembrane complex that recruit the activity of both ROCK1 and F-actin that, in turn, stimulate the phosphorylation of Smad p1/5, upstream of RUNX2, OPN and OCN expression to bring osteogenesis closer to fruition (Figure 6).

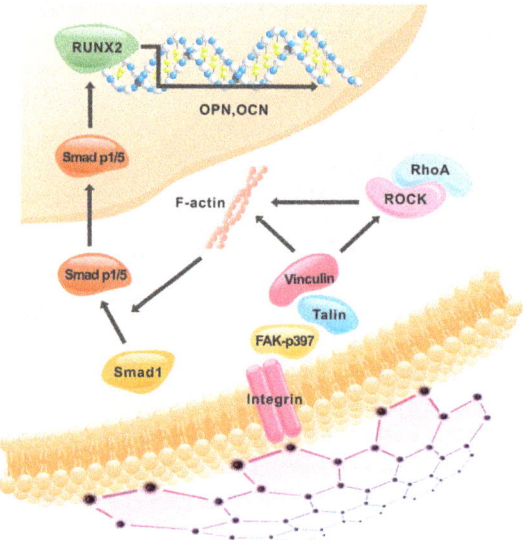

Figure 6. Mono-atomic graphene film promotes osteogenic differentiation of human MSCs via activation the integrin-FAK axis.

4. Material and Methods

4.1. Graphene Films and Scaffold Preparation

Graphene samples were produced by CVD using a custom-built furnace in a Class 1000 clean room facility at the NUS Centre for Advanced 2D Materials and Graphene Research Centre.

For single-layer graphene films, copper foils (Graphene Platform, Tokyo, Japan) were placed in a quartz tube. Thereafter, hydrogen gas (8 sccm) was introduced into the chamber and the temperature was increased to 1010 °C (25 °C/min) and maintained for 35 min. The graphene films were grown by flowing 16 sccm of CH_4 into the tube for 30 min at 500 mTorr. Finally, the tube was cooled to room temperature with H_2. The graphene-coated copper foils were spin-coated with polymethyl methacrylate (PMMA, Sigma, Saint Louis, MO, USA) and maintained at 180 °C for 10 min. The uncoated surface of the copper foil was etched with oxygen plasma (3 min, 50 W, 50 sccm, Vita-mini reactive ion etching, FEMTO Science Inc., Gyeonggi-Do, Korea). Following this, the copper was etched with 1.5% ammonium persulphate solution for 8 h and the graphene/PMMA transferred to deionized water for 2 h. Thereafter, the floating membrane was gently scooped onto the target polydimethylsiloxane (PDMS) substrates and the PMMA dissolved in acetone for 12 h. Finally, the graphene-coated samples were washed in isopropyl alcohol for 3 h.

The graphene scaffold was grown using nickel templates (density 320 ± 25 g/m^2, pore size ~500 µm, Alantum Advanced Technology Materials, Seongnam, Korea) with the same CVD protocol. Thereafter, the nickel was removed with $FeCl_3$ solution for 72 h at room temperature and washed with deionized water for 72 h.

Raman spectra of graphene films and scaffolds were obtained at room temperature, with an excitation laser source of 532 nm and power of 0.1 mW (Raman Microscope CRM 200, Witec, Ulm, Germany) and confirmed the production of single-layer films and multi-layer graphene scaffolds (Figure 7A).

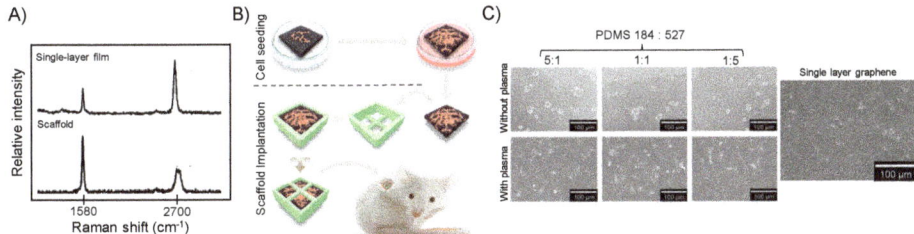

Figure 7. (**A**) Raman characterization. The Raman fingerprints confirmed the successful production of single layer graphene films and multilayer graphene scaffolds. (**B**) Graphene scaffold transplantation in vivo. MSC were seeded in the chemical vapor deposition (CVD)-grown graphene scaffolds (10 × 10 × 1 mm) and maintained in suspension plates at 37 °C for seven days to allow for attachment and proliferation. The scaffolds were placed in three-dimensionally (3D)-printed polylactide protection containers and implanted subcutaneously into immunodeficient mice for four weeks. (**C**) Cell attachment on PDMS with tunable elastic properties and graphene. Cells attached and proliferated on the surface of the oxygen plasma-etched PDMS substrates and single layer CVD-grown graphene.

4.2. Graphene Scaffold Transplantation In Vivo

The use of human cells and animals were approved by the NUS Institutional Review Board, Institutional Biosafety Committee and Institutional Animal Care and Use Committee (R17-0956, 12/12/2017).

Human dental pulp stem cells (DP003F, Alcells, Alameda, USA) were cultured in basal growth culture medium ((Dulbecco's modified Eagle's medium (DMEM, Invitrogen, Carlsbad, CA, USA), 10% fetal bovine serum (Invitrogen) and 1% penicillin/streptomycin (Invitrogen)) until 70~80% confluence and harvested (TrypLE, Invitrogen). As shown in Figure 7B, MSC (2×10^4, passage 3) were seeded in the graphene scaffold (0.8 × 0.8 × 0.2 cm) and maintained undisturbed for seven days. Following that, the scaffolds (n = 6) were placed inside 3D-printed polylactide protection containers (Cube, 3D Systems, Columbia, SC, USA) and transplanted into the subcutaneous space of immunodeficient mice (CB-17 SCID, Invivos, Singapore). After 28 days, specimens were retrieved, fixed with 4%

formaldehyde solution in phosphate-buffered saline at 4 °C for 24 h and stained with hematoxylin and eosin (H&E). Immunohistochemical analyses of the tissues formed were performed for OPN and RUNX2 (1:1000, Abcam, Cambridge, UK), and anti-human mitochondria antibodies (1:500, Abcam). Controls were tissue sections stained with an isotype-matched non-specific IgG antibody.

4.3. Role of Integrin-FAK Axis in the Graphene-Induced Osteogenic Differentiation

4.3.1. Substrate Preparation

To create the PDMS substrates with tunable modulus of elasticity, we blended Sylgard 184 and 527 (Dow Corning, Midland, TX, USA) in different proportions (5:1 = 1.3 MPa, 1:1 = 830 kPa, and 1:5 = 130 kPa [41]). Glass slides (22 mm × 22 mm) were covered with the freshly mixed PDMS and spin-coated (2000 rpm, 90 s). The PDMS substrates were maintained at 120 °C for 1 h followed by 60 °C overnight. To improve cell attachment, the PDMS substrates were treated with oxygen plasma (3 min, 5 W, 20 sccm, Vita-mini reactive ion etching, Figure 7C). Thereafter, half of the PDMS substrates were coated with a single layer of graphene (Gp) by the scooping method.

4.3.2. Inhibitor Concentration Optimization

Echistatin, Y27632 and DMH1 were used to selectively inhibit the expression of key proteins expressed in the integrin-FAK axis namely integrin, Rho-associated kinase 1 (ROCK1) and Smad.

To optimize the concentration of the inhibitors, MSC were seeded in 96-well flat-bottom plate (800/well) and maintained undisturbed in basal growth culture medium for 24 h. Thereafter, the culture medium was completely removed and 100 µL of DMEM with different concentrations of the inhibitors were added and changed every 48 h. Cell proliferation was evaluated for seven days by an MTS assay (CellTiter 96 AQueous One Solution Assay, Promega, Madison, WI, USA). The inhibitory ratio of cell proliferation was calculated based on the absorbance at 490 nm (Multiskan GO Microplate Spectrophotometer, Thermo Fisher Scientific, Waltham, MA, USA) obtained for the untreated cells (control). Statistical analyses were performed with Mann–Whitney test (α = 5%, SPSS Statistics 22, IBM, Armonk, NY, USA). The concentrations that inhibited cell proliferation by ~30% were selected for the subsequent tests.

4.3.3. Role of Integrin-FAK Axis on Graphene Osteogenic Differentiation

MSC (7×10^3) were seeded on the PDMS substrates alone (5:1, 1:1 and 5:1) or coated with graphene (Gp) and cultured for 10 days. Thereafter, the expression of integrin-FAK axis-related (FAK-p397, F-actin, ROCK, Smad p1/5) and bone-related proteins (RUNX2, OCN and OPN) were assessed by Western blot (all antibodies from Abcam) and gels imaged using ChemiDoc MP Imaging System (Bio-Rad, Hercules, CA, USA). Cells cultured with echistatin (10 nM), Y27632 (50 µM) or DMH1 (50 µM) were the controls. Three independent samples were evaluated per group. The intensity of the bands was quantified with ImageJ (NIH, Bethesfa, MD, USA). For the relative quantification, we firstly confirmed that the housekeeping protein (GAPDH) was constant across the groups. Secondly, we obtained the lane normalization factor (LNF) by dividing the housekeeping bands on the blot by the highest signal for the housekeeping protein identified. The NFs were approximately 1.0 for all groups indicating homogeneous housekeeping protein expression. Thirdly, the protein density (PD) of all the proteins of interest (e.g., FAK) was normalized to their respective housekeeping protein in each lane. Finally, the normalized signal of each PD was divided by the LNF and the relative expression was calculated as $[(Gp - PDMS)/PDMS] \times 100$.

5. Conclusions

CVD-grown graphene induces osteogenic differentiation of MSCs due to its molecular and physical properties (e.g., elastic properties, surface features) without the assistance of the conventional cocktail of osteogenic inducers considered necessary in other platforms. Here, we extended these findings by showing that graphene scaffolds also promote the osteogenic differentiation of MSCs in vivo. Osteogenesis was corroborated with the positive expression of RUNX2 and OPN, classical markers for osteogenic differentiation. Complementary in vitro tests demonstrated that the osteogenic differentiation was mediated by the activation of the mechanosensitive integrin-FAK axis. These developmental attributes appear to be inherent to the mono-atomic graphene sheet as changing the elastic modulus of the underlying PDMS had little effect on their ability to induce osteogenesis. Our findings deepen the understanding of the effects of graphene on biological systems and broaden the potential avenues for the use of this single-atom thin material to promote osteogenesis without the delivery limitations of exogenous chemical inducers.

Author Contributions: Conceptualization, V.R. and H.X.; Methodology, V.R. and H.X. Investigation, H.X.; Draft Preparation, Revision, H.X., T.C., A.F.O. and V.R.; Supervision, V.R.; Project Administration, V.R.; Funding Acquisition, V.R.

Acknowledgments: This research is supported by the National Research Foundation, Prime Minister' Office, Singapore, under its Medium Sized Centre Programme. V.R. was supported by the grants from the Singapore Ministry of Education, Singapore (Academic Research Fund Tier 1, R-221-000-091-112 and R-221-000-104-114) and the National University Health System, Singapore (NUHS Open Collaborative Research Grant NUHS O-CRG 2016 Oct-25).

Conflicts of Interest: The authors declare no conflict of interest.

References

1. Geim, A.K.; Novoselov, K.S. The rise of graphene. *Nat. Mater.* **2007**, *6*, 183–191. [CrossRef]
2. Xie, H.; Chua, M.; Islam, I.; Bentini, R.; Cao, T.; Viana-Gomes, J.C.; Castro Neto, A.H.; Rosa, V. CVD-grown monolayer graphene induces osteogenic but not odontoblastic differentiation of dental pulp stem cells. *Dent. Mater.* **2017**, *33*, e13–e21. [CrossRef] [PubMed]
3. Lee, W.C.; Lim, C.H.Y.X.; Shi, H.; Tang, L.A.L.; Wang, Y.; Lim, C.T.; Loh, K.P. Origin of enhanced stem cell growth and differentiation on graphene and graphene oxide. *ACS Nano* **2011**, *5*, 7334–7341. [CrossRef] [PubMed]
4. Xie, H.; Cao, T.; Gomes, J.V.; Castro Neto, A.H.; Rosa, V. Two and three-dimensional graphene substrates to magnify osteogenic differentiation of periodontal ligament stem cells. *Carbon* **2015**, *93*, 266–275. [CrossRef]
5. Xie, H.; Cao, T.; Rodríguez-Lozano, F.J.; Luong-Van, E.K.; Rosa, V. Graphene for the development of the next-generation of biocomposites for dental and medical applications. *Dent. Mater.* **2017**, *33*, 765–774. [CrossRef] [PubMed]
6. Li, J.; Wang, G.; Geng, H.; Zhu, H.; Zhang, M.; Di, Z.; Liu, X.; Chu, P.K.; Wang, X. CVD growth of graphene on NiTi alloy for enhanced biological activity. *ACS Appl. Mater. Interfaces* **2015**, *7*, 19876–19881. [CrossRef] [PubMed]
7. Dubey, N.; Ellepola, K.; Decroix, F.E.D.; Morin, J.L.P.; Castro Neto, A.H.; Seneviratne, C.J.; Rosa, V. Graphene onto medical grade titanium: An atom-thick multimodal coating that promotes osteoblast maturation and inhibits biofilm formation from distinct species. *Nanotoxicology* **2018**, *12*, 274–289. [CrossRef] [PubMed]
8. Nayak, T.R.; Andersen, H.; Makam, V.S.; Khaw, C.; Bae, S.; Xu, X.; Ee, P.L.; Ahn, J.H.; Hong, B.H.; Pastorin, G.; et al. Graphene for controlled and accelerated osteogenic differentiation of human mesenchymal stem cells. *ACS Nano* **2011**, *5*, 4670–4678. [CrossRef] [PubMed]
9. Dubey, N.; Bentini, R.; Islam, I.; Cao, T.; Castro Neto, A.H.; Rosa, V. Graphene: A versatile carbon-based material for bone tissue engineering. *Stem Cells Int.* **2015**, *2015*, 804213. [CrossRef]
10. Kusuma, G.D.; Carthew, J.; Lim, R.; Frith, J.E. Effect of the Microenvironment on Mesenchymal Stem Cell Paracrine Signaling: Opportunities to Engineer the Therapeutic Effect. *Stem Cells Dev.* **2017**, *26*, 617–631. [CrossRef]
11. Crowder, S.W.; Prasai, D.; Rath, R.; Balikov, D.A.; Bae, H.; Bolotin, K.I.; Sung, H.J. Three-dimensional graphene foams promote osteogenic differentiation of human mesenchymal stem cells. *Nanoscale* **2013**, *5*, 4171–4176. [CrossRef] [PubMed]
12. Lu, J.; He, Y.-S.; Cheng, C.; Wang, Y.; Qiu, L.; Li, D.; Zou, D. Self-supporting graphene hydrogel film as an experimental platform to evaluate the potential of graphene for bone regeneration. *Adv. Funct. Mater.* **2013**, *23*, 3494–3502. [CrossRef]

13. Nuttall, M.E.; Patton, A.J.; Olivera, D.L.; Nadeau, D.P.; Gowen, M. Human trabecular bone cells are able to express both osteoblastic and adipocytic phenotype: Implications for osteopenic disorders. *J. Bone Miner. Res.* **1998**, *13*, 371–382. [CrossRef]
14. Carragee, E.J.; Hurwitz, E.L.; Weiner, B.K. A critical review of recombinant human bone morphogenetic protein-2 trials in spinal surgery: Emerging safety concerns and lessons learned. *Spine J.* **2011**, *11*, 471–491. [CrossRef] [PubMed]
15. Reid, I.R. Glucocorticoid-induced osteoporosis. *Baillieres Best Pract. Res. Clin. Endocrinol. Metab.* **2000**, *14*, 279–298. [CrossRef] [PubMed]
16. Lin, F.; Du, F.; Huang, J.; Chau, A.; Zhou, Y.; Duan, H.; Wang, J.; Xiong, C. Substrate effect modulates adhesion and proliferation of fibroblast on graphene layer. *Colloids Surf. B Biointerfaces* **2016**, *146*, 785–793. [CrossRef]
17. Chandran, R.; Knobloch, T.J.; Anghelina, M.; Agarwal, S. Biomechanical signals upregulate myogenic gene induction in the presence or absence of inflammation. *Am. J. Physiol. Cell Physiol.* **2007**, *293*, C267–C276. [CrossRef]
18. Engler, A.J.; Sen, S.; Sweeney, H.L.; Discher, D.E. Matrix elasticity directs stem cell lineage specification. *Cell* **2006**, *126*, 677–689. [CrossRef]
19. Lu, Q.; Pandya, M.; Rufaihah, A.J.; Rosa, V.; Tong, H.J.; Seliktar, D.; Toh, W.S. Modulation of dental pulp stem cell odontogenesis in a tunable peg-fibrinogen hydrogel system. *Stem Cells Int.* **2015**, *2015*, 525367. [CrossRef]
20. Shih, Y.-R.V.; Tseng, K.-F.; Lai, H.-Y.; Lin, C.-H.; Lee, O.K. Matrix stiffness regulation of integrin-mediated mechanotransduction during osteogenic differentiation of human mesenchymal stem cells. *J. Bone Miner. Res.* **2011**, *26*, 730–738. [CrossRef]
21. Holle, A.W.; Engler, A.J. More than a feeling: Discovering, understanding, and influencing mechanosensing pathways. *Curr. Opin. Biotechnol.* **2011**, *22*, 648–654. [CrossRef] [PubMed]
22. Lee, J.U.; Yoon, D.; Cheong, H. Estimation of Young's modulus of graphene by Raman spectroscopy. *Nano Lett.* **2012**, *12*, 4444–4448. [CrossRef]
23. Kalbacova, M.; Broz, A.; Kong, J.; Kalbac, M. Graphene substrates promote adherence of human osteoblasts and mesenchymal stromal cells. *Carbon* **2010**, *48*, 4323–4329. [CrossRef]
24. Akhavan, O.; Ghaderi, E.; Shahsavar, M. Graphene nanogrids for selective and fast osteogenic differentiation of human mesenchymal stem cells. *Carbon* **2013**, *59*, 200–211. [CrossRef]
25. Kern, B.; Shen, J.; Starbuck, M.; Karsenty, G. Cbfa1 contributes to the osteoblast-specific expression of type I collagen genes. *J. Biol. Chem.* **2001**, *276*, 7101–7107. [CrossRef] [PubMed]
26. Sodek, J.; Chen, J.; Nagata, T.; Kasugai, S.; Todescan, R., Jr.; Li, I.W.; Kim, R.H. Regulation of osteopontin expression in osteoblasts. *Ann. N. Y. Acad. Sci.* **1995**, *760*, 223–241. [CrossRef] [PubMed]
27. Katsumi, A.; Orr, A.W.; Tzima, E.; Schwartz, M.A. Integrins in mechanotransduction. *J. Biol. Chem.* **2004**, *279*, 12001–12004. [CrossRef] [PubMed]
28. Michael, K.E.; Dumbauld, D.W.; Burns, K.L.; Hanks, S.K.; García, A.J. Focal Adhesion Kinase Modulates Cell Adhesion Strengthening via Integrin Activation. *Mol. Biol. Cell* **2009**, *20*, 2508–2519. [CrossRef]
29. Paszek, M.J.; Zahir, N.; Johnson, K.R.; Lakins, J.N.; Rozenberg, G.I.; Gefen, A.; Reinhart-King, C.A.; Margulies, S.S.; Dembo, M.; Boettiger, D.; et al. Tensional homeostasis and the malignant phenotype. *Cancer Cell* **2005**, *8*, 241–254. [CrossRef]
30. Salasznyk, R.M.; Klees, R.F.; Boskey, A.; Plopper, G.E. Activation of FAK is necessary for the osteogenic differentiation of human mesenchymal stem cells on laminin-5. *J. Cell. Biochem.* **2007**, *100*, 499–514. [CrossRef]
31. Wang, Y.K.; Yu, X.; Cohen, D.M.; Wozniak, M.A.; Yang, M.T.; Gao, L.; Eyckmans, J.; Chen, C.S. Bone morphogenetic protein-2-induced signaling and osteogenesis is regulated by cell shape, RhoA/ROCK, and cytoskeletal tension. *Stem Cells Dev.* **2012**, *21*, 1176–1186. [CrossRef] [PubMed]
32. Kumar, C.C.; Nie, H.; Rogers, C.P.; Malkowski, M.; Maxwell, E.; Catino, J.J.; Armstrong, L. Biochemical characterization of the binding of echistatin to integrin alphavbeta3 receptor. *J. Pharmacol. Exp. Ther.* **1997**, *283*, 843–853.
33. Shah, P.P.; Fong, M.Y.; Kakar, S.S. PTTG induces EMT through integrin $\alpha V\beta 3$-focal adhesion kinase signaling in lung cancer cells. *Oncogene* **2011**, *31*, 3124–3135. [CrossRef] [PubMed]
34. Hanna, S.; El-Sibai, M. Signaling networks of Rho GTPases in cell motility. *Cell. Signal.* **2013**, *25*, 1955–1961. [CrossRef] [PubMed]

35. Amano, M.; Nakayama, M.; Kaibuchi, K. Rho-Kinase/ROCK: A Key Regulator of the Cytoskeleton and Cell Polarity. *Cytoskeleton* **2010**, *67*, 545–554. [CrossRef] [PubMed]
36. Yamamoto, T.; Ugawa, Y.; Yamashiro, K.; Shimoe, M.; Tomikawa, K.; Hongo, S.; Kochi, S.; Ideguchi, H.; Maeda, H.; Takashiba, S. Osteogenic differentiation regulated by Rho-kinase in periodontal ligament cells. *Differentiation* **2014**, *88*, 33–41. [CrossRef]
37. Afzal, F.; Pratap, J.; Ito, K.; Ito, Y.; Stein, J.L.; van Wijnen, A.J.; Stein, G.S.; Lian, J.B.; Javed, A. Smad function and intranuclear targeting share a Runx2 motif required for osteogenic lineage induction and BMP2 responsive transcription. *J. Cell. Physiol.* **2005**, *204*, 63–72. [CrossRef]
38. Lee, K.S.; Hong, S.H.; Bae, S.C. Both the Smad and p38 MAPK pathways play a crucial role in Runx2 expression following induction by transforming growth factor-beta and bone morphogenetic protein. *Oncogene* **2002**, *21*, 7156–7163. [CrossRef]
39. Zouani, O.F.; Kalisky, J.; Ibarboure, E.; Durrieu, M.C. Effect of BMP-2 from matrices of different stiffnesses for the modulation of stem cell fate. *Biomaterials* **2013**, *34*, 2157–2166. [CrossRef]
40. Song, B.; Estrada, K.D.; Lyons, K.M. Smad signaling in skeletal development and regeneration. *Cytokine Growth Factor Rev.* **2009**, *20*, 379–388. [CrossRef]
41. Palchesko, R.N.; Zhang, L.; Sun, Y.; Feinberg, A.W. Development of polydimethylsiloxane substrates with tunable elastic modulus to study cell mechanobiology in muscle and nerve. *PLoS ONE* **2012**, *7*, e51499. [CrossRef]

© 2019 by the authors. Licensee MDPI, Basel, Switzerland. This article is an open access article distributed under the terms and conditions of the Creative Commons Attribution (CC BY) license (http://creativecommons.org/licenses/by/4.0/).

Article

Localized Delivery of Pilocarpine to Hypofunctional Salivary Glands through Electrospun Nanofiber Mats: An Ex Vivo and In Vivo Study

Sujatha Muthumariappan [1], Wei Cheng Ng [2], Christabella Adine [1], Kiaw Kiaw Ng [1], Pooya Davoodi [2], Chi-Hwa Wang [2] and Joao N. Ferreira [1,3,4,*]

1. Department of Oral and Maxillofacial Surgery, Faculty of Dentistry, National University of Singapore, Singapore 119085, Singapore; email2sujatha227@gmail.com (S.M.); christabella.adine@u.nus.edu (C.A.); kiawkiaw.ng@gmail.com (K.K.N.)
2. Department of Chemical and Biomolecular Engineering, National University of Singapore, Singapore 117585, Singapore; weicheng.ng@gmail.com (W.C.N.); davoodi.pooya@u.nus.edu (P.D.); chewch@nus.edu.sg (C.-H.W.)
3. Faculty of Dentistry, Chulalongkorn University, Bangkok 10330, Thailand
4. National Institute of Dental and Craniofacial Research, National Institutes of Health, Bethesda, MD 20892-4370, USA
* Correspondence: Joao.F@chula.ac.th; Tel.: +6622188816; Fax: +6622188810

Received: 18 December 2018; Accepted: 24 January 2019; Published: 28 January 2019

Abstract: Dry mouth or xerostomia is a frequent medical condition among the polymedicated elderly population. Systemic pilocarpine is included in the first line of pharmacological therapies for xerostomia. However, the efficacy of existing pilocarpine formulations is limited due to its adverse side effects and multiple daily dosages. To overcome these drawbacks, a localized formulation of pilocarpine targeting the salivary glands (SG) was developed in the current study. The proposed formulation consisted of pilocarpine-loaded Poly(lactic-*co*-glycolic acid) (PLGA)/poly(ethylene glycol) (PEG) nanofiber mats via an electrospinning technique. The nanofiber mats were fully characterized for their size, mesh porosity, drug encapsulation efficiency, and in vitro drug release. Mat biocompatibility and efficacy was evaluated in the SG organ ex vivo, and the expression of proliferation and pro-apoptotic markers at the cellular level was determined. In vivo short-term studies were performed to evaluate the saliva secretion after acute SG treatment with pilocarpine-loaded nanofiber mats, and after systemic pilocarpine for comparison purposes. The outcomes demonstrated that the pilocarpine-loaded mats were uniformly distributed (diameter: 384 ± 124 nm) in a highly porous mesh, and possessed a high encapsulation efficiency (~81%). Drug release studies showed an initial pilocarpine release of 26% (4.5 h), followed by a gradual increase (~46%) over 15 d. Pilocarpine-loaded nanofiber mats supported SG growth with negligible cytotoxicity and normal cellular proliferation and homeostasis. Salivary secretion was significantly increased 4.5 h after intradermal SG treatment with drug-loaded nanofibers in vivo. Overall, this study highlights the strengths of PLGA/PEG nanofiber mats for the localized daily delivery of pilocarpine and reveals its potential for future clinical translation in patients with xerostomia.

Keywords: electrospinning; nanomaterials; nanofibers; drug delivery; pilocarpine; salivary glands; hypofunction; dry mouth; xerostomia

1. Introduction

Dry mouth syndrome or xerostomia is a common global health problem that arises most commonly within an ageing population [1–3]. At the age of 70, 16% of men and 25% of women report xerostomia symptoms [2]. Xerostomia is mostly triggered by diuretic medication regimens, radiotherapy for head and neck cancers, auto-immune disorders, and uncontrolled diabetes, among other diseases [4,5]. Currently, muscarinic agonists like pilocarpine are the first line of pharmacological therapy for xerostomia [4]. Pilocarpine is only available in oral formulations, which upon ingestion produce significant systemic adverse effects, leading to low tolerability and reducing patient adherence to the drug [6,7]. Furthermore, pilocarpine tablets are contra-indicated in many medical conditions affecting the elderly, including cardiovascular disease, chronic obstructive pulmonary disease, and glaucoma [4,7]. Only half of the patients respond well to pilocarpine tablets, probably due to its short duration of action (3–5 h), requiring multiple daily administrations [8]. To this end, a localized pilocarpine formulation, specifically targeting the SG and bypassing systemic adsorption, is mandatory to make the saliva secretion more long-lasting. A recent double-blind randomized clinical trial showed that 0.1% pilocarpine mouthwash was not more effective than 0.9% saline after four weeks of treatment [9]. However, when pilocarpine dosing is increased to 1–2% in studies using mouthwash, there is a significant increase in salivation and high tolerability due to the absence of adverse side effects [10,11]. Several studies have tested the efficacy of topical pilocarpine administration via mouthwash or oral rinses, but they show contradicting outcomes [8,12], perhaps because pilocarpine mouthwashes only target the minor SG present in the oral mucosa, which are glands that in normal conditions do not significantly contribute to the whole saliva output (<1%) [13]. Major SG located underneath the facial skin should be targeted instead, potentially via an intradermal or transdermal route [14]. To meet this need, a new localized and SG-targeted pilocarpine drug delivery system is needed.

Poly(lactic-*co*-glycolic acid) (PLGA) and poly(ethylene glycol) (PEG) are advantageous materials for use as drug carriers due to their favorable biodegradability and biocompatibility, which ensure safe therapies [15,16]. Various carriers in the form of particles and fiber meshes fabricated in our previous works using electrohydrodynamic atomization (EHDA) techniques have been proven to work effectively for sustained and controlled drug release with hydrophilic drugs like pilocarpine [17,18]. In EHDA, the solution is subjected to high voltage, and the charged droplet disintegrates into fine jets or aerosols in a Taylor cone-like fashion, which upon immediate evaporation of the solvent leads to the formation of the desired solid carriers [19,20]. EHDA can be operated in the electrospraying or electrospinning mode to obtain different products (particles or fibers) of different sizes (microns, sub-microns, etc.) and morphologies, depending on the operating parameters (flow rates, voltage, etc.) and properties of material used (concentration, electrical conductivity, surface tension, viscosity, etc.) [21,22]. For instance, highly concentrated and viscous liquid favors the electrospinning regime to form fibers, as the liquid easily solidifies at the onset of jetting owing to a sufficiently strong elastic network that stabilizes the jet against breakup [22].

In this study, we present a new approach for the localized and controlled delivery of pilocarpine and successfully evaluated it ex vivo and in vivo. The proposed formulation was composed of homogenous pilocarpine-loaded PLGA/PEG nanofiber mats using electrohydrodynamic atomization techniques in electrospinning mode. Next, we aimed to (1) characterize the diameter distribution, porosity and in vitro drug release of this localized nanofiber mat formulation, to (2) evaluate its biocompatibility in ex vivo SG organs, and to (3) compare the in vivo saliva secretion rates of this localized formulation against systemic pilocarpine.

2. Results

2.1. Fabrication of Electrospun Nanofiber Mats

After optimization, it was found that using acetone as the solvent gave the most homogeneous and stable polymer solution/emulsion, compared to other organic solvents tested. At PLGA concentration of <30% (*w/v*), the fibers produced were intertwined with unwanted particles. Changes in the collection height and voltage did not significantly change this outcome, though increasing the PLGA concentration from 10% to 25% (*w/v*) did augment the ratio of fibers to particles. Only at 30% (*w/v*)

PLGA concentration were uniform nanofibers with no more particles consistently observed (Figure 1A), and this was accomplished with electrospinning operating parameters of 0.3 mL/h flow rate, 11.0 kV voltage, and 10 cm height of collection. Our finding was consistent with the literature in that, in general, a lower polymer concentration favors the formation of particles while a higher polymer concentration favors fiber fabrication [23].

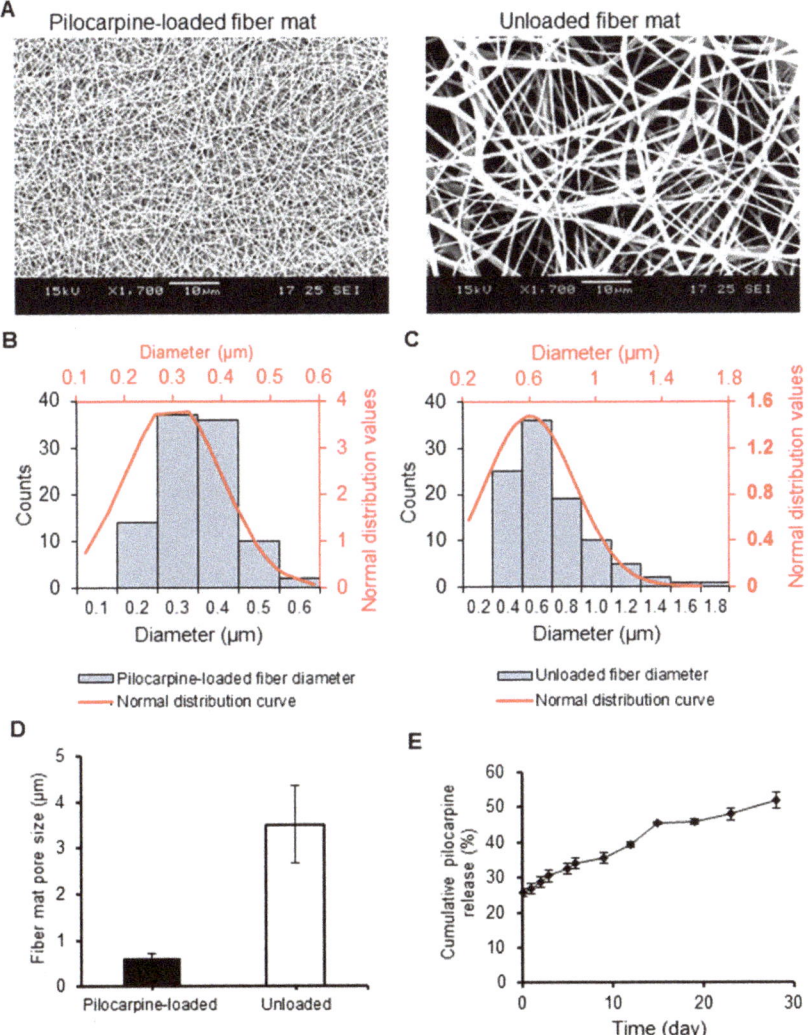

Figure 1. Structural and morphological comparison of pilocarpine-loaded and unloaded PLGA nanofiber mats, and in vitro pilocarpine release profile from pilocarpine-loaded mats. (**A**) Scanning electron microscopy (SEM) images at 1700× magnification. (**B–C**) Diameter distribution of the loaded (left) and unloaded PLGA fiber mats (right). (**D**) Mean porosity distribution of PLGA fiber mats. Error bars represent SD from $n = 10$. (**E**) Cumulative pilocarpine drug release from loaded PLGA nanofibers in the short and long term supported a steady pilocarpine release in vitro. The observed drug release profile was about 26% in the first 4.5 h, and increased steadily to 36% after 9 d, to 45% after 15 d, and to 52% after 28 d. This supports a steady pilocarpine release in vitro. Error bars represent SD from $n = 3$.

2.2. Physical and Chemical Characterization of Nanofibers

2.2.1. Scanning Electron Microscopy (SEM)

SEM images of the fabricated PLGA fiber mats (pilocarpine-loaded and unloaded) can be observed in Figure 1A. The average diameter and porosity of the fibers were determined for the loaded and unloaded formulations (Figure 1B,C, respectively). Well-distributed nano-scale fibers were observed for the pilocarpine-loaded fiber mat (diameter 384 ± 124 nm), and submicron fibers for the unloaded fiber formulation (diameter 936 ± 258 nm). The average porosity (pore size) of the pilocarpine-loaded fiber mat was significantly smaller as compared to the unloaded (611 ± 107 nm versus 3510 ± 838 nm) (Figure 1D). The porosity and diameter of the loaded nano-scale fiber mat was uniform as per its limited standard deviation.

2.2.2. Drug Loading Capacity, Encapsulation Efficiency, and Degradation

Next, the drug loading capacity of the loaded PLGA nanofiber mats as well as the encapsulation efficiency was calculated based on spectrophotometry results and standard theoretical formulas mentioned above. The drug loading capacity of the nanofibers was determined to be 0.84 ± 0.09% and the encapsulation efficiency was 81.1 ± 8.4%. For degradation studies of pilocarpine solution in different diluents (MilliQ water and 1× PBS), the optical densities of pilocarpine across a period from 4.5 h to 2 d did not change and were comparable for both diluents (results not shown), which indicates that no drug degradation occurred.

2.3. Short- and Long-Term In Vitro Pilocarpine Release

To determine the pilocarpine release from the PLGA nanofiber mats through time, we measured the pilocarpine released into a shaking PBS buffer from 4.5 h up to 28 d. In Figure 1E, the release curve showed an initial burst release of one-fourth of the encapsulated pilocarpine (25.7%) after 4.5 h, followed by a rather gradual and steady release of pilocarpine over time. After 15 d, almost 50% of encapsulated pilocarpine was released. Then, a slight plateau was noted from day 15 to day 19. After day 19, there was only a 6% increase up to day 28.

2.4. Ex Vivo Studies in Hypofunctional Salivary Glands

2.4.1. Cytotoxicity and ATP Activity of Pilocarpine

In order to test the cytotoxicity of pilocarpine in the SG organ, we cultured salivary glands ex vivo with media supplemented with this drug (Figure 2A) at a range of clinically relevant therapeutic concentrations, according to previous reports [24,25]. All pilocarpine concentrations, ranging from 0.1 to 20 µg/mL, supported the glandular growth through 48 h (Figure 2B and Figure S1), as well as the cellular metabolism as per stable ATP activity (Figure 2C). This indicated that no major organ cytotoxicity is induced by the pilocarpine solution within this clinical range. Therefore, the loaded nanofibers were fabricated to maximize the amount of pilocarpine up to 20 µg/mL.

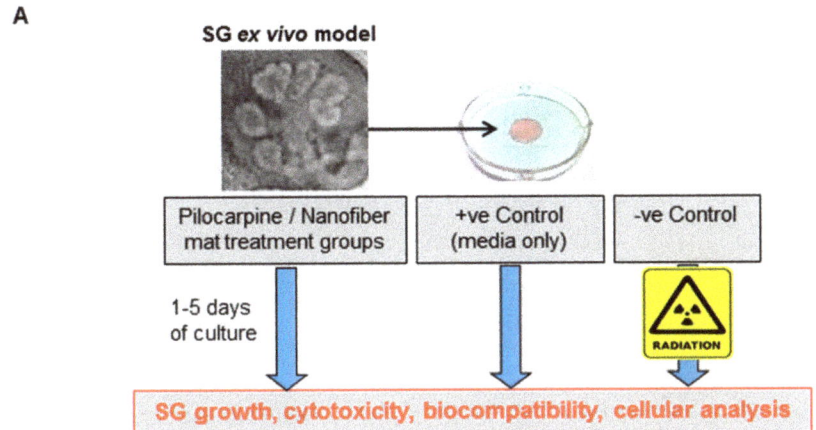
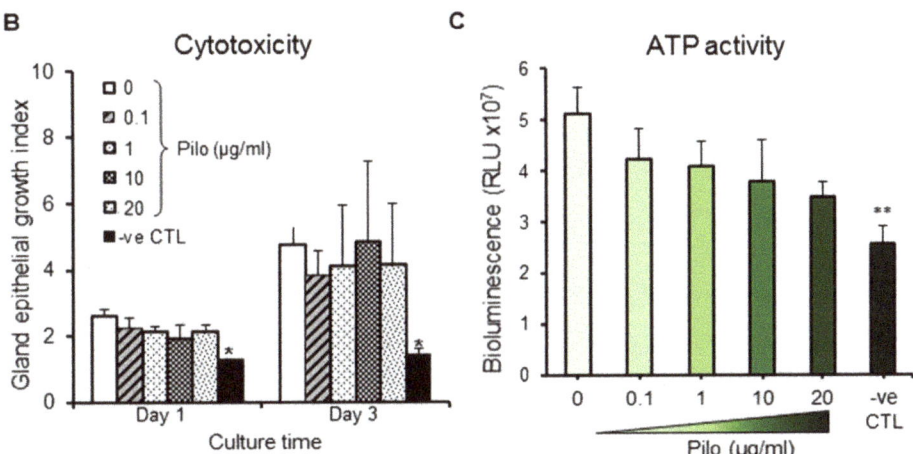

Figure 2. Biological effects in an ex vivo SG model. (**A**) Ex vivo SG model experimental set up used for testing the cytotoxicity of pilocarpine and biocompatibility of different nanofiber mats. (**B**) Glands were treated with therapeutic concentrations of pilocarpine (Pilo) ranging from 0.1 to 20 µg/mL. All gland growth index values at day 1 and 3 were normalized to time 0 (baseline). (**C**) ATP activity (a readout for gland viability) three days after glands were treated with therapeutic concentrations of pilocarpine (0.1–20 µg/mL). Error bars represent SD from $n = 4$. * $p < 0.05$ and ** $p < 0.01$ when compared to positive control without pilocarpine (0 µg/mL). Negative control (-ve CTL) represent glands damaged by gamma radiation.

2.4.2. Biocompatibility of Nanofiber Mats (Loaded and Unloaded)

To test the biocompatibility of the nanofibers, three different nanofiber discs were produced with different diameters and with a maximized concentration of pilocarpine (20 µg/mL), as shown in Figure S1. Then, ex vivo SG were cultured with media supplemented with pilocarpine-loaded and unloaded discs, and gland viability (a readout of epithelial growth) was quantified in the short and long term (Figure 3A,B). There was a steady increase in SG viability and growth in all treatment groups throughout the 5 d of culture. However, the 2 mm pilocarpine-loaded nanofiber mats produced a statistically significant lower gland viability across all culture days as compared to the gland with only media and no treatment provided (+ve CTL).

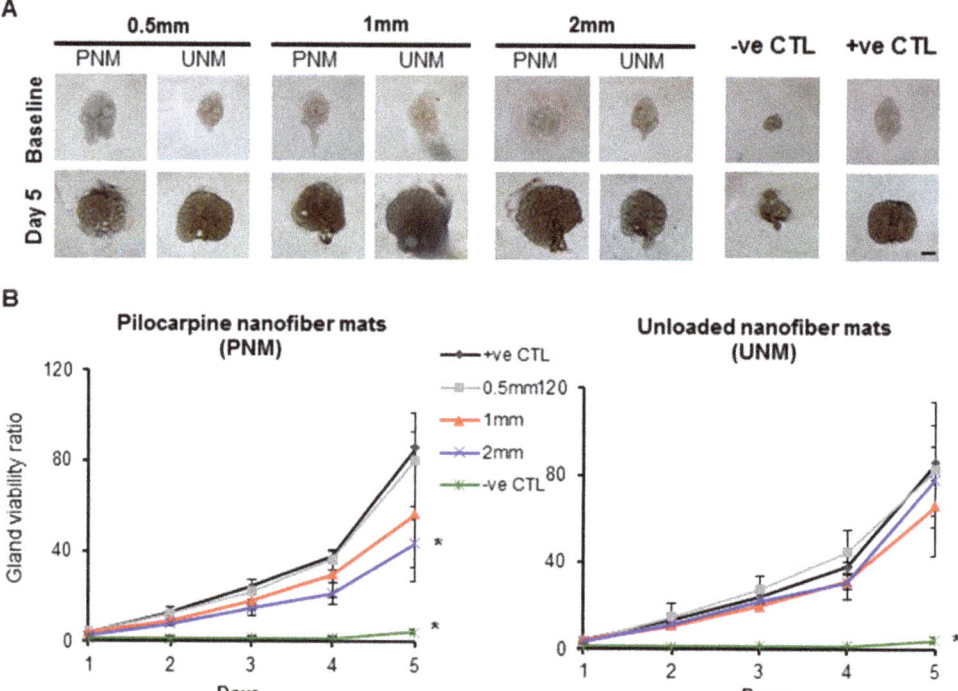

Figure 3. Biocompatibility of pilocarpine-loaded and unloaded nanofiber mats (PNM and UNM, respectively) in the ex vivo SG culture model. (A) Bright field images of the growing SG glands at 3.2× magnification. Scale bar: 400 μm. (B) SG epithelial viability (readout for organ biocompatibility) was supported by unloaded nanofibers and by 0.5 mm and 1 mm loaded nanofibers. Y axis is a ratio of epithelial bud number at a specific culture time relative to baseline. Error bars represent SD from n = 4–12. * $p < 0.05$, when compared to positive control without pilocarpine (+ve CTL) at every culture day; ns: not significant when compared to "+ve CTL". Negative control (-ve CTL) represent glands damaged by gamma radiation. PNM: Pilocarpine nanofiber mats. UNM: unloaded nanofiber mats.

2.4.3. Biological Effects of Pilocarpine-Loaded Nanofiber Mats in SG Cellular Compartments

Next, ex vivo salivary glands treated with pilocarpine-loaded nanofiber mats for 5 d were analyzed for the expression of cellular proliferation/pro-mitotic (Ki67) and pro-apoptotic (cleaved Caspase 3) markers, as seen in Figures 4 and 5, respectively. The Ki67 expression confirmed that the epithelial proliferation significantly decreased, in both the acinar branches and ductal networks, with 2 mm loaded discs (Figures 4 and 5). These cellular-based findings confirmed the outcomes from the previous organ toxicity experiments. As for cleaved Caspase 3, the expression of this pro-apoptotic marker was low ($\leq 0.01\%$), and no statistical differences were observed between the loaded discs. This suggests that salivary gland pro-apoptosis is grossly absent in the presence of the pilocarpine-loaded mats ranging from 0.5 mm to 2 mm, regardless of the pilocarpine amount in such mats.

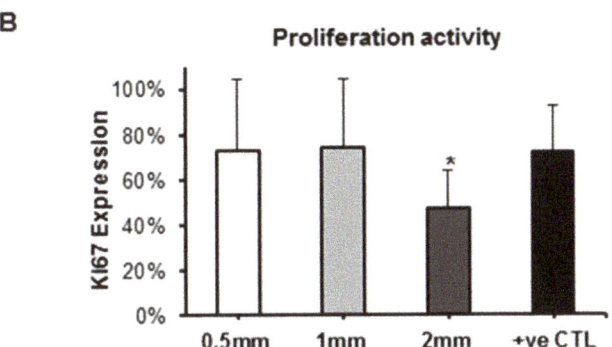

Figure 4. Expression of proliferation protein marker (Ki67) in SG after treatment with pilocarpine nanofiber mats after five culture days in the ex vivo SG model. (**A**) Fluorescence imaging after whole gland immunofluorescence staining showing expression of Ki67 in green (pro-mitotic marker), PNA in pink (peanut agglutinin staining the gland epithelial acini and ductal branched network) and nuclei in blue. (**B**) Expression of proliferation/pro-mitotic activity by quantification of Ki67 fluorescence after normalizing with nuclear counts. Error bars represent SD from $n = 4$. * $p < 0.05$ when compared to positive control. Positive control (+ve CTL) was not treated with nanofiber mats.

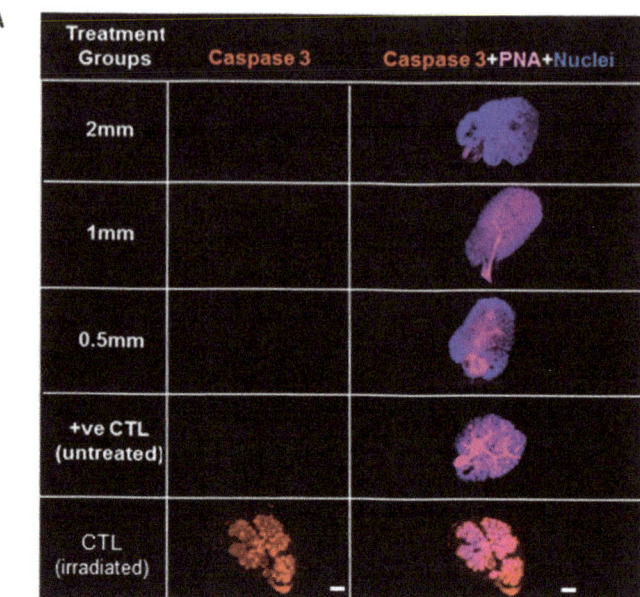

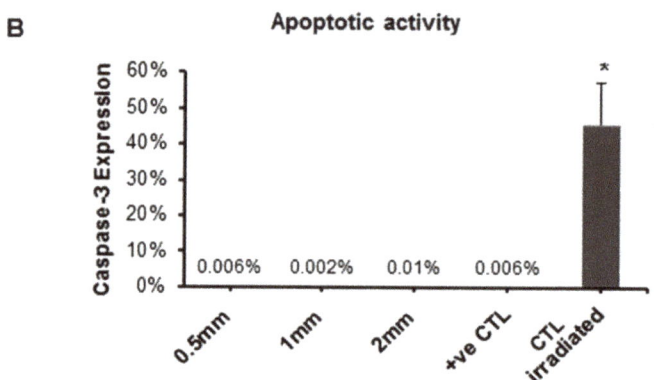

Figure 5. Expression of apoptotic protein marker (Caspase-3) in SG after treatment with pilocarpine nanofiber mats after five culture days in the ex vivo SG model. (**A**) Fluorescence imaging after whole gland immunofluorescence staining showing expression of Caspase-3 in red (apoptotic marker), PNA in pink (staining the gland epithelial acini and ductal branched network) and nuclei in blue. (**B**) Apoptotic activity by quantification of Caspase-3 fluorescence after normalizing with nuclear counts. Error bars represent SD from $n = 4$. * $p < 0.05$ when compared to positive control. Positive control (+ve CTL) was not treated with nanofiber mat and only had growth media. CTL (irradiated): gamma radiation treatment was used as a control for Caspase-3 staining since it induces apoptotic damage to the gland.

2.5. In Vivo Saliva Secretion after Intradermal Treatment with Pilocarpine-Loaded Nanofiber Mats versus Systemic Pilocarpine

Lastly, in vivo hypofunctional SG in an acute model (Figure 6A) were treated with intradermal applications of pilocarpine-loaded nanofiber mats (0.5 mm-diameter mats were selected based on high biocompatibility rates ex vivo) and compared with systemic pilocarpine (oral pilocarpine formulation could be used as per IACUC and mouse model limitations), and the salivary flow rate was determined to test our novel intradermal application formulation on a daily basis for 24 h (Figure 6B). After

4.5 h, SG with intradermal pilocarpine-loaded mats significantly increased the saliva secretion when compared to the systemic pilocarpine (Figure 6B). After 24 h, no difference was noticed between these two treatment formulations (Figure 6B). Furthermore, the whole gland weight was comparable between the two formulations indicating no gross changes in SG composition and cellular content (Figure 6C). Also, no histological differences were found between treatment groups.

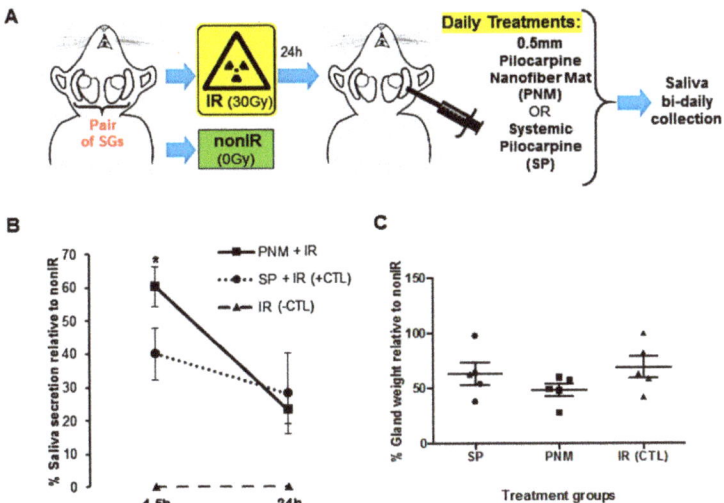

Figure 6. Treatment outcomes during daily intradermal applications of 0.5 mm pilocarpine nanofiber mats versus systemic pilocarpine in an acute in vivo SG model of SG hypofunction. (**A**) Schematic drawing of the in vivo SG hypofunction model to induce acute dry mouth for the daily proposed intradermal treatment with pilocarpine nanofiber mat and conventional systemic administration. (**B**) Saliva secretion rate during the first 24 h after daily intradermal application of 0.5 mm-diameter pilocarpine nanofiber mats (PNM) when compared to systemic pilocarpine (SP). (**C**) Salivary gland weight remained unchanged after intradermal application of 0.5 mm pilocarpine nanofiber mats. Error bars represent SEM from n = 4–5. * $p < 0.05$ when compared to irradiated group with systemic pilocarpine only (SP, which represents the positive CTL). IR: irradiated (negative CTL). nonIR: non-irradiated control group. SGs: salivary glands.

3. Discussion

This study developed a novel localized pilocarpine intradermal drug delivery system supporting the treatment of xerostomia at a greater extent than systemic pilocarpine. Our electrohydrodynamic atomization technique supported the production of a homogenous porous mats with a well-distributed network of electrospun PLGA/PEG nanofibers. Previous ophthalmic pilocarpine formulations used similar ingredients but a double emulsion fabrication method [24]. This latter method facilitated the fabrication of loaded nanoparticles, which possessed a pilocarpine encapsulation efficiency of only 57% as compared to 81.1% with our nanofibers. For double emulsion fabrication method, the chaotic liquid flow in the stirring region often results in spatially non-uniform stress and poor monodispersity, which is known to lead to low encapsulation efficiency [26]. In contrast, such stirring is not required in EHDA, and the fabrication is a one-step process as opposed to two-stage, hence minimizing the loss of drugs. This is also one key feature of the EHDA technique. Unfortunately, the oral/ophthalmic pilocarpine formulations could not be used in our mouse SG hypofunctional model as per local IACUC policies and unpredictable survival outcomes.

The nanofiber mats fabricated herein had a homogenous morphology, consistent diameter and porosity, alike other reports using electrospun fibers (Figure 1A–D) [27–29]. Other methodologies like the dropping method in a study by Kao and colleagues [25] revealed the difficulties in controlling the size of pilocarpine-loaded nanoparticles composed of chitosan and carbopol. In contrast, the nano-scale diameter of our fabricated pilocarpine-loaded mats was achievable with lesser difficulties probably due to salts present in the pilocarpine hydrochloride solution used. These salts might have enhanced the overall solution conductivity leading to nano-scale fiber formation when subjected to high voltage, as compared to the sub-micron scale of unloaded fiber mats [30].

As for the drug release profile (Figure 1E), a biphasic pilocarpine release in our nanofiber mats delivery system could be observed. The first phase consisted of an initial burst release of more than 25% within 4.5 h. This initial burst could be associated with the significant amount of drug just covering the outer surface of the nanofiber mats but not thoroughly encapsulated within, which can be easily solubilized in a favorable hydrophilic media like PBS. This initial burst may facilitate short-term localized and targeted formulations such as transdermal or topical (intra-oral as oral rinses). After 4.5 h, there was a slow and controlled drug release phase for up to 28 d of about 52% of pilocarpine. This secondary release phase may favor pilocarpine applications requiring a longer time in the mouth such as gel formulations to incorporate over stents or dentures.

Outcomes from the ex vivo study (Figures 3–5) revealed that when exposed to fiber mats, the gland cellular compartments were all found at a proliferative/pro-mitotic state with negligible pro-apoptotic activity, indicating that the toxicity of our fiber mats is low and that there is no impairment to the biocompatibility of the SG organ and its cellular compartment. Hence, this had a positive biological impact on the epithelial growth of our SG ex vivo model. These low cytotoxicity findings are also corroborated by other reports using similar materials [25,31].

Current pilocarpine systemic formulations require multiple intakes per day (three times at least) leading to poor patient adherence to this drug formulation [6,8,12]. On the other hand, our localized pilocarpine-loaded nanofiber mats could potentially be used as a single-dose daily formulation containing the therapeutic amount of pilocarpine to enhance saliva secretion when administered locally via an intradermal, as well as through subcutaneous, transdermal or topical applications (e.g., oral rinses or gels for dentures). Our in vivo study is the first comparing localized and systemic pilocarpine formulations for xerostomia (Figure 6). Applying these pilocarpine-loaded nanofiber mats locally via an intradermal route over the in vivo gland could greatly increase the saliva flow rate earlier on (4.5 h) when compared to systemic pilocarpine only (Figure 6B). Furthermore, both in vivo and ex vivo studies confirmed the biocompatibility of the nanofiber mats and supported the gland homeostasis and secretion. These outcomes make this pilocarpine-loaded nanofiber mats promising for clinical use once a day since they can readily increase saliva flow well before 24 h (4.5 h collection time). The long-term effects of these pilocarpine mats are to be explored in different in vivo SG hypofunction models using a clinically relevant radiotherapy fractionated dose regimen. Despite this, acute single-dose radiotherapy SG models, like the one used herein, can better induce epithelial damage promptly within 48 h [32], and are pertinent to study localized short-term effects for our proposed once daily pilocarpine intradermal mat administration. Pre-clinical trials determining the pilocarpine release pharmacokinetics are the next step if intradermal applications are to be tested. If transdermal applications are to be investigated, then further studies are necessary with skin adhesive patches to assess pilocarpine penetration, partitioning, diffusion and permeation properties [14].

4. Materials and Methods

4.1. Fabrication of Electrospun Nanofiber Mats

PLGA (50:50) (Lactel absorbable polymers, Birmingham, MI, USA) with an inherent viscosity of 0.26–0.54 dL/g was dissolved in acetone (VWR Chemicals, Fontenay-sous-Bois cedex, France) at 30% (w/v) overnight. A pilocarpine 2% solution (Alcon, Vilvoorde, Belgium) and polyethylene glycol-400 (PEG-400, Sigma-Aldrich, St. Louis, MO, USA) were added to the PLGA solution at PLGA: pilocarpine: PEG-400 volume ratio of 1: 0.1: 0.002 and the mixture was then sonicated at 30% amplitude (Vibra-cell

VCX 130, Sonics & Materials, Inc., Newtown, CT, USA) for a minute. For blank (unloaded) formulation, pilocarpine solution was replaced with distilled water. (Note: this final formulation was chosen after numerous rounds of optimization, using different solvents (dichloromethane (DCM), acetone, and ethanol) and mixing ratios/concentrations of each ingredients, to determine the most homogeneous and stable final polymeric solution/emulsion that does not phase separate. PEG known as a good surfactant was added to the solution to help increase the homogeneity and stability of emulsion.

After homogenization, solutions were loaded into a 5-mL syringe and the syringe was connected to the EHDA electrospinning setup to fabricate the fibers. The apparatus setup consisted of a stainless-steel needle/nozzle of inner diameter 0.72 mm (Popper and Sons, Lake Success, NY, USA), a syringe pump (KD Scientific, Holliston, MA, USA) to deliver polymer solution to the nozzle, a high voltage generator (Glassman High Voltage Inc., High Bridge, NJ, USA) to supply voltage to the nozzle via a crocodile clip, and a collecting stage at ground voltage. Upon steady state condition (i.e., the formation of Taylor cone at the tip of the nozzle), fiber sheets were collected in a glass Petri dish that was placed on the collecting stage. The dishes were subsequently wrapped with aluminum foil (since the drug is light-sensitive) and dried overnight in a vacuum oven to remove the residual solvent in the fibers. Next day, single-layer fiber sheets or mats were removed from the dishes and stored at 4 °C until use. Single-layer fiber mats were produced with different diameters (0.5, 1 and 2 mm) using a calibrated cylindrical puncher (Integra Miltex, Dutchess, NY, USA) as displayed in Table S1. This puncher allowed us to uniformly cut the mats to test the ex vivo biological effects with the different amounts/concentrations of biomaterials/pilocarpine, without changing their physical structure, fiber diameter, porosity and thickness.

4.2. Physical and Chemical Characterization of Nanofiber Mats

4.2.1. Scanning Electron Microscopy (SEM)

For SEM imaging, SEM metal stubs were placed directly under the Taylor-cone jet during electrospinning to collect representative fiber mat samples. The metal stubs were then sputter coated with Platinum (JFC-1300, JEOL, Tokyo, Japan) in vacuum at current intensity of 40 mA for 40 s. The coated samples were then viewed under the SEM (JEOL JSM 5600LV, Tokyo, Japan) to visualize the fibers morphology, distribution and pore size. To determine the diameter and porosity of the fibers, Image J software (NIH, Bethesda, MD, USA) was used.

4.2.2. Drug Loading Capacity and Encapsulation Efficiency

For this purpose, 10 mg of both unloaded and drug-loaded fiber mats were separately dissolved with 200 µL of DCM. Fibers were then vortexed for 30 s until full dissolution. Next, 5 mL of distilled water was added to the dissolved fibers, vortexed for 30 s, and centrifuged at 10,000 rpm for 10 min. Afterwards, the fibers were left for 1.5 h to allow for the separation of the organic and aqueous phases. Then, 3 mL of samples were collected from the aqueous phase. The samples were then analyzed for pilocarpine concentration by UV-Vis spectrophotometry (UV 1800, Shimadzu, Japan) at 215 nm (greatest sensitivity and maximum absorbance values; results not shown), with a standard curve having a good linear fitting (R^2 = 0.9965) within the pilocarpine concentration range of 5–40 µg/mL [33]. Unloaded fibers were used as the blank/reference in this analysis. These experiments were conducted in triplicate. Drug loading capacity was calculated using the following formula: (drug weight/(drug weight + PLGA weight + PEG-400 weight)) × 100%. The encapsulation efficiency (*EE*) was quantified using the following formula as per previous report (Kao, Lin et al. 2006): *EE* = [(Total pilocarpine-Free pilocarpine amount)/Total pilocarpine amount] × 100%.

4.2.3. Degradation Studies for the Pilocarpine Solution

The pilocarpine solution was tested for degradation in diluents such as 1× phosphate buffered saline (PBS) and ultra-pure water (MilliQ water, Sigma-Aldrich, Saint Louis, MO, USA) as this is essential to validate the drug release experiments in vitro [33,34]. For this purpose, pilocarpine solution was dissolved in 5 mL MilliQ water and 1× PBS to have a final concentration of 40 µg/mL. Solutions were vortexed for 30 s and placed in the rotary incubator at 37 °C, 100 rpm. At pre-determined time intervals (4.5 h, 1 d, 2 d), 3 mL of the drug solutions (in the different diluents) were drawn and analyzed within a wavelength ranging from 200 nm to 300 nm in the UV-Vis spectrophotometer. Reference samples of pure MilliQ water and 1× PBS only were used to blank the pilocarpine containing solutions.

4.3. Pilocarpine Drug Release Analysis from Loaded Nanofiber Mats

Fabricated nanofiber mats (pilocarpine-loaded and unloaded) weighing 20 mg each were used to analyze the drug release in vitro in the short and long term. Then, 20 mg fibers were placed in 4 mL of 1× PBS (pH 7.4) in 15-mL Falcon tubes. Tubes were vortexed for 30 s and placed in an orbital shaker bath (GFLVR 1092, Burgwedel, Germany) at 37 °C, 100 rpm. At pre-determined time intervals (4.5 h, and 1, 2, 3, 5, 6, 9, 12, 15, 19, 23, and 28 d), the tubes were subjected to centrifugation at 10,000 rpm for 10 min. After centrifugation, the supernatant (3.3 mL) was carefully collected from each tube into a new tube and stored at -20 °C for further pilocarpine quantification. Fresh 1× PBS (3.3 mL) was then added to the original sample tubes, vortexed for 30 s and placed back in the incubator. This procedure was repeated for all collection time points. Three biological replicates were run for both unloaded and loaded nanofibers for each time point. The pilocarpine concentration in the collected samples was again measured by UV-Vis spectrophotometer as described above. For each collection time point, the average absorbance readings of the unloaded nanofiber were subtracted from each of the loaded nanofibers, and then these were added up to calculate the cumulative drug release since the first collection time at 4.5 h.

4.4. Ex Vivo SG Organ Culture Studies

4.4.1. SG Organ Culture Model

Pregnant ICR mice were used as per the approved National University of Singapore IACUC protocol no. 2014-00306. Mice at embryonic day 15, after epithelial differentiation occurs to become an adult organ, were selected. Major salivary glands including submandibular and sublingual glands were isolated from ICR mouse embryos using a microdissection technique under a stereozoom microscope (M80, Leica, Wetzlar, Germany) as per previous reports [35,36]. The glands were then placed onto polycarbonate filter papers (Whatman™ Nuclepore™ Track-Etched Polycarbonate Membrane Filter, Sigma-Aldrich, Saint Louis, MO, USA). A growth medium was added (200 µL per culture plate), which was composed of DMEM/F12 media, penicillin-streptomycin (1% v/v), vitamin C (0.2% v/v), and transferrin (0.02% v/v) (Thermo Fisher Scientific, Waltham, MA, USA). Four whole salivary glands were cultured per treatment group and incubated at 37 °C with 5% CO_2.

4.4.2. Cytotoxicity of Pilocarpine

The gland growth media of the ex vivo organ culture was supplemented with clinically relevant concentrations of pilocarpine ranging from 0.1 to 25 µg/mL. The cultures were then incubated as previously. Epithelial cellular growth was analyzed at baseline (t_0), day 1 and day 3 by taking bright field microscopy images at 3.2× magnification with the above stereozoom microscope set up. The total number of epithelial buds per gland was quantified from those images using Image J (NIH). The growth index for each gland was determined by normalizing the number of epithelial buds at day 1 and day 3 to the ones at baseline [36]. The growth index was used as a readout for organ cytotoxicity. The positive control glands were those grown with the growth media only (without pilocarpine supplementation). The negative control glands were glands subjected to cytotoxic damage by gamma rays (7 Gy dosage) at baseline.

4.4.3. Biocompatibility of Nanofiber Mats

Punched nanofiber mats measuring 0.5, 1, and 2 mm were sterilize d under UV light for at least 6 h before utilization. To test the biocompatibility, mats were added to the growth media in the ex vivo SG organ cultures. These cultures were incubated for 5 d, and at each day, the viability index that measures epithelial cellular growth was determined for each gland as described above. Same positive and negative controls glands were used as before. Unloaded discs were used to make sure the unloaded nanofibers by itself are not causing any organ cytotoxicity. Three independent biological experiments were performed and four whole glands were used for each experiment.

4.4.4. Salivary Gland Cellular Analysis for Proliferation and Apoptotic Activity

Whole-mount immunofluorescence staining was used to evaluate and quantify the proliferative and apoptotic cell subpopulations in the ex vivo SG organ culture according to previous protocols [25,26]. Briefly, whole SG treated for 5 d with loaded nanofibers discs (0.5, 1, 2 mm in diameter) from the previous experiments were fixed in 4% paraformaldehyde followed by $1\times$ PBS washing steps. Then, fixed glands were permeabilized, and blocked using 10% donkey serum and 5% bovine serum albumin for 2 h. This was followed by overnight incubation at 4 °C with primary antibodies anti-Ki67 (1:200 dilution, Cat no. 556003, BD Pharmingen, San Jose, CA, USA) and anti-cleaved Caspase 3 (1:200 dilution, Cat no. 9664S, Cell Signalling, Danvers, MA, USA). After washing, incubation with respective secondary antibodies followed (1:200 dilution, Alexa Fluor 488 and Alexa Fluor 635, Thermo Fisher Scientific). Rhodamine-labelled peanut agglutinin (1:200 dilution, Vector Lab, Burlingame, CA, USA) was used to stain the acinar and ductal branched epithelial structures of each gland. A nuclear fluorescent dye (Hoechst 33342, Invitrogen, Carlsbad, CA, USA) was used to counterstain the nuclei. Whole glands were then mounted on glass slides in resin mounting media using spacers to avoid gland crushing by glass slide. Slides were visualized, scanned and analyzed with Leica DMI-8 fluorescence microscope with a z-motorized axis and Leica LAS-X software (Leica Microsystems, Wetzlar, Germany) to obtain a z-stack of images for each gland with 20 μm z-steps. Maximum intensity projections were further produced at $10\times$ and $20\times$ magnifications. Automated cell counts for each gland z-stack were done using Image J (NIH, USA) for immuno-fluorescently labeled Ki-67+ and Caspase 3+ cells, and for all cells stained with the nuclear dye. Ki-67+ and Caspase 3+ cell counts for each gland were normalized to the total number of cells (stained with the nuclear dye).

4.5. In Vivo SG Hypofunctional Model

To generate hypofunctional SG with xerostomia measurable signs, a single acute dose of gamma radiation (30 Gy) was given to eight-week-old C57BL/6 mice as per a published protocol [37]. This protocol was approved by the National University of Singapore IACUC application no. 2014-00306. All in vivo experiments complied with the ARRIVE guidelines and were carried out in accordance with the National Institutes of Health guide for the care and use of laboratory animals (NIH Publications No. 8023, revised 1978). Animals were housed in a climate- and light-controlled environment and allowed free access to food and water at the bottom of the cage. One day after radiation was delivered, our localized pilocarpine nanofiber formulation was administered over the submandibular glands (via an intradermal route) and compared with systemic pilocarpine (via an intraperitoneal route, Sigma). The oral pilocarpine clinical formulation (available in the market) could not be given according to the IACUC committee, due to the unpredictability and life-threatening adverse side effects in rodents. The following experimental mouse groups were tested on a daily basis as per the acute SG dry mouth model: (A) Non-irradiated control (nonIR, $n = 3$); (B) Irradiated control (IR, $n = 5$); (C) Irradiated and treated daily with 0.5 mm diameter pilocarpine-loaded nanofibers (PNM, $n = 5$), the ones that performed effectively in the ex vivo biocompatibility tests; (D) Irradiated and treated daily with systemic pilocarpine as a positive/conventional control treatment (SP, $n = 5$). Intraperitoneal ketamine (10 mg/mL) and xylazine (1 mg/mL) were administered before the treatment was delivered to minimize animal discomfort. As for the nanofiber composition and formulation, the sterilized ones used in the abovementioned ex vivo studies were administered here: first these were placed in a syringe vessel with a 21G needle, and a sterile PBS 1× solution was added as a vehicle.

Whole stimulated saliva was collected at 4.5 h and 24 h after nanofiber administration. Saliva was extracted by capillarity every 3 min for 15 min using a 75-mm hematocrit tube (Drummond, Scientific Company, Broomall, PA, USA) [38]. These tubes were then placed in 1.5-mL pre-weight Eppendorf tubes, which were then weighed to determine the saliva secretion rate. To determine the percentage of secreted saliva relative to the control (NonIR), the saliva secretion rate was normalized to the average of saliva secreted by the NonIR group. Submandibular glands were surgically removed and freshly dissected in a stereozoom microscope and their weight was measured before fixation. Glands were then fixed with paraformaldehyde 4% overnight at 4 °C in a rotating shaker, followed by several washing steps with PBS and incubation with 30% cold sucrose (w/v) solution in 0.01 M PBS for 3–4 d. Then, after paraffin embedding, 4-µm gland sections were made and stained with hematoxylin and eosin and visualized with bright-field microscopy at 20–40× magnification.

4.6. Statistical Analysis

Data are presented as mean ± standard deviation (SD) from 3–12 independent samples, except for the in vivo experiments, where the standard error of the mean (SEM) was used. All statistical analyses were conducted using Prism version 6 (GraphPad Software, Inc., San Diego, CA, USA). All datasets were tested for normality. Student *t*-tests were performed for two-group comparison: unpaired with Welch correction for in vivo study outcomes, or paired for time-dependent experiments for drug release profile experiments. A one-way ANOVA with Dunnet post hoc tests for comparison of three or more experimental groups with a positive control for the case of ex vivo studies. The significance level for all experiments was set at $p < 0.05$.

5. Conclusions

In this study, the ultimate aim was to determine whether an intradermal pilocarpine-loaded PLGA/PEG nanofiber mat can be utilized to release pilocarpine while supporting the SG organ viability and stimulating saliva secretion.

The intradermal pilocarpine-loaded nanofiber mats ranging from 0.5 to 1 mm diameter induced the highest SG growth and cell proliferation with negligible cytotoxicity (apoptosis) in ex vivo SG models. In the acute dry mouth in vivo model, the daily intradermal application of 0.5 mm diameter pilocarpine-loaded nanofiber mats stimulated a higher saliva secretion before 24 h as compared to the conventional systemic pilocarpine. After 24 h, saliva secretion was comparable to systemic pilocarpine.

Thus, this pilocarpine delivery system is promising for potential clinical use as an intradermal formulation for the early and prompt treatment of xerostomia before and after 24 h of application. These formulations will be particularly useful in patients who have poor saliva secretion with conventional saliva stimulants.

Supplementary Materials: The following are available online at http://www.mdpi.com/1422-0067/20/3/541/s1.

Author Contributions: Conceptualization, P.D., C.-H.W. and J.N.F.; Methodology, W.C.N., P.D. and J.F; Validation, P.D., C.-H.W. and J.N.F.; Formal Analysis, S.M. and W.N.; Investigation, S.M., C.A., and K.K.N.; Resources, C.H.W. and J.N.F.; Data Curation, J.N.F.; Writing—Original Draft Preparation, S.M., W.C.N. and J.N.F.; Writing—Review & Editing, C.A., K.K.N., P.D., C.-H.W. and J.N.F.; Visualization, S.M., W.C.N., and J.N.F.; Supervision, C.A., K.K.N., P.D., and J.N.F.; Project Administration, J.N.F.; Funding Acquisition, C.-H.W. and J.N.F.

Funding: This work was supported by National University of Singapore Cross Faculty Research grant CFGFY17P01, and by Chulalongkorn University Faculty of Dentistry start up funds.

Acknowledgments: We thank Bruce Baum (National Institutes of Health) for his critical reading.

Conflicts of Interest: The authors have no conflicts of interest to report.

References

1. Barbe, A.G.; Bock, N.; Derman, S.H.; Felsch, M.; Timmermann, L.; Noack, M.J. Self-assessment of oral health, dental health care and oral health-related quality of life among Parkinson's disease patients. *Gerodontology* **2017**, *34*, 135–143. [CrossRef] [PubMed]
2. Gil-Montoya, J.A.; Barrios, R.; Sanchez-Lara, I.; Carnero-Pardo, C.; Fornieles-Rubio, F.; Montes, J.; Gonzalez-Moles, M.A.; Bravo, M. Prevalence of Drug-Induced Xerostomia in Older Adults with Cognitive Impairment or Dementia: An Observational Study. *Drugs Aging* **2016**, *33*, 611–618. [CrossRef] [PubMed]
3. Pajukoski, H.; Meurman, J.H.; Halonen, P.; Sulkava, R. Prevalence of subjective dry mouth and burning mouth in hospitalized elderly patients and outpatients in relation to saliva, medication, and systemic diseases. *Oral Surg. Oral Med. Oral Pathol. Oral Radiol. Endod.* **2001**, *92*, 641–649. [CrossRef] [PubMed]
4. Gil-Montoya, J.A.; Silvestre, F.J.; Barrios, R.; Silvestre-Rangil, J. Treatment of xerostomia and hyposalivation in the elderly: A systematic review. *Med. Oral Patol. Oral Cir. Bucal* **2016**, *21*, e355–e366. [CrossRef] [PubMed]
5. Napenas, J.J.; Brennan, M.T.; Fox, P.C. Diagnosis and treatment of xerostomia (dry mouth). *Odontology* **2009**, *97*, 76–83. [CrossRef] [PubMed]
6. Noaiseh, G.; Baker, J.F.; Vivino, F.B. Comparison of the discontinuation rates and side-effect profiles of pilocarpine and cevimeline for xerostomia in primary Sjogren's syndrome. *Clin. Exp. Rheumatol.* **2014**, *32*, 575–577. [PubMed]
7. Hendrickson, R.G.; Morocco, A.P.; Greenberg, M.I. Pilocarpine toxicity and the treatment of xerostomia. *J. Emerg. Med.* **2004**, *26*, 429–432. [CrossRef]
8. Davies, A.N.; Thompson, J. Parasympathomimetic drugs for the treatment of salivary gland dysfunction due to radiotherapy. *Cochrane Database Syst. Rev.* **2015**, *10*, CD003782. [CrossRef]
9. Kim, J.H.; Ahn, H.J.; Choi, J.H.; Jung, D.W.; Kwon, J.S. Effect of 0.1% pilocarpine mouthwash on xerostomia: Double-blind, randomised controlled trial. *J. Oral Rehabil.* **2014**, *41*, 226–235. [CrossRef]
10. Bernardi, R.; Perin, C.; Becker, F.L.; Ramos, G.Z.; Gheno, G.Z.; Lopes, L.R.; Pires, M.; Barros, H.M. Effect of pilocarpine mouthwash on salivary flow. *Braz. J. Med. Biol. Res.* **2002**, *35*, 105–110. [CrossRef]
11. Tanigawa, T.; Yamashita, J.; Sato, T.; Shinohara, A.; Shibata, R.; Ueda, H.; Sasaki, H. Efficacy and safety of pilocarpine mouthwash in elderly patients with xerostomia. *Spec. Care Dentist.* **2015**, *35*, 164–169. [CrossRef] [PubMed]
12. Spivakovsky, S.; Spivakovsky, Y. Parasympathomimetic drugs for dry mouth due to radiotherapy. *Evid. Based Dent.* **2016**, *17*, 79. [CrossRef] [PubMed]
13. Atkinson, J.C.; Baum, B.J. Salivary enhancement: Current status and future therapies. *J. Dent. Educ.* **2001**, *65*, 1096–1101.
14. Alkilani, A.Z.; McCrudden, M.T.; Donnelly, R.F. Transdermal Drug Delivery: Innovative Pharmaceutical Developments Based on Disruption of the Barrier Properties of the stratum corneum. *Pharmaceutics* **2015**, *7*, 438–470. [CrossRef] [PubMed]
15. Dhar, S.; Gu, F.X.; Langer, R.; Farokhzad, O.C.; Lippard, S.J. Targeted delivery of cisplatin to prostate cancer cells by aptamer functionalized Pt(IV) prodrug-PLGA-PEG nanoparticles. *Proc. Natl. Acad. Sci. USA* **2008**, *105*, 17356–17361. [CrossRef] [PubMed]
16. Bi, Y.; Liu, L.; Lu, Y.; Sun, T.; Shen, C.; Chen, X.; Chen, Q.; An, S.; He, X.; Ruan, C.; et al. T7 Peptide-Functionalized PEG-PLGA Micelles Loaded with Carmustine for Targeting Therapy of Glioma. *ACS Appl. Mater. Interfaces* **2016**, *8*, 27465–27473. [CrossRef]
17. Ranganath, S.H.; Fu, Y.; Arifin, D.Y.; Kee, I.; Zheng, L.; Lee, H.S.; Chow, P.K.; Wang, C.H. The use of submicron/nanoscale PLGA implants to deliver paclitaxel with enhanced pharmacokinetics and therapeutic efficacy in intracranial glioblastoma in mice. *Biomaterials* **2010**, *31*, 5199–5207. [CrossRef]
18. Davoodi, P.; Ng, W.C.; Yan, W.C.; Srinivasan, M.P.; Wang, C.H. Double-Walled Microparticles-Embedded Self-Cross-Linked, Injectable, and Antibacterial Hydrogel for Controlled and Sustained Release of Chemotherapeutic Agents. *ACS Appl. Mater. Interfaces* **2016**, *8*, 22785–22800. [CrossRef]
19. Feng, F.; Neoh, K.Y.; Davoodi, P.; Wang, C.H. Coaxial double-walled microspheres for combined release of cytochrome c and doxorubicin. *J. Control Release* **2017**, *259*, e30–e31. [CrossRef]
20. Davoodi, P.; Srinivasan, M.P.; Wang, C.H. Effective Co-delivery of Nutlin-3a and p53 genes via Core-shell Microparticles for Disruption of MDM2-p53 Interaction and Reactivation of p53 in Hepatocellular Carcinoma. *J. Mater. Chem. B* **2017**, *5*, 5816–5834. [CrossRef]

21. Xie, J.; Jiang, J.; Davoodi, P.; Srinivasan, M.P.; Wang, C.H. Electrohydrodynamic atomization: A two-decade effort to produce and process micro-/nanoparticulate materials. *Chem. Eng. Sci.* **2015**, *125*, 32–57. [CrossRef] [PubMed]
22. Davoodi, P.; Feng, F.; Xu, Q.; Yan, W.C.; Tong, Y.W.; Srinivasan, M.P.; Sharma, V.K.; Wang, C.H. Coaxial electrohydrodynamic atomization: Microparticles for drug delivery applications. *J. Control. Release* **2015**, *205*, 70–82. [CrossRef] [PubMed]
23. Xie, J.; Tan, R.S.; Wang, C.H. Biodegradable microparticles and fiber fabrics for sustained delivery of cisplatin to treat C6 glioma in vitro. *J. Biomed. Mater. Res. A* **2008**, *85*, 897–908. [CrossRef] [PubMed]
24. Nair, K.L.; Vidyanand, S.; James, J.; Kumar, G.S.V. Pilocarpine-loaded poly (DL-lactic-co-glycolic acid) nanoparticles as potential candidates for controlled drug delivery with enhanced ocular pharmacological response. *J. Appl. Polym. Sci.* **2012**, *124*, 2030–2036. [CrossRef]
25. Kao, H.J.; Lin, H.R.; Lo, Y.L.; Yu, S.P. Characterization of pilocarpine-loaded chitosan/Carbopol nanoparticles. *J. Pharm. Pharmacol.* **2006**, *58*, 179–186. [CrossRef] [PubMed]
26. Yuan, Q.C.; Williams, R.A. Precision emulsification for droplet and capsule production. *Adv. Powder Technol.* **2014**, *25*, 122–135. [CrossRef]
27. Chua, K.N.; Chai, C.; Lee, P.C.; Tang, Y.N.; Ramakrishna, S.; Leong, K.W.; Mao, H.Q. Surface-aminated electrospun nanofibers enhance adhesion and expansion of human umbilical cord blood hematopoietic stem/progenitor cells. *Biomaterials* **2006**, *27*, 6043–6051. [CrossRef]
28. Blakney, A.K.; Little, A.B.; Jiang, Y.; Woodrow, K.A. In vitro-ex vivo correlations between a cell-laden hydrogel and mucosal tissue for screening composite delivery systems. *Drug. Deliv.* **2016**, *24*, 582–590. [CrossRef] [PubMed]
29. Zeng, J.; Xu, X.; Chen, X.; Liang, Q.; Bian, X.; Yang, L.; Jing, X. Biodegradable electrospun fibers for drug delivery. *J. Control Release* **2003**, *92*, 227–231. [CrossRef]
30. Choi, J.S.; Lee, S.W.; Jeong, L.; Bae, S.H.; Min, B.C.; Youk, J.H.; Park, W.H. Effect of organosoluble salts on the nanofibrous structure of electrospun poly (3-hydroxybutyrate-co-3-hydroxyvalerate). *Int. J. Biol. Macromol.* **2004**, *34*, 249–256. [CrossRef]
31. Han, F.; Zhang, H.; Zhao, J.; Zhao, Y.; Yuan, X. Diverse release behaviors of water-soluble bioactive substances from fibrous membranes prepared by emulsion and suspension electrospinning. *J. Biomater. Sci. Polym. Ed.* **2013**, *24*, 1244–1259. [CrossRef] [PubMed]
32. Varghese, J.J.; Schmale, I.L.; Mickelsen, D.; Hansen, M.E.; Newlands, S.D.; Benoit, D.S.W.; Korshuno, V.A.; Ovitt, C.E. Localized Delivery of Amifostine Enhances Salivary Gland Radioprotection. *J. Dent. Res.* **2018**, *97*, 1252–1259. [CrossRef] [PubMed]
33. Agban, Y.; Lian, J.; Prabakar, S.; Seyfoddin, A.; Rupenthal, I.D. Nanoparticle cross-linked collagen shields for sustained delivery of pilocarpine hydrochloride. *Int. J. Pharm.* **2016**, *501*, 96–101. [CrossRef] [PubMed]
34. Saleh, N.; Al-Handawi, M.B.; Al-Kaabi, L.; Ali, L.; Salman Ashraf, S.; Thiemann, T.; Al-Hindawi, B.; Meetani, M. Intermolecular interactions between cucurbit [7]uril and pilocarpine. *Int. J. Pharm.* **2014**, *460*, 53–62. [CrossRef] [PubMed]
35. Adine, C.; Ng, K.K.; Rungarunlert, S.; Souza, G.R.; Ferreira, J.N. Engineering innervated secretory epithelial organoids by magnetic three-dimensional bioprinting for stimulating epithelial growth in salivary glands. *Biomaterials* **2018**, *180*, 52–66. [CrossRef] [PubMed]
36. Ferreira, J.N.; Zheng, C.; Lombaert, I.M.A.; Goldsmith, C.M.; Cotrim, A.P.; Symonds, J.M.; Patel, V.N.; Hoffman, M.P. Neurturin Gene Therapy Protects Parasympathetic Function to Prevent Irradiation-Induced Murine Salivary Gland Hypofunction. *Mol. Ther. Methods Clin. Dev.* **2018**, *23*, 172–180. [CrossRef]
37. Lin, A.L.; Johnson, D.A.; Wu, Y.; Wong, G.; Ebersole, J.L.; Yeh, C.K. Measuring short-term gamma-irradiation effects on mouse salivary gland function using a new saliva collection device. *Arch. Oral Biol.* **2001**, *46*, 1085–1089. [CrossRef]
38. Baum, B.J.; Afione, S.; Chiorini, J.A.; Cotrim, A.P.; Goldsmith, C.M.; Zheng, C. Gene Therapy of Salivary Diseases. *Methods Mol. Biol.* **2017**, *1537*, 107–123. [PubMed]

© 2019 by the authors. Licensee MDPI, Basel, Switzerland. This article is an open access article distributed under the terms and conditions of the Creative Commons Attribution (CC BY) license (http://creativecommons.org/licenses/by/4.0/).

Article

The Retentive Strength of Laser-Sintered Cobalt-Chromium-Based Crowns after Pretreatment with a Desensitizing Paste Containing 8% Arginine and Calcium Carbonate

Raphael Pilo, Sharon Agar-Zoizner, Shaul Gelbard and Shifra Levartovsky *

Department of Oral Rehabilitation, The Maurice and Gabriela Goldschleger School of Dental Medicine, Tel- Aviv University, Tel -Aviv 6997801, Israel; rafipilo@gmail.com (R.P.); sharon.agar@gmail.com (S.A.-Z.); s1gelbard@gmail.com (S.G.)
* Correspondence: shifralevartov@gmail.com; Tel.: +972-52-3515403; Fax: +972-3-6409250

Received: 25 October 2018; Accepted: 13 December 2018; Published: 17 December 2018

Abstract: The retention of laser-sintered cobalt-chromium (Co-Cr)-based crowns were examined after dentin pretreatment with desensitizing paste containing 8% arginine and calcium carbonate (DP-ACC). Forty lower first molars were prepared using a standardized protocol. The Co-Cr crowns were produced using selective laser melting. The teeth were either pretreated with the desensitizing paste or not pretreated. After one week, each group was cemented with glass ionomer cement (GIC) or zinc phosphate cement (ZPC). Surface areas of the teeth were measured before cementation. After aging, a universal testing machine was used to test the retentive strength of the cemented crown-tooth assemblies. The debonded surfaces of the teeth and crowns were examined at 2.7× magnification. Pretreating the dentin surfaces with the desensitizing paste before cementation with GIC or ZPC did not affect the retention of the Co-Cr crowns. The retention of the GIC group (6.04 ± 1.10 MPa) was significantly higher than that of the ZPC group (2.75 ± 1.25 MPa). The predominant failure mode for the ZPC and the nontreated GIC group was adhesive cement-dentin failure; for the treated GIC group, it was adhesive cement-crown failure. The desensitizing paste can be safely used to reduce post-cementation sensitivity without reducing the retentive strength of Co-Cr crowns cemented with GIC or ZPC.

Keywords: desensitizing paste; dentin; retention; cements; cobalt-chromium

1. Introduction

Dentin hypersensitivity following tooth preparation and cementation of fixed partial dentures (FPDs) has been a common phenomenon [1]. Post-cementation complaints from patients are received for 20 to 30% of crowns inserted [2], and this rate remains at 6% and 3% after two and three years, respectively [3]. There are several explanations for this postoperative sensitivity. One outcome of aggressive tooth preparation is an increased number of opened and expanded dentinal tubules [4,5]. This condition is further aggravated by inadequate provisional restorations and removal of the smear layer due to acid etching induced by the cements [6]. Porcelain fused to metal (PFM) restorations, for example, are most commonly luted with zinc phosphate or glass ionomer cements (GICs), which are acidic in nature [7].

In an effort to control postoperative sensitivity, various desensitizing agents have been used to seal dentinal tubules before crown cementation; however, the literature is inconsistent regarding the effects of these agents on the retentive strength of FPDs. Sailer et al. [8,9] and Stawarczyk et al. [10,11] showed that glutaraldehyde/HEMA pretreatment and resin sealing of dentin following tooth preparation had a beneficial effect on the shear bond strength of self-adhesive resin cement. Other studies have reported that dentin desensitizing by means of glutaraldehyde-containing primers or dentin sealing by means of bonding agents did not affect the bond strength of the cements tested [12,13]. On the other hand, several studies have demonstrated that these agents decrease crown retention to some extent [14,15]. Aranha et al. [16] showed that specimens treated with dentin desensitizers (except Gluma) yielded significantly lower mean bond strengths than nontreated control specimens.

Recently, a new in-office Colgate Sensitive Pro-Relief Desensitizing Paste containing 8% arginine and calcium carbonate (DP-ACC) was shown to provide immediate and lasting relief from dentin hypersensitivity [17–21]. No significant difference in the bonding strength of composites to enamel or dentin pretreated with this desensitizing paste has been reported [22–24]. In addition, pretreating dentin surfaces with DP-ACC prior to cementation did not affect the retention of complete cast metal crowns luted with a glass ionomer cement (GIC) [25] or the retention of zirconium oxide crowns luted with a resin-modified GIC or a self-adhesive resin cement [26].

Recently, a new additive manufacturing technology operated by computer-aided design and computer-aided manufacturing (CAD/CAM), referred to as selective laser melting (SLM) technology, has been introduced for fabricating cobalt-chromium (Co-Cr) frameworks for PFM crowns. Co-Cr crowns produced with SLM exhibit a marginal and internal accuracy that is comparable to that of conventional production procedures but save time and facilitate laboratory procedures [27–29].

No information has been found in the literature concerning the influence of pretreating dentin with DP-ACC on the retentive strength of Co-Cr crowns produced by the SLM technology and cemented by zinc phosphate cement (ZPC) and glass ionomer cement (GIC), which are the most frequently used cements for luting metal-based restorations.

The aim of this in vitro study was to evaluate the effect of the pretreatment of dentin with DP-ACC on the retentive strength of SLM Co-Cr copings cemented by ZPC and GIC. The null hypotheses were as follows: (1) the retentive strength of the SLM Co-Cr copings cemented by ZPC and GIC to human extracted teeth is not affected by DP-ACC, and (2) the retentive strength of the two cements is similar.

2. Results

The retentive strength (mean, SD) of the treated and untreated cementation groups are presented in Table 1. Pretreating the dentin surfaces with DP-ACC prior to cementation with either GIC or ZPC did not affect the retentive strength of the Co-Cr copings ($p = 0.780$). The retention obtained with Fuji I capsules (GIC) was significantly ($p = 0.001$) higher than that obtained with Harvard Cement OptiCaps (ZPC). The interaction between cement and dentin treatment was not significant ($p = 0.208$).

Table 1. Mean (SD) retentive strength (MPa) of the cobalt-chromium-based crown for all cementation groups.

Cement Type	Treatment	Sample No.	Mean Retentive Value (MPa)	Standard Deviation
GIC	1	10	6.39	1.06
	2	10	5.73	1.10
	Total	20	6.04	1.10
ZPC	1	10	2.39	0.99
	2	10	3.10	1.44
	Total	20	2.75	1.25
Total	1	20	4.29	2.27
	2	20	4.41	1.83
	Total	40	4.36	2.03

Treatment: 1, Without pretreatment with DP-ACC (control); 2, With pretreatment with DP-ACC.

Examination by magnifying glasses of the failure mode after the dislodgment of the crown revealed that for ZPC, the predominant failure mode was adhesive cement-dentin failure. In 85% of the surfaces, all (53%) or part (32%) of the surface of the crown was covered with cement, and the rest were detected on the dentin (Figure 1). This mode of failure was consistent regardless of whether the dentin was pretreated with DP-ACC or not. In the GIC group, the predominant failure mode was adhesive cement-crown failure. In 62% of the surfaces, all (40%) or part (22%) of the surface of the dentin was covered with cement, and the rest were detected on the crown (Figure 1). This failure mode was inconsistent between the groups. In the nontreated group, more surfaces exhibited the adhesive cement-dentin mode of failure; in the treated group, more surfaces exhibited the adhesive cement-crown mode of failure (Figure 1). Cohesive cement failure was barely seen in all groups, while cohesive dentin failure did not occur.

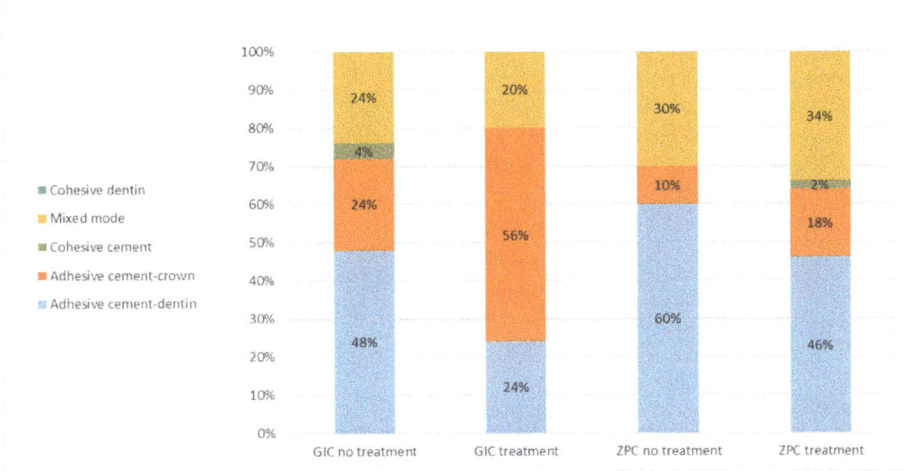

Figure 1. Distribution of failure modes (number of surfaces) for each cementation group.

Scanning electron microscopy (SEM) images of the dentin surfaces illustrating the various modes of failure are presented for the ZPC (Figure 2a–c) and GIC (Figure 3a,b) groups. Figure 2a–c illustrates the mixed mode of failure, whereas most of the dentin surface exhibits longitudinal striations of the bur, with a small part covered with the ZPC. This type of failure was consistent in the untreated (A,B), as well as the treated (C) ZPC groups.

Figure 3a,b illustrates the adhesive crown cement mode of failure, with most of the dentin surface covered with cement. This mode of failure was typical of the treated GIC group.

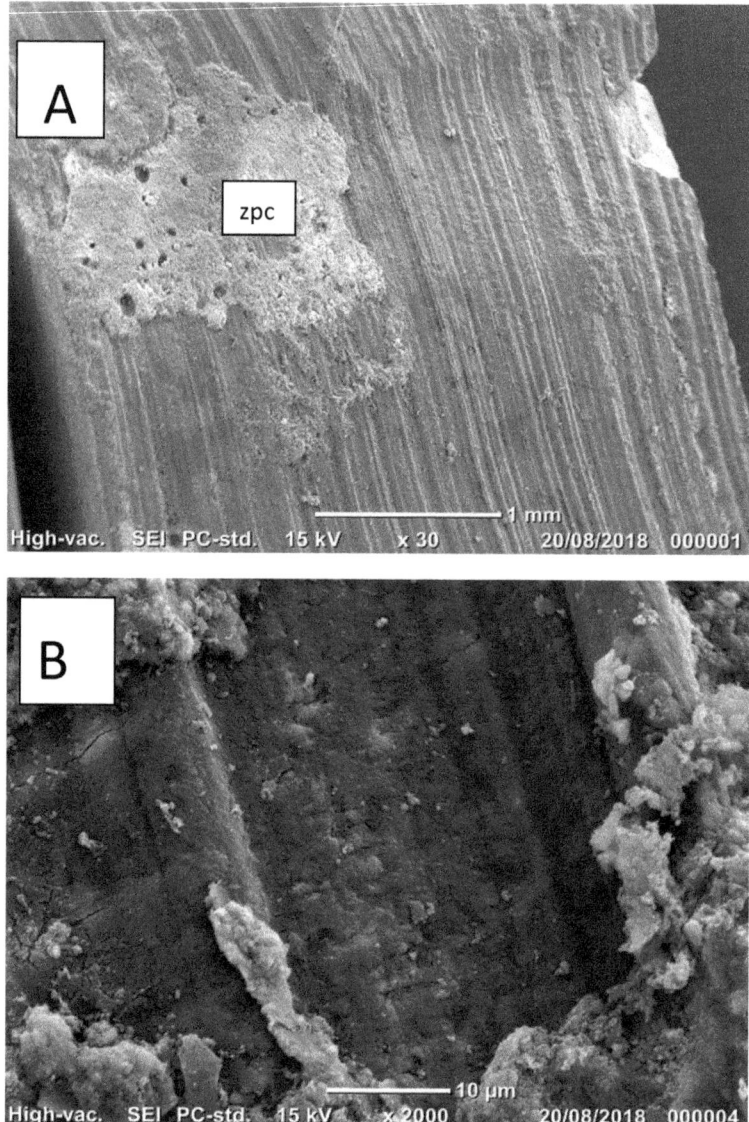

Figure 2. *Cont.*

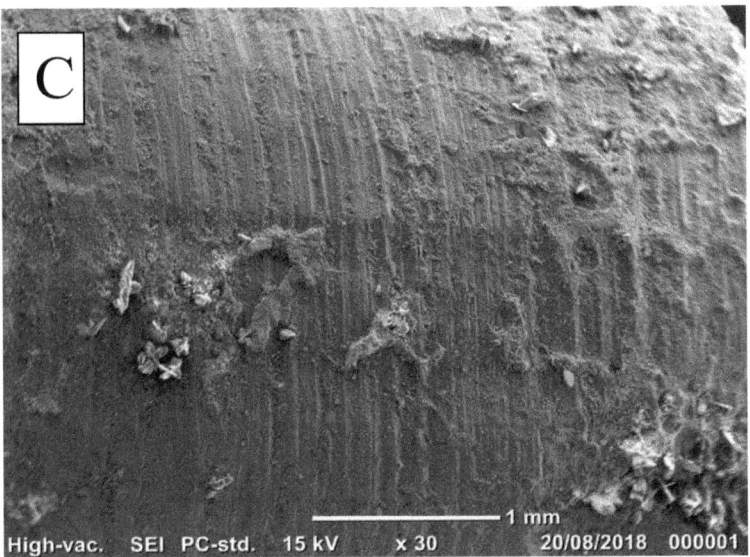

Figure 2. **A–C** (ZPC group). Scanning electron microscopy image of the untreated (**A,B**) and treated (**C**) dentin surfaces after failure illustrating the mixed mode of failure; the majority exhibited adhesive cement-dentin failure, while most of the dentin surface exhibited longitudinal striations of the bur, with only a small part covered with cement (ZPC). This type of predominantly adhesive cement-dentin failure was typical of the ZPC group.

Figure 3. *Cont.*

Figure 3. A,B (GIC group). Scanning electron microscopy images of the treated dentin surface after failure. (**A**) Most of the dentin surface is covered with cement (30×), and only a small part of the dentin is exposed (d), demonstrating the striations of the bur. (**B**) At greater magnification (170×), craze lines in the cement layer caused by the dehydration process are evident.

3. Discussion

Colgate Sensitive Pro-Relief Desensitizing Paste contains arginine, bicarbonate, and calcium carbonate and is highly effective in occluding dentin tubules, as was previously demonstrated by confocal laser scanning microscopy (CLSM) and SEM [30]. This paste has been shown to physically plug and seal exposed dentin tubules and to effectively provide dentin hypersensitivity relief [31], which has been reported to last for up to 28 days and to reduce the post-preparation and post-cementation sensitivity of vital teeth that serve as abutments for FPDs; however, this treatment may be advocated only if retentive strength is not affected [20].

In the current study, the retention of cobalt-chromium-based (Co-Cr) copings was tested one week after dentin pretreatment with DP-ACC in order to resemble a period of reevaluation prior to final cementation. The luting agents used in this study were GIC and ZPC. Both cements are acid-base materials associated with post-cementation sensitivity, which can last for a week [32]. GIC and ZPC are popular choices for luting metal-based restorations. GIC relies on both the mechanical retention to surface irregularities and the chelation to calcium in the tooth structure, while ZPC relies only on mechanical retention to both dentin and crown surfaces [33,34]. GICs are naturally adhesive to dentin; initially due to the presence of polyacrylic acid, forming hydrogen bonds between the free carboxyl groups and the strongly bound water layers residing on the surface of the dentin. Subsequently they are gradually replaced by ionic bonds involving mainly calcium in the mineral phase at the surface of the dentin and carboxylate groups in the cement. Imaging of dentin surfaces treated with DP-ACC reveals surface coverage including the tubule orifices. This surface coverage may thus interfere with the interaction of the cement with the mineral phase of the surface dentin as well as block the longitudinal striations of the bur and, thus, impair both the micromechanical and chemical mode of action of the cements, affecting the bond strength/retention.

The current results support the first null hypothesis of the study, implying that pretreatment with DP-ACC would have no effect on the retentive strength of SLM Co-Cr crown copings cemented to human extracted teeth with GIC or ZPC. Our results are in agreement with those of another study that demonstrated that pretreating dentin surfaces with DP-ACC prior to cementation did not affect the retention of complete cast metal crowns luted with GIC [25]. Moreover, in the aforementioned study, pretreatment with DP-ACC showed the best retention of complete cast metal crowns compared to all other dentin desensitizers tested. Pilo et al. [26] showed the same results with zirconium oxide (Y-TZP) crowns luted with either resin-modified glass ionomer cement (RMGIC) or self-adhesive resin cement (SARC).

Our second null hypothesis was rejected because the retentive strength of GIC was significantly higher than that of ZPC. This conclusion is in accordance with a previous study of Wiskott et al. [35] demonstrating that crowns luted with resin composite cement and GIC were more resistant to dynamic lateral loading than those luted using ZPC. On the other hand, Gorodovsky et al. [36] reported no significant difference between the retention of zinc phosphate and that of glass ionomer. A review of the research of different cements did not reveal a consistent conclusion about the retentive strength of GIC in comparison with that of ZPC. Some studies showed a higher retentive strength for ZPC, and others reported a higher retentive strength for GIC; some showed no significant difference between the two cements [33]. However, it should be noted that all the aforementioned ZPC studies used the classic Harvard Cement normal setting, whereas in the current study, the newer Harvard Cement OptiCaps was used. The latter is the capsulated version intended to overcome mistakes in mixing and dosing. It has been shown that Vickers hardness increases with more powder with a rise in mean values following an exponential curve ranging between 34 and 66 MPa [37]. Comparison studies between hand-mixed and capsulated ZPC have not been reported yet.

The lack of effect on the retentive strength of ZPC from DP-ACC was also verified by an absence of change in failure mode, which was mainly adhesive cement-dentin failure. This finding implies that all or most of the crown was covered by ZPC, probably due to surface irregularities of the intaglio of the Co-Cr copings. In the GIC group, the predominant failure mode was adhesive cement-crown failure. This finding implies that all or most of the dentin was covered by GIC, probably due to the chemical interaction between the calcium in the mineral phase at the surface of the dentin and carboxylate groups in the cement, implying that in spite of the surface coverage by the DP-ACC, many calcium ions are still available for bonding. Although the DP-ACC did not affect the retentive strength, the failure mode varied between the groups; in the nontreated group, most of the cement remained on the crown, while in the treated group, most of the dentin was covered with cement. These differences might be explained by chelation between the polyalkenoic chains in GIC also with the calcium carbonate contained in the DP-ACC, which physically plugs and seals exposed dentin tubules.

This is an in vitro study, hence, the subjective desensitizing effect of the DP-ACC on vital teeth prepared for FPD must be validated in clinical studies before being recommended for use.

4. Materials and Methods

The study sample comprised forty freshly extracted, caries-free, intact lower first molars that were extracted for periodontal reasons (age range 40–60). Approval from the Ethics Committee of Tel Aviv University was obtained (#21-08-16), and all individuals signed an informed consent.

The teeth were stored in a germ-free 0.1% thymol tap water solution at room temperature for a maximum of two weeks until experimentation. Each tooth was suspended in the middle of an aluminum ring and was mounted 2 mm apical to the cementoenamel junction (CEJ) in poly(methyl methacrylate) resin (Quick resin, Ivoclar, Schaan, Liechtenstein) after notching the roots for retention purposes. The mounted teeth were stored in tap water at room temperature at all times.

A standardized protocol yielding an axial height of 5 mm and a 10° taper was followed for preparation. The occlusal surface was sectioned perpendicular to the long axis with a water-cooled precision saw (Isomet Plus, Buehler, IL, USA). A 0.4-mm, 360° chamfer finish line located 1 mm above the CEJ with a 10° taper preparation was obtained by a rigidly secured, high-speed handpiece equipped with a diamond bur (C1-Strauss, Ra'anana, Israel) mounted on a custom-designed, surveyor-like apparatus. A new diamond bur was used for each tooth.

The prepared teeth were digitally scanned by a laboratory scanner (Series 7, Dental Wing, Letourneux, Montreal, Canada) operated by blue light and equipped with five axes of rotation, and STL files were obtained. Forty Co-Cr copings were produced using an SLM system (Eosint M 280, EOS, Krailling, Germany) at a commercial dental laboratory (MS Systems, Or Yehuda, Israel). The CAD-CAM Co-Cr cores were 1.0 mm thick with a 50 μm virtual cement spacer layer. To facilitate tensile loading, an occlusal loop (4-mm outer diameter and 2-mm inner diameter) was designed extending coronally from the occlusal surface [38]. The teeth were randomly assigned to two groups (2 × 20). In one group, the dentin surfaces were pretreated with DP-ACC using prophy cups under light pressure according to the manufacturer's recommendations. In the second group (the control group), the dentin surfaces were not pretreated. After placing the Co-Cr copings on each tooth, they were stored at 37 °C under 100% humidity for one week, resembling a period of reevaluation prior to final cementation. Two luting cements were evaluated (2 × 10) in each group: a GIC (GC Fuji I Capsule, GC, Tokyo, Japan) and a ZPC (Harvard Cem OptiCaps, Harvard Dental International, GmbH). The areas of the axial and occlusal surfaces of each prepared tooth were measured prior to cementation, as previously described [38]. The cements were used according to the manufacturer's recommendations. Cementing each crown to its tooth was conducted in a standardized manner under a constant load of 50 N (Force gauge, FG 20, Lutron, Taiwan) for 10 min and then allowed to set for 24 h.

The cemented crown-tooth assemblies were stored in tap water at 37 °C for two weeks, followed by thermal cycling between water temperatures of 5 and 55 °C for 5000 cycles with a 10 s dwell time (Y. Manes, Tel-Aviv, Israel). After thermal cycling, the crown-tooth assemblies were subjected to dislodgment forces through a 1.2-mm diameter metal cable entangled through the occlusal loop along the apico-occlusal axis using a universal testing machine (Instron, Model 4502, Instron Corp., Buckinghamshire, UK) at a crosshead speed of 1 mm/min until failure. The force at dislodgment was recorded and divided by the total surface area of each prepared sample to yield the retention value (Pa).

The debonded surfaces of the teeth and crowns were examined with magnifying glasses at 2.7× magnification (Orascoptic, Middleton, WI, USA). Each surface of the dentin-crown interface was analyzed separately (five surfaces per tooth). Failure was classified based on the criteria presented in Table 2.

Table 2. Classification of failure criteria.

Classification	Description	Criteria
1	Cement principally on crown surface	Adhesive cement-dentin
2	Cement principally on dentin surface	Adhesive cement-crown
3	Cement equally distributed on dentin & crown surfaces	Cohesive cement
4	Mixed mode	Adhesive & cohesive cement
5	Fracture of the tooth	Cohesive dentin

A separate analysis was performed for each matched Co-Cr tooth surface (buccal, lingual, mesial, distal, and occlusal). For each category, the number of surfaces was counted and presented as a percentage of all the surfaces for the specific cement.

To analyze the dentin surfaces, the debonded surfaces of some teeth representing different failure categories from each group were examined under an SEM (Quantum 2000) in high vacuum mode following gold sputter-coating. The acquisition conditions were as follows: 25 kV, 90 μA and 40–1000× magnification.

Statistical Analysis

Retentive strength was evaluated using two-way analysis of variance (ANOVA) with repeated measures; cement ($n = 2$) and pretreatment ($n = 2$) were the independent variables. The level of significance was 0.05.

5. Conclusions

An 8.0% arginine and calcium carbonate in-house desensitizing paste can be safely used on dentin to reduce post-cementation sensitivity without compromising the retention of SLM Co-Cr crowns cemented with either ZPC or GIC.

Author Contributions: Conceptualization: R.P., S.G. and S.L.; investigation: S.A.-Z.; methodology: R.P. and S.L.; supervision: S.L.; writing—original draft: R.P. and S.L.; writing—review and editing: R.P.

Funding: This research received no external funding.

Conflicts of Interest: The authors declare no conflict of interest.

References

1. Rosenstiel, S.F.; Rashid, R.G. Postcementation hypersensitivity: Scientific data versus dentists' perceptions. *J. Prosthodont.* **2003**, *12*, 73–81. [CrossRef]
2. Johnson, G.H.; Powell, L.V.; DeRouen, T.A. Evaluation and control of post-cementation pulpal sensitivity: Zinc phosphate and glass ionomer luting cements. *J. Am. Dent. Assoc.* **1993**, *124*, 38–46. [CrossRef]
3. Thylstrup, A.; Bille, J.; Qvist, V. Radiographic and observed tissue changes in approximal carious lesions at the time of operative treatment. *Caries Res.* **1986**, *20*, 75–84. [CrossRef] [PubMed]
4. Bernier, J.L.; Knapp, M.J. A new pulpal response to high-speed dental instruments. *Oral Surg. Oral Med. Oral Pathol. Oral Radiol.* **1958**, *11*, 167–183. [CrossRef]
5. Richardson, D.; Tao, L.; Pashley, D.H. Dentin permeability: Effects of crown preparation. *Int. J. Prosthodont.* **1991**, *4*, 219–225.
6. Langeland, K.; Langeland, L.K. Pulp reactions to crown preparation, impression, temporary crown fixation, and permanent cementation. *J. Prosthet. Dent.* **1965**, *15*, 129–143. [CrossRef]
7. Zaimoglu, A.; Aydin, A.K. An evaluation of smear layer with various desensitizing agents after tooth preparation. *J. Prosthet. Dent.* **1992**, *68*, 450–457. [CrossRef]
8. Sailer, I.; Tettamanti, S.; Stawarczyk, B.; Fischer, J.; Hammerle, C.H. In vitro study of the influence of dentin desensitizing and sealing on the shear bond strength of two universal resin cements. *J. Adhes. Dent.* **2010**, *12*, 381–392. [CrossRef]
9. Sailer, I.; Oendra, A.E.; Stawarczyk, B.; Hammerle, C.H. The effects of desensitizing resin, resin sealing, and provisional cement on the bond strength of dentin luted with self-adhesive and conventional resincements. *J. Prosthet. Dent.* **2012**, *107*, 252–260. [CrossRef]
10. Stawarczyk, B.; Hartmann, R.; Hartmann, L.; Roos, M.; Ozcan, M.; Sailer, I.; Hammerle, C.H. The effect of dentin desensitizer on shear bond strength of conventional and self-adhesive resin luting cements after aging. *Oper. Dent.* **2011**, *36*, 492–501. [CrossRef]
11. Stawarczyk, B.; Hartmann, L.; Hartmann, R.; Roos, M.; Ender, A.; Ozcan, M.; Sailer, I.; Hammerle, C.H. Impact of gluma desensitizer on the tensile strength of zirconia crowns bonded to dentin: An in vitro study. *Clin. Oral Investig.* **2012**, *16*, 201–213. [CrossRef] [PubMed]
12. Cobb, D.S.; Reinhardt, J.W.; Vargas, M.A. Effect of HEMA-containing dentin desensitizers on shear bond strength of a resin cement. *Am. J. Dent.* **1997**, *10*, 62–65. [PubMed]
13. Soeno, K.; Taira, Y.; Matsumura, H.; Atsuta, M. Effect of desensitizers on bond strength of adhesive luting agents to dentin. *J. Oral Rehabil.* **2001**, *28*, 1122–1128. [CrossRef] [PubMed]
14. Yim, N.H.; Rueggeberg, F.A.; Caughman, W.F.; Gardner, F.M.; Pashley, D.H. Effect of dentin desensitizers and cementing agents on retention of full crowns using standardized crown preparations. *J. Prosthet. Dent.* **2000**, *83*, 459–465. [CrossRef]

15. Huh, J.B.; Kim, J.H.; Chung, M.K.; Lee, H.Y.; Choi, Y.G.; Shim, J.S. The effect of several dentin desensitizers on shear bond strength of adhesive resin luting cement using self-etching primer. *J. Dent.* **2008**, *36*, 1025–1032. [CrossRef] [PubMed]
16. Aranha, A.C.; Siqueira Junior Ade, S.; Cavalcante, L.M.; Pimenta, L.A.; Marchi, G.M. Microtensile bond strengths of composite to dentin treated with desensitizer products. *J. Adhes. Dent.* **2006**, *8*, 85–90. [PubMed]
17. Hamlin, D.; Williams, K.P.; Delgado, E.; Zhang, Y.P.; DeVizio, W.; Mateo, L.R. Clinical evaluation of the efficacy of a desensitizing paste containing 8% arginine and calcium carbonate for the in-office relief of dentin hypersensitivity associated with dental prophylaxis. *Am. J. Dent.* **2009**, *22*, 16A–20A.
18. Schiff, T.; Delgado, E.; Zhang, Y.P.; Cummins, D.; DeVizio, W.; Mateo, L.R. Clinical evaluation of the efficacy of an in-office desensitizing paste containing 8% arginine and calcium carbonate in providing instant and lasting relief of dentin hypersensitivity. *Am. J. Dent.* **2009**, *22*, 8A–15A.
19. Ayad, F.; Ayad, N.; Zhang, Y.P.; DeVizio, W.; Cummins, D.; Mateo, L.R. Comparing the efficacy in reducing dentin hypersensitivity of a new toothpaste containing 8.0% arginine, calcium carbonate, and 1450 ppm fluoride to a commercial sensitive toothpaste containing 2% potassium ion: An eight-week clinical study on Canadian adults. *J. Clin. Dent.* **2009**, *20*, 10–16.
20. Docimo, R.; Montesani, L.; Maturo, P.; Costacurta, M.; Bartolino, M.; DeVizio, W.; Zhang, Y.P.; Cummins, D.; Dibart, S.; Mateo, L.R. Comparing the efficacy in reducing dentin hypersensitivity of a new toothpaste containing 8.0% arginine, calcium carbonate, and 1450 ppm fluoride to a commercial sensitive toothpaste containing 2% potassium ion: An eight-week clinical study in Rome, Italy. *J. Clin. Dent.* **2009**, *20*, 17–22.
21. Nathoo, S.; Delgado, E.; Zhang, Y.P.; DeVizio, W.; Cummins, D.; Mateo, L.R. Comparing the efficacy in providing instant relief of dentin hypersensitivity of a new toothpaste containing 8.0% arginine, calcium carbonate, and 1450 ppm fluoride relative to a benchmark desensitizing toothpaste containing 2% potassium ion and 1450 ppm fluoride, and to a control toothpaste with 1450 ppm fluoride: A three-day clinical study in New Jersey, USA. *J. Clin. Dent.* **2009**, *20*, 123–130. [PubMed]
22. Garcia-Godoy, A.; Garcia-Godoy, F. Effect of an 8.0% arginine and calcium carbonate in-office desensitizing paste on the shear bond strength of composites to human dental enamel. *Am. J. Dent.* **2010**, *23*, 324–326. [PubMed]
23. Wang, Y.; Liu, S.; Pei, D.; Du, X.; Ouyang, X.; Huang, C. Effect of an 8.0% arginine and calcium carbonate in-office desensitizing paste on the microtensile bond strength of self-etching dental adhesives to human dentin. *Am. J. Dent.* **2012**, *25*, 281–286. [PubMed]
24. Canares, G.; Salgado, T.; Pines, M.S.; Wolff, M.S. Effect of an 8.0% arginine and calcium carbonate desensitizing toothpaste on shear dentin bond strength. *J. Clin. Dent.* **2012**, *23*, 68–70. [PubMed]
25. Chandavarkar, S.M.; Ram, S.M. A comparative evaluation of the effect of dentin desensitizers on the retention of complete cast metal crowns. *Contemp. Clin. Dent.* **2015**, *6*, S45–S50. [CrossRef] [PubMed]
26. Pilo, R.; Harel, N.; Nissan, J.; Levartovsky, S. The retentive strength of cemented zirconium oxide crowns after dentin pretreatment with desensitizing paste containing 8% arginine and calcium carbonate. *Int. J. Mol. Sci.* **2016**, *17*, 426. [CrossRef] [PubMed]
27. Quante, K.; Ludwig, K.; Kern, M. Marginal and internal fit of metal-ceramic crowns fabricated with a new laser melting technology. *Dent. Mater.* **2008**, *24*, 1311–1315. [CrossRef]
28. Ucar, Y.; Akova, T.; Akyil, M.S.; Brantley, W.A. Internal fit evaluation of crowns prepared using a new dental crown fabrication technique: Laser-sintered Co-Cr crowns. *J. Prosthet. Dent.* **2009**, *102*, 253–259. [CrossRef]
29. Tamac, E.; Toksavul, S.; Toman, M. Clinical marginal and internal adaptation of CAD/CAM milling, laser sintering, and cast metal ceramic crowns. *J. Prosthet. Dent.* **2014**, *112*, 909–913. [CrossRef]
30. Lavender, S.A.; Petrou, I.; Heu, R.; Stranick, M.A.; Cummins, D.; Kilpatrick-Liverman, L.; Sullivan, R.J.; Santarpia, R.P., III. Mode of action studies on a new desensitizing dentifrice containing 8.0% arginine, a high cleaning calcium carbonate system and 1450 ppm fluoride. *Am. J. Dent.* **2010**, *23*, 14A–19A.
31. Panagakos, F.; Schiff, T.; Guignon, A. Dentin hypersensitivity: Effective treatment with an in-office desensitizing paste containing 8% arginine and calcium carbonate. *Am. J. Dent.* **2009**, *22*, 3A–7A. [PubMed]
32. Shetty, R.M.; Bhat, S.; Mehta, D.; Srivatsa, G.; Shetty, Y.B. Comparative analysis of postcementation hypersensitivity with glass ionomer cement and a resin cement: An in vivo study. *J. Contemp. Dent. Pract.* **2012**, *13*, 327–331. [PubMed]
33. Rosenstiel, S.F.; Land, M.F.; Crispin, B.J. Dental luting agents: A review of the current literature. *J. Prosthet. Dent.* **1998**, *80*, 280–301. [CrossRef]

34. Pattanaik, B.K.; Nagda, S.J. An evaluation of retention and marginal seating of Ni-Cr alloy cast restorations using three different luting cements: An in vitro study. *Indian. J. Dent. Res.* **2012**, *23*, 20–25. [CrossRef] [PubMed]
35. Wiskott, H.W.; Nicholls, J.I.; Belser, U.C. The relationship between abutment taper and resistance of cemented crowns to dynamic loading. *Int. J. Prosthodont.* **1996**, *9*, 117–139. [PubMed]
36. Gorodovsky, S.; Zidan, O. Retentive strength, disintegration, and marginal quality of luting cements. *J. Prosthet. Dent.* **1992**, *68*, 269–274. [CrossRef]
37. Behr, M.; Rosentritt, M.; Loher, H.; Kolbeck, C.; Trempler, C.; Stemplinger, B.; Kopzon, V.; Handel, G. Changes of cement properties caused by mixing errors: The therapeutic range of different cement types. *Dent. Mater.* **2008**, *24*, 1187–1193. [CrossRef] [PubMed]
38. Pilo, R.; Lewinstein, I.; Ratzon, T.; Cardash, H.S.; Brosh, T. The influence of dentin and/or metal surface treatment on the retention of cemented crowns in teeth with an increased taper. *Dent. Mater.* **2008**, *24*, 1058–1064. [CrossRef] [PubMed]

© 2018 by the authors. Licensee MDPI, Basel, Switzerland. This article is an open access article distributed under the terms and conditions of the Creative Commons Attribution (CC BY) license (http://creativecommons.org/licenses/by/4.0/).

Article

Influence of Implant Material and Surface on Differentiation and Proliferation of Human Adipose-Derived Stromal Cells

Susanne Jung, Lauren Bohner, Marcel Hanisch, Johannes Kleinheinz and Sonja Sielker *

Department of Cranio-Maxillofacial Surgery, Research Unit Vascular Biology of Oral Structures (VABOS), University Hospital Muenster, 48149 Muenster, Germany; Susanne.Jung@ukmuenster.de (S.J.); Lauren.Oliveiralimabohner@ukmuenster.de (L.B.); Marcel.Hanisch@ukmuenster.de (M.H.); Johannes.Kleinheinz@ukmuenster.de (J.K.)
* Correspondence: Sonja.Sielker@ukmuenster.de; Tel.: +49-251-83-47007

Received: 20 November 2018; Accepted: 12 December 2018; Published: 13 December 2018

Abstract: For the guided regeneration of periimplant hard and soft tissues, human adipose-derived stromal cells (hADSC) seem to be a promising source for mesenchymal stromal cells. For this, the proliferation and differentiation of hADSC were evaluated on titanium and zirconia dental implants with different surface treatments. Results were compared to edaphic cells as human osteoblasts (hOB) and human gingival fibroblasts (HGF). Primary cells were cultured on (1) titanium implants with a polished surface (Ti-PT), (2) sandblasted and acid-etched titanium (Ti-SLA), (3) sandblasted and alkaline etched zirconia (ZrO_2-ZLA) and (4) machined zirconia (ZrO_2-M). The cell proliferation and differentiation on osteogenic lineage were assessed after 1, 7 and 14 days. Statistical analysis was performed by one-way ANOVA and a modified Levene test with a statistical significance at $p = 0.05$. PostHoc tests were performed by Bonferroni-Holm. Zirconia dental implants with rough surface (ZrO_2-ZLA) showed the highest proliferation rates ($p = 0.048$). The osteogenic differentiation occurred early for zirconia and later for titanium implants, and it was enhanced for rough surfaces in comparison to polished/machined surfaces. Zirconia was more effective to promote the proliferation and differentiation of hADSCs in comparison to titanium. Rough surfaces were able to improve the biological response for both zirconia and titanium.

Keywords: hADSC; tissue regeneration; implants; titanium; zirconia

---

1. Introduction

The treatment of periimplant bone defects usually requires a reconstructive surgical procedure, with autogenous bone or biomaterial-based bone grafting to stimulate bone regeneration [1–3]. Despite the favorable clinical results showed by these techniques, disadvantages arise, such as high morbidity, limited availability and risk of infection [2–4].

Bone regeneration pathway involves the proliferation and differentiation of mesenchymal stem cells into the osteogenic lineage [5]; thus, bone tissue engineering makes it a suitable option to treat periimplant defects. The application of osteogenic cells in combination with physiological components involved on bone remodeling have shown improved outcomes in comparison to biomaterials alone [3].

A promising source for the stem cells is the adipose tissue, since it is easily collected using a low invasive approach. Human adipose-derived cells (hADSCs) differentiate into osteogenic and angiogenic lineage and represent for potential bone regeneration [6]. Additionally, extrinsic and intrinsic osteogenic inducers are recommended to optimize the culture environment and increase the cell regeneration potential [7,8]. Nonetheless, the imperative need of this osteogenic inducer is unclear and the efficacy of osteogenic differentiation without the use of an osteogenic medium has been assessed [9].

The cell biological response in periimplant defects is influenced by the dental implant material and surface [10–13]. Hempel et al. [10] showed a greater cell proliferation and differentiation on zirconia than titanium. A rough surface improved and accelerated the osteogenic response, by increasing the surface area to mimic the bone morphology [11–15].

Previous studies showed the osteoinduction and osteoconduction capability of hADSCs [16,17], however, the influence of the dental implant material and surface treatment has not yet been well established. The present study evaluated the proliferation and differentiation of hADSCs over titanium and zirconia with various surface treatments. The hADSCs were cultured without inducing factors and their biological response was evaluated using human primary osteoblasts (hOBs) and human gingiva fibroblasts (HGF).

2. Results

2.1. Cell Viability

Cell viability among experimental and control groups after 14 days is shown in Figure 1. Zirconia dental implants with a rough surface showed the highest vitality ($p = 0.022$) and proliferation rates ($p = 0.041$). Rough surfaces showed better results compared to polished or machined surfaces. Titanium implants showed higher cytotoxicity than zirconia dental implants ($p = 0.0016$), although minor, and did not affect the cell viability.

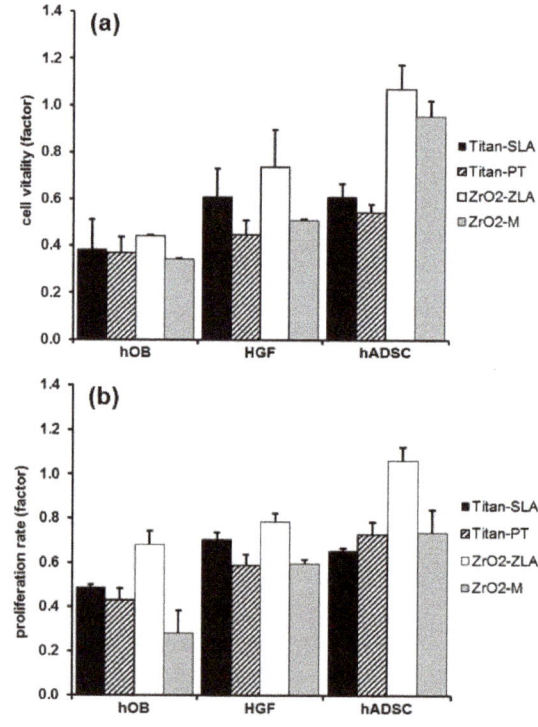

Figure 1. *Cont.*

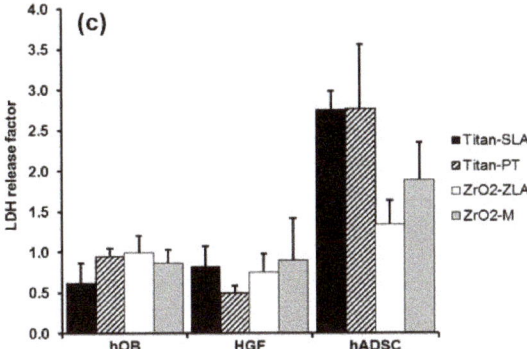

Figure 1. Cell Viability at day 14 correlated to control for human osteoblasts (hOB), human gingival fibroblasts (HGF), and human adipose-derived stromal cells (hADSC) (**a**) cell vitality $p = 0.021$; (**b**) proliferation rate, $p = 0.041$, (**c**) LDH release factor, $p = 0.0016$; standard abbreviation as error marks.

2.2. Expression of Stem Cell and Osteogenic Marker

Figure 2 shows the change in stem cells markers compared to day 1 ($p = 0.04$). Gene expression of stem cell markers differed strongly between zirconia and titanium implants (PostHoc analysis $p = 0.017$). For zirconia, gene expression was lower and decreased from day 7 to day 14. In contrast, for titanium and the control group, gene expression was higher and increased from day 7 to day 14 (Figure 2).

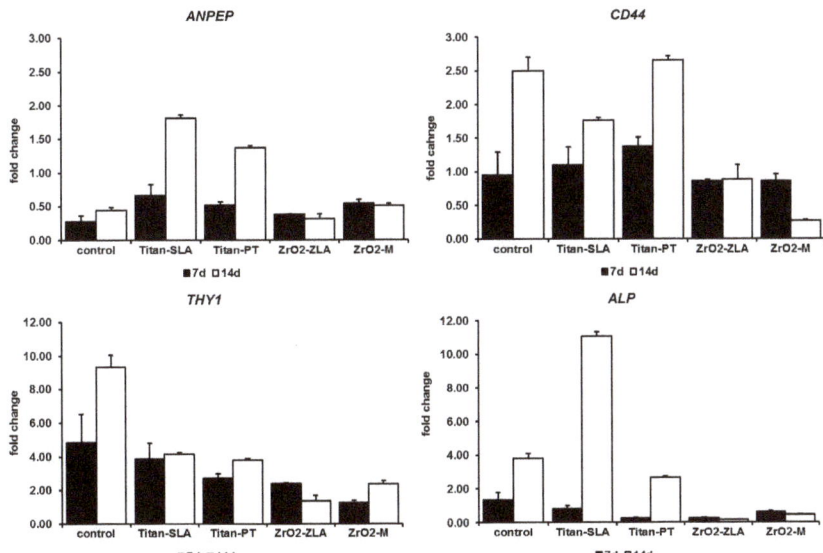

Figure 2. Fold change of stem cell marker compared to day 1 (*ANPEP* $p = 0.038$, *CD44* $p = 0.18$, *THY1* $p = 0.004$, *ALP* $p = 0.17$; standard abbreviation as error marks).

Figure 3 shows the change of osteogenic marker compared to day 1 for hADSC ($p = 0.0083$) and hOB ($p = 0.022$). Gene expression of *RANKL* was not detectable in any sample. Expression of osteogenic markers did not show a consistent picture, unlike the stem cell marker. In hOB, expression of *RUNX2* and osteoprotegerin increased from day 7 to day 14 with a low fold change, and expression of osteopontin remained constant (Figure 3). In hADSC, expression of osteogenic marker increased strongly and fold change was stronger on rough surfaces except for *RUNX2* on zirconia dental implants. Changes in expression of osteocalcin were remarkable, because no dexamethasone was added to hADSC. Expression of osteocalcin increased strongly, whereas the expression in hOB remained constant (Figure 3).

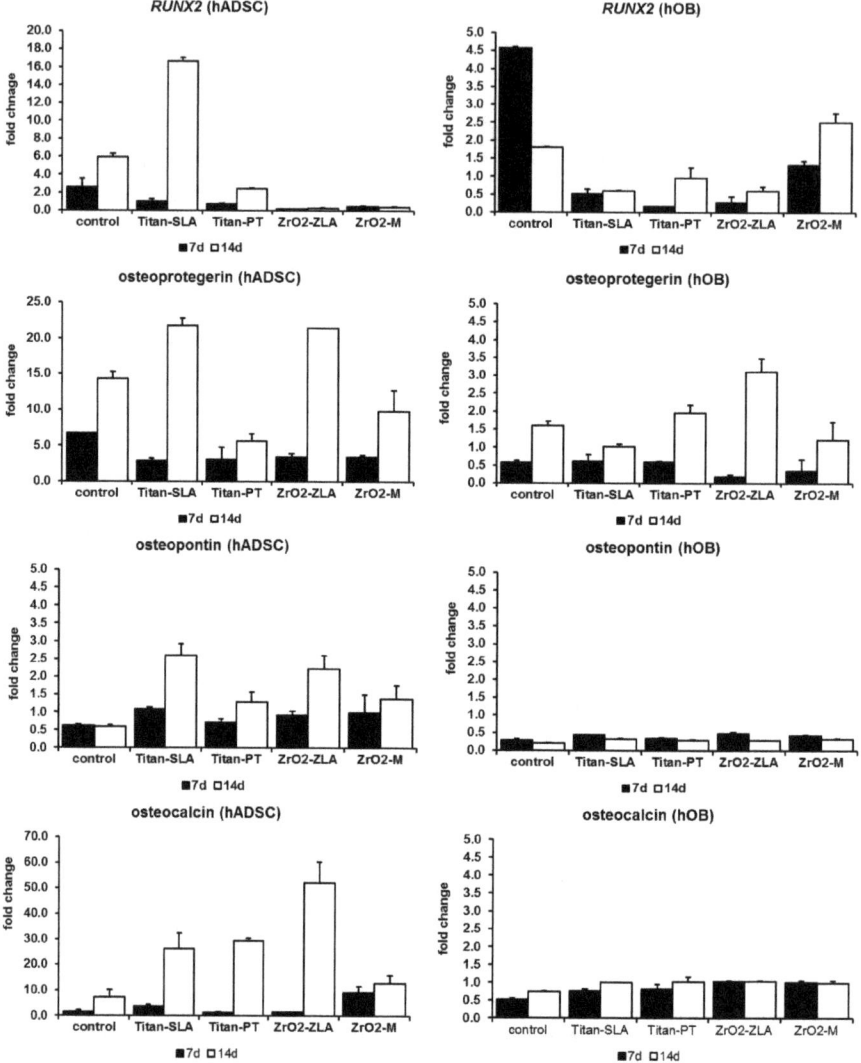

Figure 3. Fold change of osteogenic markers compared to day 1 (*RUNX2* $p = 021$ for hADSC, $p = 0.135$ for hOB; osteoprotegerin $p = 0.011$ for hADSC, $p = 0.008$ for hOB; osteopontin $p = 0.07$ for hADSC, $p = 0.024$ for hOB, osteocalcin $p = 0.024$ for hADSC, $p = 0.25$ for hOB; standard abbreviation as error marks).

3. Discussion

Zirconia implants showed higher cell viability than titanium implants. From all groups, zirconia dental implants with a machined surface showed the greatest metabolic activity and differentiation. The potential of zirconia to stimulate the osteogenic differentiation was confirmed by the decrease of stem cell markers over time, on rough and machined surfaces, which indicated the trans-differentiation into osteogenic lineage. This effect was not observed for the titanium implants. Similar findings have been reported by Hempel et al. [10], who showed that the metabolic activity was significantly greater for zirconia than for titanium implants 24 and 48 h after plating. Zirconia dental implants seem to be capable of inducing cell proliferation and accelerating the differentiation toward osteogenesis.

The findings of the present study showed that *RUNX2* (the master regulator of osteoblast differentiation) was early expressed in hADSCs over zirconia surfaces. Expression of *RUNX2* takes place at an early stage of osteogenic differentiation, followed by the inhibition of the process at later stages [18]. As the expression of *RUNX2* was not significant for zirconia dental implants after 14 days, it was assumed that the differentiation occurred predominantly during the first 24 h, and remained constant on the following days.

According to Carinci et al. [19], zirconia materials are correlated with an early gene expression, which occurs within the first 24 h of stimulation. Similar findings were shown by Altmann et al. [20], whose osteogenic marker *RUNX2* had higher levels on zirconia after 1 and 7 days. The authors claimed that this response was not only time-dependent, but a function of biomaterial. This hypothesis was confirmed in the present study, since the differentiation of hADSCs was different between titanium and zirconia implants. The titanium implants showed a late response for cell differentiation, resulting in a higher change on the expression of osteogenic markers after 14 days. The late response of titanium to cell viability and differentiation in comparison to zirconia implants had already been shown in previous studies [10,21].

The same hypothesis may be applied to the surface treatment [22]. In this study, rough surfaces showed an early expression of osteogenic markers compared to smooth surfaces. The expression of osteoprotegerin, a receptor related with the osteoclastogenesis [23], was higher on rough surfaces and lower on machined and polished surfaces. Osteogenic markers showed similar results.

Regarding to the use of inducing factors, the expression of osteocalcin did not significantly change over time for osteoblasts with inducing media, whereas for hADSCs, it decreased after 14 days without the use of inducing factors. Thus, albeit the expression of osteogenic markers may be accelerated when an inducing media is used, it seems not to be essential for the osteogenic differentiation.

Previous studies have been controversial regarding the role of inducing factors on osteogenesis. Cecchinato et al. [14] showed the osteogenic differentiation of hADSCs cells on titanium surfaces after 2 weeks; however, Abagnale et al. [21] showed no differentiation of hADSCs without osteogenic medium. For Faia-Torres et al. [9], the osteogenic differentiation only occurred with rough surfaces [9]. Thus, although the findings of this study suggested that material and surface treatment alone may induce the osteogenic differentiation, additional research is required to confirm this hypothesis.

4. Materials and Methods

4.1. Study Design and Ethical Approval

The study evaluated the cell viability of hADSCs, primary human osteoblasts (hOB) and human gingiva fibroblasts (HGF) on titanium and zirconia with polished/machined and rough surfaces. The differentiation into osteogenic lineage was assessed for hADSCs and hOBs. The cell viability was assessed based on cell vitality, proliferation and cytotoxicity, whereas the differentiation in osteogenic lineage was measured using on the protein expression analysis. The experiment was designed according to the "Declaration of Helsinki" and approved by the Ethics Committee of the Faculty of Medicine, University of Muenster (#2016-624-f-S, 07 December 2016). Previous to the cell isolation, a written informed consent was obtained from all donors.

4.2. Isolation of Primary Human Cell Cultures

All cell colleting procedures were performed anonymously and under sterile conditions. Human fat tissue was collected from leftover tissue of patients who had undergone elective abdominal surgery at the General and Visceral Surgery, University Hospital, Muenster. Oncological surgeries were not included. Human spongiosa and gingiva samples were obtained from the leftover tissue of patients treated at the Department of Craniofacial Surgery, University Hospital Muenster.

Isolation and culture techniques of human fat tissue, primary human osteoblast and human gingiva fibroblasts cells were performed as described previously [6]. Details about the culture medium are described in Table 1. The cells were cultivated in a 5% CO_2 humidified atmosphere at 37 °C, being fed every 2–3 days and passaged with 10,000 cells/cm^2 after reaching 90% of confluence.

Table 1. Culture medium.

Cells	Medium	Culture Formula
Human adipose-derived stromal cells (hADSC)	minimum essential medium–Alpha Eagle (α-MEM) (Lonza, Walkersville, MD, USA)	10% fetal bovine serum, 1% amphotericin B (250 µg/mL), 1% glutamine (200 Mm), 1% penicillin (10,000 U/mL)/streptomycin (10,000 µg/mL) (Biochrom Merck, Berlin, Germany)
Primary human osteoblast (hOBs)	Primary human osteoblast (hOBs)	12% fetal bovine serum, 1% amphotericin B (250 µg/mL), 1% glutamine (200 mM), 1% penicillin (10000 U/mL)/streptomycin (10,000 µg/mL) (Biochrom Merck, Germany). For osteogenic differentiation, 16 ng/mL dexamethasone (Merck Pharma, Darmstadt, Germany) was added to the medium.
Primary human gingiva fibroblasts (HGF)	DMEM medium (high glucose and L-glutamine; Gibco, USA) with ¼ Ham's F12 nutrient mixtures (Sigma, Hamburg, Germany)	10% fetal bovine serum, 1% amphotericin B (250 µg/mL), 1% penicillin (10,000 U/mL)/streptomycin (10,000 µg/mL) (all Biochrom Merck, Berlin, Germany)

4.3. Main Cell Culture

Cells were cultured on various titanium and zirconia discs (Straumann, Basel, Switzerland) measuring 5 mm diameter, the groups were: (1) titanium implants with polished surface (Ti-PT); (2) sandblasted and acid-etched titanium (Ti-SLA); (3) sandblasted and alkaline etched zirconia (ZrO_2-ZLA) and (4) machined zirconia (ZrO_2-M). For hADSCs, no inducing factor was used, whereas the hOB culture was supplemented by an osteogenic medium. The samples were placed in 48-well cell culture plates (Greiner Bio One; Bad Nenndorf, Germany) with a 5% CO_2 humidified atmosphere at 37 °C. Culturing medium was replaced every 2–3 days during cell culture study. In addition, cells for control group were cultivated as a monolayer in 48-well cell culture plates. Samples were analyzed 1, 7 and 14 days after the beginning of the experiment. Cell culture part was repeated three times.

4.4. Cell Viability

The living cell count was performed with the LUNA II system (Logos Biosystems Inc., Villeneuve d'Ascq, France). Proliferation rate was estimated with an in-house MTT assay, which determines the metabolic activity of vital cells. The conversion of the yellow thiazolyl blue tetrazolium bromide (0.5 mg/mL; Sigma-Aldrich, Hamburg, Germany) to the purple formazan was measured at a wavelength of 570 nm. Cytotoxic effects were determined with the Pierce™ LDH Cytotoxicity Assay (ThermoFisher Scientific; Wesel, Germany). All assays were performed according to manufacture protocols and done in triplicates.

4.5. Protein Expression Analysis

For protein expression analysis, cells were lysed with the Pierce™ IP Lysis Buffer (ThermoFisher Scientific, Waltham, MA, USA) according to the manufacture's protocol. The supernatant was frozen at −80 °C for subsequent assays. To determinate secreted proteins, part of the culturing medium was taken before lysis and frozen at −80 °C. Quantification protein determination was performed with the Pierce™ BCA Protein Assay (ThermoFisher Scientific, Wesel, Germany) according to the manufacture's protocol. The µQuant reader (BioTek, Bad Friedrichshall, Germany) was used for protein determination and ALP assay. Protein expression analysis was evaluated with enzyme-linked immunosorbent assay (ELISA). The ELISAs are listed in Table 2 (abcam, Cambridge, England). ELISAs were performed according to the manufacture's protocol. Absorbance was measured at a wavelength of 450 nm with the µQuant reader (BioTek, Bad Friedrichshall, Germany). Protein expression was normalized in two steps. The first step was a correlation to whole protein, and the second step was a correlation to day 1. Changes in expression related to day 1 are shown.

Table 2. Used and primer (Qiagen, Hilden, Germany) and enzyme-linked immunosorbent assay (ELISA)-Kits (abcam, Cambridge, England).

Gene/Protein	Primer	Protein Assay/ELISA
Stem Cell Marker		
ANPEP/CD13	PPH05672A	
CD44	PPH00114A	
THY1/CD90	PPH02406G	
alkaline phosphatase	PPH01311F	
Osteogenic Marker		
RUNX2	PPH01897C	
TNFSF11/RANKL	PPH01048F	
osteoprotegerin		ab189580
osteopontin		ab192143
osteocalcin		ab195214
Housekeeping Genes		
RPLP0	PPH21138F	
B2M	PPH01094E	
GAPDH	PPH00150F	
HPRT1	PPH01018C	
ACTB	PPH00073G	

4.6. RNA Extraction and Real-Time qPCR

For RNA isolation and purification, an RNeasy Micro Kit (Qiagen, Hilden, Germany) was used and performed according to the manufacture's protocol. Purity and concentration of the isolated RNA was determined by a spectrophotometric reading (NanoDrop™ 2000; ThermoFisher Scientific, Wesel, Germany). For RT-qPCR, a custom-designed RT2 PCR Array (Qiagen, Hilden, Germany) was used (Table 2). For each sample 50 ng RNA was transcripted in complementary DNA (cDNA) with the RT2 First Strand Kit (Qiagen, Hilden, Germany). A DNase treatment to eliminate existent genomic DNA was part of this step. RT-qPCR was performed with the RT2 SYBR Green ROX pPCR Mastermix (Qiagen, Hilden, Germany) according to the manufacture's protocol in Eppendorf mastercycler ep realplex 4S (Eppendorf, Hamburg, Germany). One µL cDNA was used as template for each reaction. The PCR protocol was: activation of hot start Taq-polymerase for 10 min at 95 °C, followed by 40 cycles with a denaturation step for 15 sec at 95 °C and an annealing/elongation step for 60 s at 60 °C. A melting curve between 65 and 95 °C with a time ramp of 2 min for 1 °C was connected. The $\Delta\Delta CT$ method was used for data analysis with the RT2 Profiler PCR array data analysis web portal (https://www.qiagen.com/de/shop/genes-and-path-ways/data-analysis-center-overview-page/custom-rt2-pcr-arrays-data-analysis-center, accessed on 28 March 2018). Fold changes related to day 1 are shown.

4.7. Statistical Analysis

Statistical analysis of expression factors and protein expression factors was carried out by one-way ANOVA and a modified Levene test with a statistical significance at $p = 0.05$. PostHoc tests were performed by Bonferroni-Holm test (Daniel's XL Toolbox version 6.53; https://www.xltoolbox.net/).

5. Conclusions

Zirconia promoted the proliferation and differentiation of hADSCs more than titanium. Rough surfaces improve the biological response for both zirconia and titanium.

Author Contributions: Conceptualization, S.J., J.K. and S.S.; methodology, S.S.; validation, S.S.; formal analysis, S.S.; investigation, S.S.; writing—original draft preparation, L.B. and S.S.; writing—review and editing, S.J., M.H., L.B. and S.S.; visualization, S.S..; supervision, J.K.; project administration, S.J. and J.K.; funding acquisition, S.J. and J.K. All authors gave final approval and agree to be accountable for all aspects of the work.

Funding: This work was supported by an ITI Research Grant (application no. 1195_2016) of the International Team for Implantology, Basel, Switzerland.

Acknowledgments: This study was supported by Professor Senninger, the head of the department of Visceral Surgery of the University Hospital Muenster. Special thanks go to L.B. for her work in manuscript drafting (ITI Scholar 2017–2018).

Conflicts of Interest: The authors declare no conflict of interest. The funders had no role in the design of the study; in the collection, analyses, or interpretation of data; in the writing of the manuscript, or in the decision to publish the results.

Abbreviations

hADSC   Human adipose-derived stromal cells
hOB   Human osteoblasts
HGF   Human gingival fibroblasts

References

1. Mercado, F.; Hamlet, S.; Ivanovski, S. Regenerative surgical therapy for peri-implantitis using deproteinized bovine bone mineral with 10% collagen, enamel matrix derivative and Doxycycline-A prospective 3-year cohort study. *Clin. Oral Implants Res.* **2018**, *29*, 583–591. [CrossRef] [PubMed]
2. Renvert, S.; Roos-Jansaker, A.M.; Persson, G.R. Surgical treatment of peri-implantitis lesions with or without the use of a bone substitute—A randomized clinical trial. *J. Clin. Periodontol.* **2018**, *45*, 1266–1274. [CrossRef] [PubMed]
3. Shanbhag, S.; Pandis, N.; Mustafa, K.; Nyengaard, J.R.; Stavropoulos, A. Bone tissue engineering in oral peri-implant defects in preclinical in vivo research: A systematic review and meta-analysis. *J. Tissue Eng. Regen. Med.* **2018**, *12*, e336–e349. [CrossRef] [PubMed]
4. Maglione, M.; Salvador, E.; Ruaro, M.E.; Melato, M.; Tromba, G.; Angerame, D.; & Bevilacqua, L. Bone regeneration with adipose derived stem cells in a rabbit model. *J. Biomed. Res.* **2018**. [CrossRef]
5. Valenti, M.T.; Dalle Carbonare, L.; Mottes, M. Osteogenic Differentiation in Healthy and Pathological Conditions. *Int. J. Mol. Sci.* **2016**, *18*, 41. [CrossRef]
6. Jung, S.; Kleineidam, B.; Kleinheinz, J. Regenerative potential of human adipose-derived stromal cells of various origins. *J. Craniomaxillofac. Surg.* **2015**, *43*, 2144–2151. [CrossRef] [PubMed]
7. Chen, Q.; Yang, Z.; Sun, S.; Huang, H.; Sun, X.; Wang, Z.; Zhang, Y.; Zhang, B. Adipose-derived stem cells modified genetically in vivo promote reconstruction of bone defects. *Cytotherapy* **2010**, *12*, 831–840. [CrossRef]
8. Shiraishi, T.; Sumita, Y.; Wakamastu, Y.; Nagai, K.; Asahina, I. Formation of engineered bone with adipose stromal cells from buccal fat pad. *J. Dent. Res.* **2012**, *91*, 592–597. [CrossRef]
9. Faia-Torres, A.B.; Charnley, M.; Goren, T.; Guimond-Lischer, S.; Rottmar, M.; Maniura-Weber, K.; Spencer, D.D.; Reis, R.; Textor, M. Osteogenic differentiation of human mesenchymal stem cells in the absence of osteogenic supplements: A surface-roughness gradient study. *Acta Biomater.* **2015**, *28*, 64–75. [CrossRef]
10. Hempel, U.; Hefti, T.; Kalbacova, M.; Wolf-Brandstetter, C.; Dieter, P.; Schlottig, F. Response of osteoblast-like SAOS-2 cells to zirconia ceramics with different surface topographies. *Clin. Oral Implants Res.* **2010**, *21*, 174–181. [CrossRef]
11. Hirano, T.; Sasaki, H.; Honma, S.; Furuya, Y.; Miura, T.; Yajima, Y.; Yoshinari, M. Proliferation and osteogenic differentiation of human mesenchymal stem cells on zirconia and titanium with different surface topography. *Dent. Mater. J.* **2015**, *34*, 872–880. [CrossRef] [PubMed]
12. Ito, H.; Sasaki, H.; Saito, K.; Honma, S.; Yajima, Y.; Yoshinari, M. Response of osteoblast-like cells to zirconia with different surface topography. *Dent. Mater. J.* **2013**, *32*, 122–129. [CrossRef] [PubMed]

13. Pandey, A.K.; Pati, F.; Mandal, D.; Dhara, S.; Biswas, K. In vitro evaluation of osteoconductivity and cellular response of zirconia and alumina based ceramics. *Mater. Sci. Eng. C Mater. Biol. Appl.* **2013**, *33*, 3923–3930. [CrossRef] [PubMed]
14. Abagnale, G.; Steger, M.; Nguyen, V.H.; Hersch, N.; Sechi, A.; Joussen, S.; Denecke, B.; Merkel, R.; Hoffmann, B.; Dreser, A.; et al. Surface topography enhances differentiation of mesenchymal stem cells towards osteogenic and adipogenic lineages. *Biomaterials* **2015**, *61*, 316–326. [CrossRef] [PubMed]
15. Faia-Torres, A.B.; Guimond-Lischer, S.; Rottmar, M.; Charnley, M.; Goren, T.; Maniura-Weber, K.; Spencer, N.D.; Reis, R.L.; Textor, M.; Neves, N.M. Differential regulation of osteogenic differentiation of stem cells on surface roughness gradients. *Biomaterials* **2014**, *35*, 9023–9032. [CrossRef] [PubMed]
16. Lee, H.R.; Kim, H.J.; Ko, J.S.; Choi, Y.S.; Ahn, M.W.; Kim, S.; Do, S.H. Comparative characteristics of porous bioceramics for an osteogenic response in vitro and in vivo. *PLoS ONE* **2013**, *8*, e84272. [CrossRef] [PubMed]
17. Lu, W.; Ji, K.; Kirkham, J.; Yan, Y.; Boccaccini, A.R.; Kellett, M.; Jin, Y.; Yang, X.B. Bone tissue engineering by using a combination of polymer/Bioglass composites with human adipose-derived stem cells. *Cell Tissue Res.* **2014**, *356*, 97–107. [CrossRef] [PubMed]
18. Bruderer, M.; Richards, R.G.; Alini, M.; Stoddart, M.J. Role and regulation of RUNX2 in osteogenesis. *Eur. Cell Mater.* **2014**, *28*, 269–286. [CrossRef] [PubMed]
19. Carinci, F.; Pezzetti, F.; Volinia, S.; Francioso, F.; Arcelli, D.; Farina, E.; Piattelli, A. Zirconium oxide: Analysis of MG63 osteoblast-like cell response by means of a microarray technology. *Biomaterials* **2004**, *25*, 215–228. [CrossRef]
20. Altmann, B.; Rabel, K.; Kohal, R.J.; Proksch, S.; Tomakidi, P.; Adolfsson, E.; Bernsmann, F.; Palermo, P.; Fürderer, T.; Steinberg, T. Cellular transcriptional response to zirconia-based implant materials. *Dent. Mater.* **2017**, *33*, 241–255. [CrossRef]
21. Cecchinato, F.; Karlsson, J.; Ferroni, L.; Gardin, C.; Galli, S.; Wennerberg, A.; Zavan, B.; Andeersson, M.; Jimbo, R. Osteogenic potential of human adipose-derived stromal cells on 3-dimensional mesoporous TiO$_2$ coating with magnesium impregnation. *Mater. Sci. Eng. C Mater. Biol. Appl.* **2015**, *52*, 225–234. [CrossRef] [PubMed]
22. Miranda, R.B.P.; Grenho, L.; Carvalho, A.; Fernandes, M.H.; Monteiro, F.J.; Cesar, P.F. Micropatterned Silica Films with Nanohydroxyapatite for Y-TZP Implants. *J. Dent. Res.* **2018**, *97*, 1003–1009. [CrossRef] [PubMed]
23. Liu, W.; Zhang, X. Receptor activator of nuclear factor-kappaB ligand (RANKL)/RANK/osteoprotegerin system in bone and other tissues (review). *Mol. Med. Rep.* **2015**, *11*, 3212–3218. [CrossRef] [PubMed]

© 2018 by the authors. Licensee MDPI, Basel, Switzerland. This article is an open access article distributed under the terms and conditions of the Creative Commons Attribution (CC BY) license (http://creativecommons.org/licenses/by/4.0/).

Article

Studies on the Curing Efficiency and Mechanical Properties of Bis-GMA and TEGDMA Nanocomposites Containing Silver Nanoparticles

Izabela Barszczewska-Rybarek [1],* and Grzegorz Chladek [2]

[1] Department of Physical Chemistry and Technology of Polymers, Silesian University of Technology, 44-100 Gliwice, Poland
[2] Institute of Engineering Materials and Biomaterials, Silesian University of Technology, 44-100 Gliwice, Poland; Grzegorz.Chladek@polsl.pl
* Correspondence: Izabela.Barszczewska-Rybarek@polsl.pl; Tel.: +48-32-237-1509

Received: 31 October 2018; Accepted: 5 December 2018; Published: 7 December 2018

Abstract: Bioactive dimethacrylate composites filled with silver nanoparticles (AgNP) might be used in medical applications, such as dental restorations and bone cements. The composition of bisphenol A glycerolate dimethacrylate (Bis-GMA) and triethylene glycol dimethacrylate (TEGDMA) mixed in a 60/40 wt% ratio was filled from 25 to 5000 ppm of AgNP. An exponential increase in resin viscosity was observed with an increase in AgNP concentration. Curing was performed by way of photopolymerization, room temperature polymerization, and thermal polymerization. The results showed that the polymerization mode determines the degree of conversion (*DC*), which governs the ultimate mechanical properties of nanocomposites. Thermal polymerization resulted in a higher *DC* than photo- and room temperature polymerizations. The *DC* always decreased as AgNP content increased. Flexural strength, flexural modulus, hardness, and impact strength initially increased, as AgNP concentration increased, and then decreased at higher AgNP loadings. This turning point usually occurred when the *DC* dropped below 65% and moved toward higher AgNP concentrations, according to the following order of polymerization methods: photopolymerization < room temperature polymerization < thermal polymerization. Water sorption (WS) was also determined. Nanocomposites revealed an average decrease of 16% in *WS* with respect to the neat polymer. AgNP concentration did not significantly affect WS.

Keywords: silver nanocomposite; dimethacrylate; molecular structure; flexural properties; water sorption

---

1. Introduction

Recently, polymeric biomaterials with microbiological activity have increasingly gained attention in the potential treatment of various types of inflammation or infection [1]. In dentistry and orthopedy, biomaterials based on dimethacrylates have a superior status, serving as restorative dental materials [2] and bone cements [3–5]. The most commonly used monomers of this type are: 2,2'-bis-[4-(2-hydroxy-3-methacryloyloxy propoxy)phenyl]propane (Bis-GMA) and triethylene glycol dimethacrylate (TEGDMA)—a diluting monomer (Scheme 1) [2–5]. Their polymerization results in a highly crosslinked composite matrix [6,7]. One of the negative consequences of this process is polymerization shrinkage, which causes interfacial gap formation. This phenomenon is responsible for the microleakage and accumulation of bacteria beneath and around the reconstruction, causing tissue inflammation and recurrent caries [8–10].

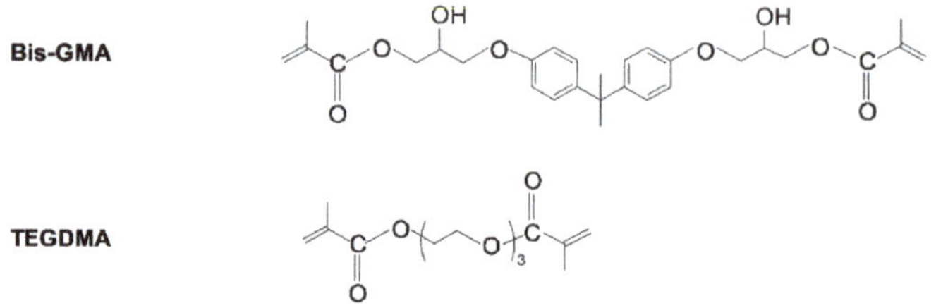

Scheme 1. The Bis-GMA and TEGDMA chemical structure.

It is widely recognized that the interaction of a restorative material with microorganisms is important for its longevity and effectiveness [9]. It might be improved by the incorporation of antimicrobial substances into tooth reconstruction or bone cement [10–12]. The group of *Streptococcus mutans* has been described as the most important bacteria related to the formation of dental caries [10,13,14]. Complications of bone replacements are, in turn, primarily caused by staphylococci, in particular, *Staphylococcus aureus* [15]. Faced with such problems, the need to further develop dental restorative materials, adhesives, and bone cements containing antimicrobial agents has arisen.

A variety of nanoparticles are currently under investigation in order to improve the antibacterial properties of biomaterials. Polymeric nanocomposites filled with nanoparticles, such as calcium phosphate [16,17], zinc oxide [13,18,19], titanium dioxide [20], gold [21,22], and silver (AgNP) [13,16,23–28] have shown promising capabilities in reducing bacterial proliferation, thereby preventing the degradation of tissue and restoration.

AgNP advantages, such as a wide spectrum and long-term antibacterial activity, low toxicity, and good biocompatibility with human cells have led to its application in medical products, such as wound dressings, catheters, and prostheses [28]. AgNP has also been applied in several areas of dentistry [24–29] and orthopedy [15,23,26]. Literature provides much evidence for the antibacterial action of AgNP-containing dental materials against oral streptococci [13,26–28]. On the other hand, bone cements filled with AgNP exhibited antimicrobial activity against staphylococci, such as *Staphylococcus aureus*, *Staphylococcus epidermidis*, and *Acinetobacter baumannii* [23,26]. The decrease in the development of recurrent caries and formation of bacterial biofilm on tooth restoration [16,27,30] and bone cementation [23] was achieved by the introduction of AgNP at the concentration of 500 to 5000 ppm.

Despite evident benefits of nanocomposites with microbiological activity, their potential is limited by the nanofiller agglomeration process at a high nanofiller volume fraction. If the nanofiller content exceeds a critical value, the polymerization extent decreases, which leads to a decrease in polymer network homogeneity [31]. The number of pendant groups, pendant chains, or even free monomer molecules within the poly(dimethacrylate) matrix increases. Consequently, macroscopic properties of the composite are negatively influenced [32]. Additionally, the elution of unreacted monomer from a reconstruction may negatively influence the biological compatibility of the applied biomaterial. The literature has revealed that the greater the content of nanoparticles in the composite, the greater the amount of monomers in the eluate [29].

In view of the above, knowledge regarding structure-property relationships within nanocomposites represents a crucial aspect. The literature, dealing with dimethacrylate systems filled with AgNP, is still lacking a thorough study on this issue.

The purpose of this study was to characterize polymer nanocomposites obtained by photochemical, room temperature, and thermal polymerizations of Bis-GMA/TEGDMA 60/40 wt% composition enriched with silver nanoparticles (AgNP). The AgNP content was increased from 25 to 5000 ppm. The work focused on differences induced by the polymerization initiation methods and AgNP content on the degree of conversion, mechanical properties, and water sorption. Mechanical examination of nanocomposites included the determination of the modulus of elasticity, flexural strength, hardness, and impact resistance.

2. Results

In the present study, the mixture of Bis-GMA and TEGDMA at 60/40 wt% ratio was loaded with AgNP in the quantity of 25 to 5000 ppm. As shown in Table 1, eight Bis-GMA/TEGDMA/AgNP resin compositions were produced using the solvent casting method. n-Hexane was used as a solvent.

Table 1. The AgNP concentrations in Bis-GMA/TEGDMA/AgNP compositions, AgNP concentrations in AgNP/hexane colloids, and colloid amounts used in nanocomposite manufacturing.

AgNP Concentration in Bis-GMA/TEGDMA/AgNP Compositions (ppm)	AgNP Concentration in hexane Colloid (ppm)	The Amount of AgNP/hexane Colloid (g)
25	100	25.0
50	100	50.0
100	100	100.0
150	100	150.0
250	100	250.0
500	100	500.0
1500	500	300.4
5000	1000	502.5

Before polymerization, Bis-GMA/TEGDMA/AgNP compositions, as well as the neat Bis-GMA/TEGDMA mixture, were tested for viscosity. An exponential increase in viscosity with AgNP concentration was observed (Figure 1).

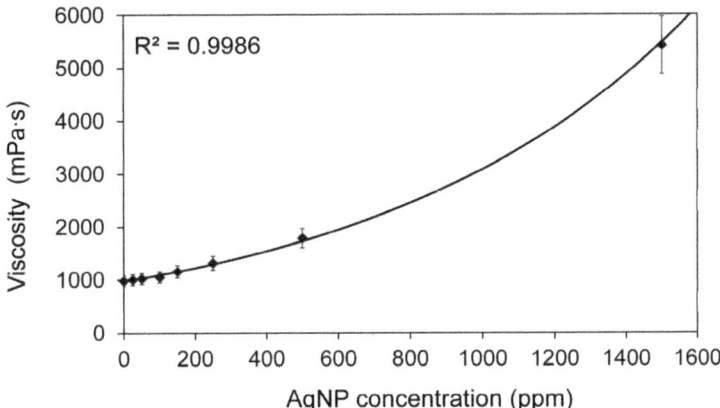

Figure 1. The relationship between viscosity and AgNP concentration in Bis-GMA/TEGDMA/AgNP compositions. All results were statistically significant ($p < 0.05$).

As shown in Figure 1, by increasing the AgNP concentration from 0 to 1500 ppm, the viscosity increased from 980 to 5420 mPa·s. Above 1500 ppm of AgNP, the viscosity increased dramatically, which necessitated the use of another torque in further studies. Consequently, 98 mPa·s was determined for the sample filled with 5000 ppm of AgNP. This value could not be compared with the previous results, found using the torque designed for lower viscosities. Thus, this value was not taken into account when the correlation between viscosity and AgNP concentration was constructed.

The monomer mixtures were then activated for polymerization in three ways:

1. Photopolymerization—by the addition of CQ (camphorquinone) 0.4 wt.% and DMAEMA (N,N-dimethylaminoethyl methacrylate) 1 wt.%;

2. Room temperature polymerization—by the addition of BPO (benzoyl peroxide) 0.5 wt% and DMPT (N,N-dimethyl-p-toluidine) 0.05 wt%;
3. Thermal polymerization—by the addition of BPO 1 wt%.

Macroscopic observations led to the first conclusion that the polymerization initiation mechanism restricts the AgNP capacity in the nanocomposite.

Photopolymerization was limited by 250 ppm of AgNP. A higher AgNP content resulted in a poor sample quality, which was mechanically weak and brittle. The absorption of UV-VIS radiation was suspected as the key factor responsible for this limitation. To check this hypothesis, UV-VIS analysis was performed on the neat Bis-GMA/TEGDMA composition, as well as its AgNP loaded modifications. As shown in Figure 2, all samples were likely to absorb UV-VIS radiation within the range of 190 to 300 nm, with the maximum at around 215 nm. However, the absorption intensity of nanocomposites was higher than the absorption intensity of the neat network. The higher the AgNP quantity, the higher the absorption intensity. This region corresponds to the CQ absorption in the UV region [33]. Additionally, CQ absorbs electromagnetic radiation in the VIS region from 400 to 550 nm, with the maximum absorption intensity at 467 nm [33]. This part of the spectra remained free from absorption bands of Bis-GMA/TEGDMA/AgNP (Figure 2b).

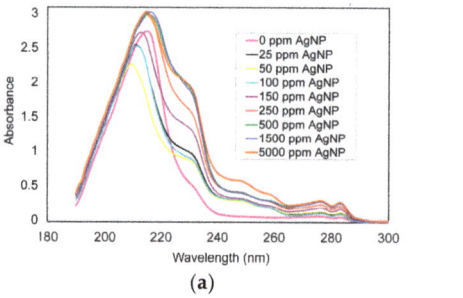 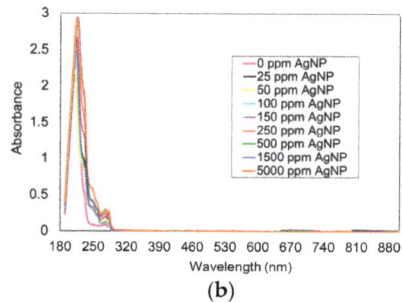

(a) (b)

Figure 2. UV/VIS spectra of the pristine Bis-GMA/TEGDMA and its AgNP loaded modifications in the range of: (a) 190–300 nm; (b) 190–900 nm.

In the case of room temperature polymerization, 1500 ppm of AgNP was the maximum concentration that could be employed to maintain a good visual quality of the sample. The introduction of 5000 ppm of AgNP caused an enormous increase in the viscosity of the resinous system. This prevented the escape of air bubbles before the polymer was hardened. Consequently, a significant weakening of the nanocomposite was observed.

Thermal initiation was successfully adopted to produce nanocomposites of a high quality and durability over the whole AgNP concentration range.

Following the initial sample quality assessment, the AgNP concentration range was narrowed individually for each group of nanocomposites according to the polymerization technique. Thus, the following maximum AgNP concentrations were set up in a sample preparation for further studies: 250 ppm—photopolymerization, 1500 ppm—room temperature polymerization, and 5000 ppm—thermal polymerization. The list of tested samples and their names are presented in Table 2.

Table 2. The Bis-GMA/TEGDMA/AgNP nanocomposite samples studied and their names: Ph (photopolymerization), RT (room temperature polymerization), T (thermal polymerization).

AgNP Concentration in Nanocomposite (ppm)	Sample Name		
	Photopolymerization	Room Temperature Polymerization	Thermal Polymerization
0	Ph0	RT0	T0
25	Ph25	RT25	T25
50	Ph50	RT50	T50
100	Ph100	RT100	T100
150	Ph150	RT150	T150
250	Ph250	RT250	T250
500	-	RT500	T500
1500	-	RT1500	T1500
5000	-	-	T5000

In Table 3, the results for the degree of conversion (DC) and polymerization shrinkage (S) are summarized.

Table 3. The polymerization shrinkage and degree of conversion in studied Bis-GMA/TEGDMA/AgNP nanocomposites.

Sample	Polymerization Shrinkage (%)	Degree of Conversion (%)
Ph0	8.37 ± 0.58 [1]	69.8 ± 6.0 [31,32]
Ph25	8.22 ± 0.60 [2,6]	67.2 ± 7.0 [33]
Ph50	8.01 ± 0.70 [3,7]	64.8 ± 4.6 [34,35,f]
Ph100	7.41 ± 0.62 [4,6]	60.2 ± 4.2 [g]
Ph150	6.95 ± 0.96 [5,7,a,b]	54.8 ± 5.6 [31,34,h]
Ph250	5.44 ± 0.53 [1–5,c,d]	50.2 ± 7.5 [32,33,35,i,j]
RT0	8.39 ± 0.60 [8,9]	69.6 ± 5.4 [36,39,k]
RT25	8.13 ± 0.70 [10]	68.0 ± 4.9 [37,40]
RT50	7.99 ± 0.62 [11]	66.8 ± 4.7 [41,l]
RT100	7.84 ± 0.44 [12]	65.4 ± 7.2 [38,42]
RT150	7.77 ± 0.78 [13,a]	61.4 ± 7.2 [m]
RT250	7.66 ± 0.73 [14,c]	60.2 ± 6.3 [j,n]
RT500	7.41 ± 0.62 [8,15]	56.2 ± 7.7 [36–38,o]
RT1500	5.34 ± 0.59 [9–15,e]	52.0 ± 9.0 [39–42,p]
T0	8.50 ± 0.50 [16,23]	78.2 ± 5.5 [43,k]
T25	8.37 ± 0.43 [17,24]	78.0 ± 5.7 [44]
T50	8.28 ± 0.57 [18,25]	78.2 ± 5.4 [45,f,l]
T100	8.19 ± 0.48 [19,26]	77.4 ± 7.0 [46,g]
T150	8.13 ± 0.69 [20,27,b]	76.8 ± 7.0 [47,h,m]
T250	8.11 ± 0.51 [21,28,d]	76.6 ± 7.2 [48,i,n]
T500	8.01 ± 0.70 [22,29]	73.8 ± 4.9 [49,o]
T1500	7.15 ± 0.74 [16–22,30,e]	70.4 ± 7.4 [50,p]
T5000	5.01 ± 0.99 [23–30]	43.6 ± 7.8 [43–50]

[1–50] $p < 0.05$—statistically significant results within the nanocomposite series. [a–p] $p < 0.05$—statistically significant results between nanocomposites, having the same AgNP content, but obtained in different polymerization modes.

The DC in the neat networks produced by photo- and room temperature polymerizations was almost the same and equaled 70%. The DC in the thermally polymerized neat network was higher and equaled 78%. However, a statistically significant difference was only found for the RT0 and T0 couple. The average polymerization shrinkage of the neat networks was 8.4% and there was no statistically significant difference between the Ph0, RT0, and T0 samples.

The analysis of results for nanocomposites showed that the introduction of AgNP into the Bis-GMA/TEGDMA matrix resulted in a decrease of the DC and S (polymerization shrinkage).

The higher the AgNP concentration, the lower the *DC* and *S*. It is worth noting that the *DC* in T5000 did not achieve a satisfactory degree of conversion, which was 43.6%. Statistically significant decreases in the *DC*, with respect to the neat networks, were found, beginning from the following AgNP concentrations: 150 ppm—for photopolymerization, 500 ppm—for room temperature polymerization, and 5000 ppm—for thermal polymerization. An analogous analysis of the results for *S* allowed for the specification of the following AgNP concentrations, which resulted in statistically significant decreases in *S*, with respect to the neat networks, respectively: 250 ppm, 500 ppm, and 1500 ppm.

A detailed analysis of the *DC* and *S* led to the conclusion that both parameters increased by the initiation mechanism according to the following order: photopolymerization < room temperature polymerization < thermal polymerization. The *DC* in nanocomposites obtained by thermal polymerization was usually statistically significantly higher than the *DC* in nanocomposites obtained by photopolymerization and room temperature polymerization.

In Table 4, the results for flexural properties are summarized. It may be seen that the flexural strength (σ) and flexural modulus (*E*) depend on the AgNP concentration.

Table 4. The flexural properties of studied Bis-GMA/TEGDMA/AgNP nanocomposites.

Sample	Flexural Strength (MPa)	Flexural Modulus (MPa)
Ph0	88.6 ± 9.1 [1,5]	3819.4 ± 255.8 [40]
Ph25	92.4 ± 11.7 [2,6,a]	3848.5 ± 210.5 [41,45]
Ph50	98.3 ± 11.7 [3,7,b]	4039.2 ± 361.9 [42,46]
Ph100	97.4 ± 8.7 [4,8,c]	3897.2 ± 314.3 [43,47]
Ph150	74.4 ± 7.7 [1–4,9,d,f]	3179.6 ± 343.6 [44,45–47,p]
Ph250	57.6 ± 6.8 [5–9,e,g]	1808.2 ± 254.9 [40–44,q,r]
RT0	79.8 ± 10.2 [10,h]	3610.5 ± 316.0 [48,54]
RT25	88.2 ± 9.8 [11,16,i]	3665.6 ± 248.6 [49,55]
RT50	95.5 ± 9.0 [12,17,j]	3822.3 ± 292.7 [50,56,61]
RT100	98.5 ± 10.4 [13,18,k]	4001.6 ± 442.3 [51,57,62]
RT150	96.7 ± 10.8 [10,14,19,a,f,l]	3746.8 ± 290.3 [52,58]
RT250	89.6 ± 5.9 [15, 20,c,g,m]	3336.5 ± 307.4 [53,59,61,62,q,s]
RT500	73.1 ± 7.8 [11–15,n]	2676.8 ± 227.8 [47–53,60,t]
RT1500	68.1 ± 7.1 [16–20,o]	1905.9 ± 289.8 [54–60,u]
T0	100.3 ± 6.8 [21,25,h]	3872.5 ± 297.9
T25	109.0 ± 11.0 [26,27,33,a,i]	4034.0 ± 437.6
T50	112.0 ± 12.5 [28,34,b,j]	4066.4 ± 415.0
T100	120.16 ± 8.4 [21,29,35,c,k]	4044.4 ± 272.6 [61]
T150	124.8 ± 6.3 [22,30,36,d,l]	4066.5 ± 407.3 [p]
T250	129.3 ± 15.8 [23,26,31,37,e,m]	4153.5 ± 303.4 [62,r,s]
T500	129.2 ± 11.2 [24,32,38,n]	4210.6 ± 396.6 [63,t]
T1500	96.6 ± 7.2 [27–32,39,o]	4596.1 ± 590.6 [64,u]
T5000	58.7 ± 6.0 [25,33–39]	3368.9 ± 479.4 [61–64]

[1–64] $p < 0.05$—statistically significant results within the nanocomposite series. [a–u] $p < 0.05$—statistically significant results between nanocomposites, having the same AgNP content, but obtained in different polymerization modes.

The flexural strength of the neat networks derived from room temperature polymerization (79.8 MPa) and photopolymerization (88.6 MPa) was lower than that from thermal polymerization (100.3 MPa). The results for only the RT0 and T0 couple were statistically significant. The elasticity of the pristine networks was similar ($p \geq 0.05$) and it was characterized by the average *E* value of 3767.5 MPa. By filling Bis-GMA/TEGDMA with AgNP, increases in both elastic properties were observed. The flexural strength of nanocomposites produced by photopolymerization increased by 11%, whereas the modulus increased by 6%. σ of samples produced by room temperature polymerization increased by 23%, whereas *E* increased by 11%. Thermally activated polymerizations resulted in the greatest increases in σ and *E*, which correspond, respectively, to 29 and 19%. The detailed analysis of results for σ and *E* within particular series, led to the observation that their values increased with an increasing AgNP content at the beginning (low AgNP concentrations), but then decreased with an increasing AgNP content at high AgNP concentrations. The maximum values of σ were noted for the following nanocomposites: Ph50, RT100, and T250, whereas the maximum values for E were found for: Ph50, RT100, and T1500. Statistically significant drops in σ with respect to the pristine networks were recorded for the first time for Ph150 and T5000. σ of none of the chemically cured nanocomposites revealed statistically significant decreases with respect to RT0. The first statistically significant drop in E, with respect to Ph0, was recorded for Ph250, whereas with respect to RT0, it was recorded for RT1500. None of the thermally cured samples showed statistically significant decreases in E with respect T0. This result may be due to the fact that in thermally activated samples, the DC was relatively high (in most cases it was higher than 70%) compared to Ph or RT samples.

When comparing the elastic properties of nanocomposites by the polymerization method, the following order of increasing σ and E might be constructed: photopolymerization < room temperature polymerization < thermal polymerization. Statistical analysis showed that thermally polymerized materials had a significantly higher bending strength than products of photo- and room temperature polymerizations. The same tendency could be found by comparing results for modulus; however, they did not usually show statistical significance.

The decreases in σ and E, observed with increasing AgNP concentration in nanocomposites, coincide with the decreasing DC. The comprehensive analysis of changes in the DC, σ, and E led to the conclusion that AgNP in the Bis-GMA/TEGDMA/AgNP nanocomposites can play a reinforcing role, by increasing σ and E, when the DC is at a minimum of about 65%.

In Table 5, the results for the hardness and impact strength of studied materials are summarized.

Table 5. Hardness and impact resistance of studied Bis-GMA/TEGDMA/AgNP nanocomposites.

Sample	Hardness (N/mm^2)	Impact Resistance (kJ/m^2)
Ph0	115.1 ± 9.5 [1]	4.57 ± 0.35 [26,30]
Ph25	117.9 ± 6.1 [2]	4.77 ± 0.36 [27,31]
Ph50	123.3 ± 16.4 [3,6]	4.82 ± 0.47 [28,32]
Ph100	122.8 ± 10.6 [4,7]	4.21 ± 0.45 [29,33,g]
Ph150	100.9 ± 12.1 [5,6,7,a,b]	3.41 ± 0.35 [26–29,34,h]
Ph250	81.4 ± 7.3 [1–5,c,d]	2.48 ± 0.31 [30–34,i,j]
RT0	110.2 ± 7.6 [8,9]	4.07 ± 0.37 [35,40,k]
RT25	117.7 ± 4.2 [10]	4.24 ± 0.44 [36,41,l]
RT50	121.3 ± 6.5 [11]	4.25 ± 0.42 [37,42,m]
RT100	128.5 ± 13.1 [12]	4.27 ± 0.49 [38,43,n]
RT150	130.2 ± 15.1 [8,13,16,a]	4.03 ± 0.44 [39,44,o]
RT250	122.7 ± 14.5 [14,c]	3.67 ± 0.36 [45,j,p]
RT500	113.1 ± 11.6 [15,16,e]	3.34 ± 0.28 [35–39,46,q]
RT1500	83.94 ± 6.7 [9–15,f]	2.26 ± 0.15 [40–46,r]
T0	121.8 ± 11.6 [17,19,22]	5.20 ± 0.28 [47,50,56,k]
T25	126.4 ± 12.1 [18,20,23]	5.38 ± 0.51 [51,57,l]
T50	130.5 ± 15.5 [21,24]	5.47 ± 0.56 [48,52,58,m]
T100	139.2 ± 18.3	5.54 ± 0.44 [49,53,59,64,f,n]
T150	143.2 ± 21.7 [b]	5.30 ± 0.45 [54,60,h,o]
T250	142.8 ± 20.8 [25,d]	4.73 ± 0.45 [55,61,64,i,p]
T500	148.5 ± 13.0 [17,18,e]	4.42 ± 0.50 [47–49,62,q]
T1500	149.1 ± 7.1 [19–21,f]	4.06 ± 0.41 [50–55,63,r]
T5000	158.9 ± 13.5 [22–25]	2.26 ± 0.15 [56–63]

[1–64] $p < 0.05$—statistically significant results within the nanocomposite series. [a–r] $p < 0.05$—statistically significant results between nanocomposites, having the same AgNP content, but obtained in different polymerization modes.

The hardness (H) of pristine networks did not differ significantly and the average value equaled 115.7 N/mm^2. In the composite series, the values of H increased to a certain AgNP limiting content and afterwards, they decreased. Thus, the following nanocomposites were characterized by the highest hardness in the series: Ph50, RT150, and T5000, which correspond to the following percentage increases, respectively: 7%, 18%, and 30%. Statistically significant decreases in *H*, with respect to pristine polymers, were found for Ph250 and RT1500, i.e., for nanocomposites with the highest AgNP loadings in the photo- and room temperature polymerized series. The hardness of the thermally polymerized samples increased over the whole concentration range. Moreover, the T5000 nanocomposite was characterized by the highest hardness noted in this study.

When comparing the hardness of nanocomposites according to the initiation mechanism, the following increasing order can be constructed: photopolymerization < room temperature polymerization < thermal polymerization. However, these results were usually not statistically significant.

When comparing the impact strength of pristine polymers by the polymerization technique, the values might be ordered accordingly: room temperature polymerization (4.07 kJ/m^2) < photopolymerization (4.57 kJ/m^2) < thermal polymerization (5.20 kJ/m^2). Statistically significant results were produced for the RT0 and T0 couple. The same order of polymerization techniques might be constructed if a_n of nanocomposites was analyzed. The results for nanocomposites obtained by thermal polymerization were statistically significantly higher compared to results for their counterpart samples produced in photo- or chemically initiated processes.

By comparing results in the nanocomposite series, it can be seen that the impact strength increased as the AgNP content increased up to a certain limit and then decreased, with a further increase of AgNP concentration. The impact resistance of photopolymerized samples reached the maximum in Ph50, which corresponded to 5.5% income with respect to Ph0. The impact resistance (a_n) of samples polymerized on the chemical route increased by 4.9%, which was recorded for RT100. Thermal polymerization resulted in a 6.5% increase in a_n, which was achieved for T100. All these increases were statistically insignificant with respect to the pristine polymers. Statistically significant drops in a_n of nanocomposites, with respect to the pristine polymers, were noted for Ph250, RT500, and T500.

In Figure 3, the results for water sorption (WS) are shown. It may be seen that AgNP incorporation caused the decrease in the values by around 16.5%. The average WS of nanocomposites was 33.2 µg/mm^3, whereas the WS of pristine polymers equaled 37.7 µg/mm^3. The polymerization technique did not significantly influence WS ($p \geq 0.05$).

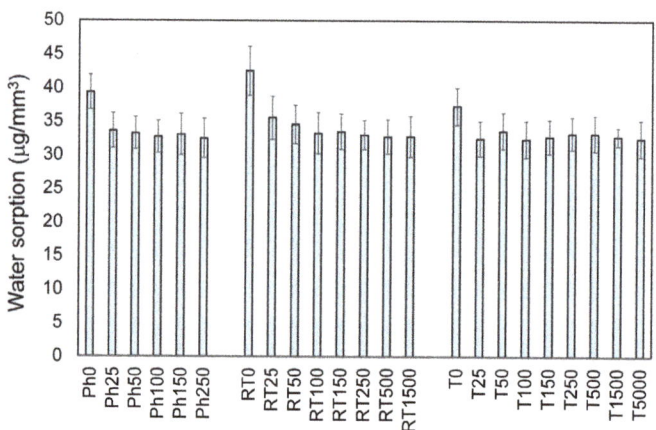

Figure 3. Water sorption of Bis-GMA/TEGDMA/AgNP nanocomposites. The statistically significant results were obtained between pristine polymers and nanocomposites in each series ($p < 0.05$).

3. Discussion

The present paper describes research regarding the effect of AgNP concentration in the Bis-GMA/TEGDMA 60/40 wt% on the viscosity of its liquid form, as well as structural, physical, and mechanical properties of hardened materials.

The results showed that AgNP concentration influences Bis-GMA/TEGDMA/AgNP resin viscosity, which increases as the AgNP content increases. An exponential increase in viscosity resulted in the enormously limited flowability of the mixture filled with 5000 ppm of AgNP. This means that the concentration of AgNP in the Bis-GMA/TEGDMA/AgNP mixture should not be higher than 1500 ppm. A higher AgNP concentration may considerably restrict resin flowability, which gives rise to air bubble trapping and can exclude the possibility of introducing other fillers, typically used in dental composites and bone cements to improve their mechanical characteristics.

The structure-property relationships were tested for Bis-GMA/TEGDMA/AgNP nanocomposites obtained by radical polymerization of the dimethacrylate matrix, which was initiated according to three mechanisms:

1. Photopolymerization—commonly used for direct curing of dental restorations [2];
2. Room temperature chemical polymerization—commonly used in bone cementation [3];
3. Thermal polymerization—commonly used in industrial processes for the manufacturing of thick parts [34].

The neat networks were characterized by the DC from 70 to 78%, which is in agreement with the literature data. The polymerization of dimethacrylates is never complete. A considerable fraction of the double bonds remains unreacted because of immobilization, caused by a set of certain complex features, such as autoacceleration, autodeceleration, steric isolation, and vitrification [6,7,35–38]. This has been demonstrated in many studies on the photopolymerization [35,36], room temperature polymerization [37], and thermal polymerization [38] of Bis-GMA/TEGDMA systems. The DC of 78% was the highest measured in this study and corresponded to the thermally polymerized sample. It can be explained by the effects of temperature on the polymerization kinetics. It is well-known that the higher polymerization temperature, the higher the final conversion in the dimethacrylate network. Heating provides the energy to improve the segmental mobility of the polymer chain and makes more residual unsaturation sites accessible for polymerization [35,36]. Additionally, the thermodynamic miscibility of Bis-GMA and TEGDMA monomers is greater at elevated temperatures. They can undergo phase separation at room temperature, which can result in the formation of more heterogeneous networks [36]. In the early stage of the thermal polymerization process, the temperature was kept at 40 °C for 1 h. At this temperature, the BPO decomposition rate was slow enough that the system nearly did not polymerize [39] and the viscosity decreased sufficiently enough to allow air bubbles to escape. Due to the preheating of liquid samples, high macroscopic homogeneity of the cured materials was maintained, despite the increase in viscosity. Gradually raising the temperature to 100 °C, in combination with a 24 h total polymerization time, resulted in materials of the highest structural homogeneity and best mechanical performance. The temperature of 100 °C was higher than the glass temperature of the Bis-GMA/TEGDMA polymer network [36]. It was chosen to intensify segmental mobility during post-curing.

The results for the DC in nanocomposites showed that it depends on the AgNP concentration, as well as the polymerization initiation mechanism.

The decrease in DC with the increase of AgNP concentration is likely caused by the increasing distance between double bonds. The free volume in the nanocomposite is occupied by the AgNP agglomerates. The higher the AgNP concentration, the higher the tendency of forming larger agglomerates [31]. As the dimensions of agglomerates and their number increase, the reactive groups drift further apart and thus become less available and less reactive.

The following general order of increasing DC according to polymerization methods: photopolymerization < room temperature polymerization < thermal polymerization, results from a variety of factors. The poorest curing efficiency of photopolymerization might be explained by the presence of partially overlapping absorption of Bis-GMA/TEGDMA/AgNP and CQ in the UV region. Even though the UV/VIS spectrum was free from any absorption bands from Bis-GMA/TEGDMA/AgNP in the VIS region, where CQ absorbs in the 400–550 nm range, the curing efficiency in nanocomposites produced by photopolymerization was insufficient when AgNP content exceeded 250 ppm. The effectiveness of room temperature polymerization was most likely restricted by the excessive high viscosity of the resin system, which causes the trapping of air bubbles. The highest DC of nanocomposites manufactured by thermal polymerization can be explained by the same kinetic factors as mentioned above. However, it has to be noted that the T5000 nanocomposite was characterized by the DC of 44%. Since the DC lower than 50% gives the information of free monomer occurrence in the system, 1500 ppm might be proposed as a maximum AgNP concentration in possible potential applications of studied nanocomposites [40].

Since S is the consequence of the DC, as predicted, the same trend in S values was observed with increasing AgNP content and polymerization technique.

The results also showed that mechanical properties of Bis-GMA/TEGDMA/AgNP nanocomposites mainly depend on the DC in the dimethacrylate polymer matrix, which is governed by the initiation mechanism (Figure 4).

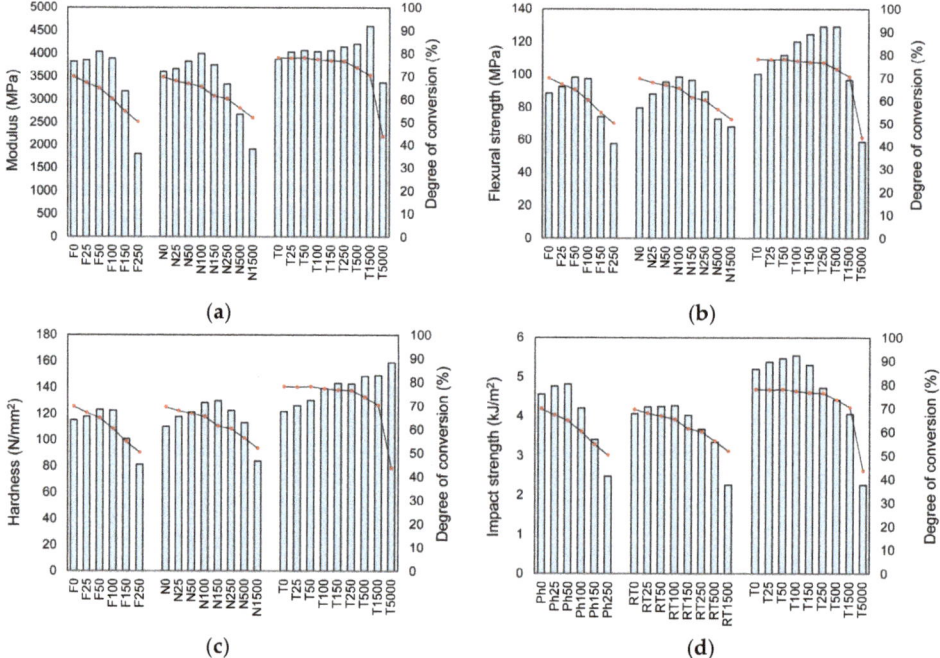

Figure 4. The comparison of mechanical properties with the degree of conversion: (**a**) modulus; (**b**) flexural strength; (**c**) hardness; (**d**) impact resistance. Bar charts correspond to mechanical properties, whereas line charts correspond to the DC.

Thermal polymerization resulted in a higher flexural strength, modulus, and hardness than photopolymerization and room temperature polymerization. Impact strength increased accordingly: room temperature polymerization < photopolymerization < thermal polymerization. The top position of thermally polymerized nanocomposites in these arrangements might be attributed to their maximum DC and the best sample quality. The lowest a_n of chemically polymerized nanocomposites might be attributed to lower DC and a larger number of air bubbles trapped in the sample. Since the room temperature polymerization begins just after mixing the components of the initiation system, the resin viscosity increases radically in a short time and perfect trapped air removal is unattainable. For this reason, bone cements typically display an overall porosity ranging from 2 to 18% [41,42]. This means that the latter property may be considerably deteriorated by air trapped bubbles, which are specific for room temperature polymerization.

As can be seen from Figure 4, mechanical properties usually increased at the beginning, at lower AgNP concentrations, and then decreased, when AgNP concentrations exceeded a certain level. The decreases in σ, E, H, and a_n, which were observed with increasing AgNP concentration in nanocomposites, coincide with the decreasing DC. This result coincides with previous findings, which show that the modulus, mechanical strength, and hardness decrease as the degree of conversion decreases [6,7]. The detailed analysis of changes in the DC, σ, and E leads to the conclusion that AgNP in the Bis-GMA/TEGDMA/AgNP nanocomposites can play a reinforcing role on the condition that the DC is at a minimum of 65%. In the series of nanocomposites obtained by thermal polymerization, hardness increased throughout the series, regardless of the degree of conversion. Therefore, hardness might be considered as the least sensitive to the DC property.

Taking the above considerations into account, the following AgNP concentrations might be recommended as the upper limits for manufacturing Bis-GMA/TEGDMA/AgNP nanocomposites with a satisfactory physico-mechanical performance: 250 ppm—for photopolymerization and 1500 ppm—for room temperature and thermal polymerizations.

The water sorption of the nanocomposites studied decreases with the mere addition of AgNP by an average of 16% and an increase in its concentration does not significantly influence the WS behavior. This result might be attributed to the hydrophobic character of AgNP [43].

4. Materials and Methods

4.1. Materials

The dimethacrylate resins: Bis-GMA and TEGDMA, components of the initiation systems: CQ (camphorquinone), DMAEMA (N,N-dimethylaminoethyl methacrylate), and DMPT (N,N-dimethyl-p-toluidine) were purchased from Sigma-Aldrich (St. Louis, MO, USA) and used as received. The BPO initiator—(benzoyl peroxide) came from POCh (Gliwice, Poland) and it was purified by dissolving in chloroform (POCh, Gliwice, Poland) and precipitated by adding methanol (POCh, Gliwice, Poland). n-Hexane was purchased from POCh (Gliwice, Poland) and was used as received.

AgNP colloid in hexane of the 1000 ppm concentration was purchased from AMEPOX Co. Ltd. (Lodz, Poland). The AgNP/hexane colloid was used as received or diluted with n-hexane to the concentrations of 500 and 100 ppm (Table 1). The mean particle size was 22.8 nm, as measured in a previous study [44].

4.2. Preparation of AgNP-Loaded Resin Compositions

The monomer blend consisted of Bis-GMA and TEGDMA in a 60/40 wt% ratio. It was loaded with AgNP. Eight concentrations of AgNP: 25, 50, 100, 150, 250, 500, 1500, and 5000 ppm, were prepared by solvent casting, with the use of n-hexane as a solvent.

First, 40 g of TEGDMA was dissolved in 100 mL of n-hexane. Next, the precisely measured amount of AgNP/hexane colloid was added to the TEGDMA solution in n-hexane (Table 1). Mixing was performed in an Erlenmayer flask on a magnetic stirrer for 10 min. Hexane was evaporated from the TEGDMA/AgNP/hexane dispersion in a rotary evaporator (IKA RV-10, Staufen, Germany) for 20 min, under the pressure of 50 mbar and then for 20 min, under the pressure of 5 mbar. The effectiveness of the evaporation procedure was confirmed in the ^{1}H NMR experiments (UNITY INOVA, Varian, 300 MHz, Palo Alto, CA, USA), which did not show the presence of peaks corresponding to n-hexane on

1H NMR spectra. TEGDMA loaded with AgNP was then blended with 60 g of Bis-GMA. Mixing was performed in a glass vessel with a mechanical stirrer at 40 °C for 20 min.

4.3. Curing Procedure

Polymerizations were carried out in 12 cm diameter Petri dishes and 10 mm diameter Teflon O-rings placed on a glass surface. The polymer amounted to the obtained thicknesses of 2 and 1 mm, respectively.

4.3.1. Photopolymerization

Bis-GMA/TEGDMA/AgNP liquid compositions, loaded with AgNP in the concentration specified in Table 1 were mixed with 0.4 wt.% of CQ (initiator) and 1 wt.% of DMAEMA (reducer) and poured into molds. Before irradiation, the sample surface was covered with a PET (poly(ethylene terephtalate)) film in order to avoid the oxygen inhibition effect. The mixtures were irradiated with a mercury UV/VIS lamp (FAMED-1, Zywiec, Poland, power 375 W) set at a distance of 10 cm for 15 min.

4.3.2. Room Temperature Polymerization (Chemical Polymerization)

Bis-GMA/TEGDMA/AgNP liquid compositions, loaded with AgNP in the concentration specified in Table 1, were divided into two equimass portions. One of them was mixed with 0.5 wt% of BPO (radical polymerization initiator) and the other with 0.05 wt% of DMPT (redox activator for BPO). To perform curing, both components were thoroughly mixed for 60 s. After stirring, the mixtures were poured into molds, covered with PET film, and left for 24 h.

4.3.3. Thermal Polymerization

Bis-GMA/TEGDMA/AgNP liquid compositions, loaded with AgNP in the concentration specified in Table 1, were mixed with 1 wt% of BPO, poured into molds, and polymerized in a drying oven under nitrogen—reducing oxygen inhibition, by raising the temperature gradually from 40 to 100 °C over 24 h [40]. First, the polymerizing mixture was reheated at 40 °C for 1 h; next, the temperature was increased to 70 °C and maintained at that temperature overnight; and finally, the temperature was increased to 100 °C and maintained for 2 h.

4.4. UV/VIS Spectroscopy

The UV/VIS spectra of pristine Bis-GMA/TEGDMA monomer composition, as well as its modifications with AgNP, were recorded over a wavelength ranging from 190 to 500 nm, utilizing a UV/VIS spectrometer (UV 2600, Shimadzu Co., Kyoto, Japan) with a 10-cm path-length quartz cuvette.

4.5. Viscosity

The viscosity (η, mPa·s) of liquid Bis-GMA/TEGDMA and its compositions with AgNP was measured by means of a rotating spindle viscometer (Brookfield Fungilab Viscometer, Visco Star Plus L, Barcelona, Spain) at 25 °C, according to PN-ISO 2555 (Plastics—Resins in the liquid state or as emulsions or dispersions—Determination of apparent viscosity by the Brookfield method). Viscosity was measured using the appropriate spindle, which allowed for recording viscosity values between 10% and 90% torque. Compositions containing 0 to 1500 ppm AgNP were tested utilizing the same spindle of the designation "1". The composition containing 5000 ppm was characterized by a significantly higher viscosity, which required the use of a spindle with the designation of "3".

4.6. Polymerization Shrinkage

The density of resinous compositions (d_m) was measured utilizing a liquid pyknometer at 25 °C according to ISO 1675 (Plastics—Liquid resins—Determination of density by the pyknometer method). Their cured forms were tested for density (d_p) according to Archimedes' principle, on the Mettler

Toledo XP Balance with a 0.01 mg accuracy (Greifensee, Switzerland) with the density determination kit at 25 °C. Water was used as the immersing liquid. The volumetric shrinkage (S) was determined by the following equation:

$$S(\%) = \frac{d_p - d_m}{d_p} \times 100 \qquad (1)$$

4.7. Degree of Conversion

The degree of conversion (DC) in the polymer matrix of studied nanocomposites was determined by using an FTIR spectrophotometer (Bio-Rad Laboratories, FTS 175C, Hercules, CA, USA). The spectra of the monomers and their polymers were recorded with 128 scans at a resolution of 1 cm^{-1}. The monomer samples were tested as very thin films on KBr pellets. The cured samples were pulverized into fine powder with a particle diameter of less than 24 µm and analyzed as KBr pellets. The DC was calculated from the decrease of the absorption band at 1637cm^{-1}, referring to the C=C stretching vibration ($A_{C=C}$), in relation to the peak at 1609 cm^{-1} assigned to aromatic stretching vibrations (A_{Ar}) [45]:

$$DC(\%) = \left(1 - \frac{(A_{C=C}/A_{Ar})_{polymer}}{(A_{C=C}/A_{Ar})_{monomer}}\right) \times 100 \qquad (2)$$

4.8. Mechanical Properties

4.8.1. Flexural Properties

The flexural modulus (E) and flexural strength (σ) were determined in accordance with ISO 178 (Plastics—Determination of flexural properties) in three-point bending tests, using a universal testing machine (Zwick Z020, Ulm, Germany). The rectangular samples (length × width × thickness: 40 mm × 10 mm × 2 mm) were tested. E and σ were calculated according to the following equations:

$$E \text{ (MPa)} = \frac{P_1 l^3}{4bd^3\delta} \qquad (3)$$

$$\sigma \text{ (MPa)} = \frac{3Pl}{2bd^2} \qquad (4)$$

where, P is the maximum load, P_1 is the load at a selected point of the elastic region of the stress–strain plot, l is the distance between supports, b is the specimen width, d is the specimen thickness, and δ is the deflection of the specimen at P_1.

4.8.2. Hardness

The ball indentation hardness (H) was determined according to ISO 2039-1 (Plastics—Determination of hardness—Part 1: Ball indentation method) on the VEB Werkstoffprüfmaschinen apparatus (Leipzig, Germany). The 4 mm-thick disc-like samples were prepared by stacking two 2 mm layers. H was calculated according to the following equation:

$$H \text{ (MPa)} = \frac{F_m\left(\frac{0.21}{h-h_r+0.21}\right)}{\pi d h_r} \qquad (5)$$

where, h_r is the reduced depth of impression (h_r = 0.25 mm), d is the diameter of the ball indenter (d = 5 mm), F_m is the test load on the indenter, and h is the depth of impression.

4.8.3. Impact Strength

The impact strength (a_n) was determined in accordance with DIN 53435 (Testing of plastics. Bending test and impact test on Dynstat test pieces) using VEB Werkstoffprüfmaschinen Dynstat apparatus on rectangular unnotched specimens (length × width × thickness: 15 mm × 10 mm × 2 mm). The following formula was applied to calculate a_n:

$$a_n \left(\frac{kJ}{m^2}\right) = \frac{A_n}{bd} \quad (6)$$

where, A_n is the impact energy required to cause a material to fracture, and b and d are the width and thickness of the specimen, respectively.

4.9. Water Sorption

Water sorption was measured according to ISO 4049 (Dentistry—Polymer-based restorative materials). Disc-like specimens (diameter × thickness: 10 mm × 1 mm) of each nanocomposite were dried in a pre-conditioning oven at 37 °C until their weight was constant. This result was recorded as m_0 (Mettler Toledo XP Balance with 0.01 mg accuracy, Greifensee, Switzerland). The specimens were then immersed in distilled water and maintained at 37 °C for a week. After this time, the samples were removed, blotted dry, and weighed (m_1). Water sorption (WS) was calculated using the following formula:

$$WS\left(\frac{\mu g}{mm^3}\right) = \frac{m_1 - m_0}{V_0} \quad (7)$$

where, m_0 is the initial sample weight, V_0 – the initial volume of the sample.

4.10. Statistical Analysis

The experimental results were analyzed using one-way analysis of variance (ANOVA). The pair-wise comparisons were conducted by means of the Student's t-test with a significance level (p) of 0.05. For each physical property, a set of five samples was tested. The results of measurements were expressed as mean values with associated standard deviations.

5. Conclusions

This study explained the AgNP influence on the degree of conversion and mechanical properties of Bis-GMA/TEGDMA/AgNP nanocomposites. It was shown that the higher the AgNP concentration, the lower the *DC*. Among the polymerization methods used, the thermal polymerization resulted in the highest level of curing.

Further analysis showed that AgNP can have a strengthening effect on mechanical properties. However, their values increased at the beginning, at lower AgNP concentrations, and then decreased, as AgNP concentrations exceeded a certain level, which was varied and dependent on the property. Hardness showed the least sensitivity to the *DC*, followed by the modulus, bending strength, and impact resistance, in that order. The latter property, besides possessing the highest sensitivity to the *DC*, was most likely deteriorated by air trapped bubbles, which are specific for chemical polymerization.

Author Contributions: Conceptualization, I.M.B.-R.; methodology and investigation, I.M.B.-R. and G.C.; writing—original draft preparation, review, and editing, I.M.B.-R.; funding acquisition, I.M.B.-R.

Funding: This research was funded by the Rector's grant for scientific research and development activities in Silesian University, grant number: 04/040/RGJ18/0075.

Conflicts of Interest: The authors declare no conflict of interest.

Abbreviations

AgNP	Silver nanoparticles
Bis-GMA	2,2′-bis-[4-(2-hydroxy-3-methacryloyloxy propoxy)phenyl]propane
BPO	Benzoyl peroxide
RT	Room temperature polymerization
CQ	Camphorquinone
DC	Degree of conversion
DMPT	N,N-dimethyl-p-toluidine
DMAEMA	N,N-dimethylaminoethyl methacrylate
E	Flexural modulus
H	Hardness
TEGDMA	Triethylene glycol dimethacrylate
Ph	Photopolymerization
S	Polymerization shrinkage
T	Thermal polymerization
WS	Water sorption

References

1. Francolini, I.; Vuotto, C.; Piozzi, A.; Donelli, G. Antifouling and antimicrobial biomaterials: An overview. *APMIS* **2017**, *125*, 392–417. [CrossRef] [PubMed]
2. Powers, J.M.; Sakaguchi, R.L. Restorative materials—Composites and polymers. In *Craig's Restorative Dental Materials*, 13th ed.; Mosby: St. Louis, MI, USA, 2013; ISBN 9780323081085.
3. Khader, B.A.; Towler, M.R. Materials and techniques used in cranioplasty fixation: A review. *Mater. Sci. Eng. C* **2016**, *66*, 315–322. [CrossRef] [PubMed]
4. Vallo, C.I.; Schroeder, W.F. Properties of acrylic bone cements formulated with Bis-GMA. *J. Biomed. Mater. Res.* **2005**, *74B*, 676–685. [CrossRef] [PubMed]
5. Otsuka, M.; Sawada, M.; Matsuda, Y.; Nakamura, T.; Kokubo, T. Antibiotic delivery system using bioactive bone cement consisting of Bis-GMA/TEGDMA resin and bioactive glass ceramics. *Biomaterials* **1997**, *18*, 1559–1564. [CrossRef]
6. Barszczewska-Rybarek, I.; Jurczyk, S. Comparative study of structure-property relationships in polymer networks based on Bis-GMA, TEGDMA and various urethane-dimethacrylates. *Materials* **2015**, *8*, 1230–1248. [CrossRef] [PubMed]
7. Barszczewska-Rybarek, I. Structure-property relationships in dimethacrylate networks based on Bis-GMA, UDMA and TEGDMA. *Dent. Mater.* **2009**, *25*, 1082–1089. [CrossRef] [PubMed]
8. Irie, M.; Suzuki, K.; Watts, D.C. Marginal gap formation of light-activated restorative materials: Effects of immediate setting shrinkage and bond strength. *Dent. Mater.* **2002**, *18*, 203–210. [CrossRef]
9. Mishra, S.K.; Chowdhary, R.; Kumari, S. Microleakage at the Different Implant Abutment Interface: A Systematic Review. *J. Clin. Diagn. Res.* **2017**, *11*, ZE10–ZE15. [CrossRef]
10. Hamouda, I.M. Current perspectives of nanoparticles in medical and dental biomaterials. *J. Biomed. Res.* **2012**, *26*, 143–151. [CrossRef]
11. Ástvaldsdóttir, Á.; Dagerhamn, J.; van Dijken, J.W.V.; Naimi-Akbar, A.; Sandborgh-Englund, G.; Tranæus, S.; Nilsson, M. Longevity of posterior resin composite restorations in adults—A systematic review. *J. Dent.* **2015**, *43*, 934–954. [CrossRef]
12. Karthikeyan, S.; Wagar, A. Nanotechnology and its applications in dentistry. In *Emerging Nanotechnologies in Dentistry*, 2nd ed.; Karthikeyan, S., Wagar, A., Eds.; William Andrew (Elsevier): Norwich, NY, USA, 2017; pp. 1–15, ISBN 9780128122914.
13. Kasraei, S.; Sami, L.; Hendi, S.; AliKhani, M.-Y.; Rezaei-Soufi, L.; Khamverdi, Z. Antibacterial properties of composite resins incorporating silver and zinc oxide nanoparticles on *Streptococcus mutans* and Lactobacillus. *Restor. Dent. Endod.* **2014**, *39*, 109–114. [CrossRef] [PubMed]

14. Beyth, N.; Yudovin-Farber, I.; Bahir, R.; Domb, A.J.; Weiss, E.I. Antibacterial activity of dental composites containing quaternary ammonium polyethylenimine nanoparticles against Streptococcus mutans. *Biomaterials* **2006**, *27*, 3995–4002. [CrossRef] [PubMed]
15. Tunney, M.; Dunne, N.; Einarsson, G.; McDowell, A.; Kerr, A.; Patrick, S. Biofilm formation by bacteria isolated from retrieved failed prosthetic hip implants in an in vitro model of hip arthroplasty antibiotic prophylaxis. *J. Orthop. Res.* **2007**, *25*, 2–10. [CrossRef] [PubMed]
16. Melo, M.A.; Cheng, L.; Zhang, K.; Weir, M.D.; Rodrigues, L.K.; Xu, H.H. Novel dental adhesives containing nanoparticles of silver and amorphous calcium phosphate. *Dent. Mater.* **2013**, *29*, 199–210. [CrossRef] [PubMed]
17. Xu, H.H.K.; Moreau, J.L.; Sun, L.; Chow, L.C. Nanocomposite containing amorphous calcium phosphate nanoparticles for caries inhibition. *Dent. Mater.* **2011**, *27*, 762–769. [CrossRef] [PubMed]
18. Aydin Sevinç, B.; Hanley, L. Antibacterial activity of dental composites containing zinc oxide nanoparticles. *J. Biomed. Mater. Res. B Appl. Biomater.* **2010**, *94*, 22–31. [CrossRef] [PubMed]
19. Tavassoli Hojati, S.; Alaghemand, H.; Hamze, F.; Babaki, F.A.; Rajab-Nia, R.; Rezvani, M.B.; Kaviani, M.; Atai, M. Antibacterial, physical and mechanical properties of flowable resin composites containing zinc oxide nanoparticles. *Dent. Mater.* **2013**, *29*, 495–505. [CrossRef]
20. Poosti, M.; Ramazanzadeh, B.A.; Zebarjad, M.; Javadzadeh, P.; Naderinasab, M.; Shakeri, M.T. Shear bond strength and antibacterial effects of orthodontic composite containing TiO_2 nanoparticles. *Eur. J. Orthod.* **2013**, *35*, 676–679. [CrossRef]
21. Russo, T.; Gloria, A.; De Santis, R.; Amora, U.D.; Balato, G.; Vollaro, A.; Oliviero, O.; Improta, G.; Triassi, M.; Ambrosio, L. Preliminary focus on the mechanical and antibacterial activity of a PMMA-based bone cement loaded with gold nanoparticles. *Bioact. Mater.* **2017**, *2*, 156–161. [CrossRef]
22. Chamundeeswari, M.; Sobhana, S.S.; Jacob, J.P.; Kumar, M.G.; Devi, M.P.; Sastry, T.P.; Mandal, A.B. Preparation, characterization and evaluation of a biopolymeric gold nanocomposite with antimicrobial activity. *Biotechnol. Appl. Biochem.* **2010**, *55*, 29–35. [CrossRef]
23. Prokopovich, P.; Köbrick, M.; Brousseau, E.; Perni, S. Potent antimicrobial activity of bone cement encapsulating silver nanoparticles capped with oleic acid. *J. Biomed. Mater. Res. B Appl. Biomater.* **2015**, *103*, 273–281. [CrossRef] [PubMed]
24. Fatemeh, K.; Mohammad Javad, M.; Samaneh, K. The effect of silver nanoparticles on composite shear bond strength to dentin with different adhesion protocols. *J. Appl. Oral Sci.* **2017**, *25*, 367–373. [CrossRef] [PubMed]
25. Chladek, G.; Kasperski, J.; Barszczewska-Rybarek, I.; Żmudzki, J. Sorption, Solubility, Bond Strength and Hardness of Denture Soft Lining Incorporated with Silver Nanoparticles. *Int. J. Mol. Sci.* **2013**, *14*, 563–574. [CrossRef] [PubMed]
26. Burduşel, A.-C.; Gherasim, O.; Grumezescu, A.M.; Mogoantă, L.; Ficai, A.; Andronescu, E. Biomedical Applications of Silver Nanoparticles: An Up-to-Date Overview. *Nanomaterials* **2018**, *8*, 681. [CrossRef] [PubMed]
27. Kassaee, M.Z.; Akhavan, A.; Sheikh, N.; Sodagar, A. Antibacterial effects of a new dental acrylic resin containing silver nanoparticles. *J. Appl. Polym. Sci.* **2008**, *110*, 1699–1703. [CrossRef]
28. Corrêa, J.M.; Mori, M.; Sanches, H.L.; da Cruz, A.D.; Poiate, I.A.V.P. Silver Nanoparticles in Dental Biomaterials. *Int. J. Biomater.* **2015**, 485275. [CrossRef]
29. Durner, J.; Stojanovic, M.; Urcan, E.; Hickel, R.; Reichl, F.X. Influence of silver nano-particles on monomer elution from light-cured composites. *Dent. Mater.* **2011**, *27*, 631–636. [CrossRef]
30. Nam, K.-Y. In vitro antimicrobial effect of the tissue conditioner containing silver nanoparticles. *J. Adv. Prosthodont.* **2011**, *3*, 20–24. [CrossRef]
31. Shubnikov, A.V. Nanomaterials. In *Advances in Nanotechnology Research and Application*, 2012 ed.; Acton, A.Q., Ed.; ScholarlyEditions™ eBook: Atlanta, GR, USA, 2012; Volume 86, pp. 4339–4579. ISBN 978-1-4649-046-5.
32. Pfeifer, C.S.; Shelton, Z.R.; Braga, R.R.; Windmoller, D.; Machado, J.C.; Stansbury, J.W. Characterization of dimethacrylate polymeric networks: A study of the crosslinked structure formed by monomers used in dental composites. *Eur. Polym. J.* **2011**, *47*, 162–170. [CrossRef]
33. Kamoun, E.A.; Winkel, A.; Eisenburger, M.; Menzel, H. Carboxylated camphorquinone as visible-light photoinitiator for biomedical application: Synthesis, characterization, and application. *Arab. J. Chem.* **2016**, *9*, 745–754. [CrossRef]

34. Abliz, D.; Duan, Y.; Steuernagel, L.; Xie, L.; Li, D.; Ziegmann, G. Curing Methods for Advanced Polymer Composites—A Review. *Polym. Polym. Compos.* **2018**, *21*, 341–348. [CrossRef]
35. Andrzejewska, E. Photopolymerization kinetics of multifunctional monomers. *Prog. Polym. Sci.* **2001**, *26*, 605–665. [CrossRef]
36. Stansbury, J.W. Dimethacrylate network formation and polymer property evolution as determined by the selection of monomers and curing conditions. *Dent. Mater.* **2012**, *28*, 13–22. [CrossRef] [PubMed]
37. Pomrink, G.J.; DiCicco, M.P.; Clineff, T.D.; Erbe, E.M. Evaluation of the reaction kinetics of CORTOSS, a thermoset cortical bone void filler. *Biomaterials* **2003**, *24*, 1023–1031. [CrossRef]
38. Jancar, J.; Wang, W.; DiBenedetto, A.T. On the heterogeneous structure of thermally cured bis-GMA/TEGDMA resins. *J. Mater. Sci. Mater. Med.* **2000**, *11*, 675–682. [CrossRef] [PubMed]
39. Newcomb, M. Small Radical Chemistry. In *Handbook of Radical Polymerization*, 1st ed.; Matyjaszewski, K., Davis, T., Eds.; John Wiley & Sons: Hoboken, NJ, USA, 2002; Volume 2, pp. 77–116. ISBN 978-0-471-39274-3.
40. Barszczewska-Rybarek, I.; Korytkowska, A.; Gibas, M. Investigations on the structure of poly(dimethacrylate)s. *Des. Monomers Polym.* **2001**, *4*, 301–314. [CrossRef]
41. Bercier, A.; Gonçalves, S.; Lignon, O.; Fitremann, J. Calcium Phosphate Bone Cements Including Sugar Surfactants: Part One—Porosity, Setting Times and Compressive Strength. *Materials* **2010**, *3*, 4695–4709. [CrossRef] [PubMed]
42. Mau, H.; Schelling, K.; Heisel, C.; Wang, J.S.; Breusch, S.J. Comparison of various vacuum mixing systems and bone cements as regards reliability, porosity and bending strength. *Acta Orthop. Scand.* **2004**, *75*, 160–172. [CrossRef] [PubMed]
43. Guo, R.; Peng, L.; Lan, J.; Jiang, S.; Yan, W. Microstructure and hydrophobic properties of silver nanoparticles on amino-functionalised cotton fabric. *Mater. Technol.* **2016**, *31*, 139–144. [CrossRef]
44. Chladek, G.; Mertas, A.; Barszczewska-Rybarek, I.; Nalewajek, T.; Żmudzki, J.; Król, W.; Łukaszczyk, J. Antifungal Activity of Denture Soft Lining Material Modified by Silver Nanoparticles—A Pilot Study. *Int. J. Mol. Sci.* **2011**, *12*, 4735–4744. [CrossRef] [PubMed]
45. Barszczewska-Rybarek, I.M. Quantitative determination of degree of conversion in photocured poly(urethane-dimethacrylate)s by FTIR spectroscopy. *J. Appl. Polym. Sci.* **2012**, *123*, 1604–1611. [CrossRef]

© 2018 by the authors. Licensee MDPI, Basel, Switzerland. This article is an open access article distributed under the terms and conditions of the Creative Commons Attribution (CC BY) license (http://creativecommons.org/licenses/by/4.0/).

Article

Human In Situ Study of the effect of Bis(2-Methacryloyloxyethyl) Dimethylammonium Bromide Immobilized in Dental Composite on Controlling Mature Cariogenic Biofilm

Mary Anne S. Melo [1,*], Michael D. Weir [1], Vanara F. Passos [2], Juliana P. M. Rolim [3], Christopher D. Lynch [4], Lidiany K. A. Rodrigues [2,*] and Hockin H. K. Xu [1,5,6,*]

1. Department of Advanced Oral Sciences and Therapeutics, University of Maryland School of Dentistry, Baltimore, MD 21201, USA; Mweir@umaryland.edu
2. Postgraduate Program in Dentistry, Faculty of Pharmacy, Dentistry and Nursing, Federal University of Ceara, Fortaleza, CE 60430-355, Brazil; vanarapassos@hotmail.com
3. Faculty of Dentistry UniChristus, Fortaleza, CE 60160-230, Brazil; julianapml@unichristus.br
4. Restorative Dentistry, University Dental School and Hospital, University College Cork, Wilton T12 K8AF, Ireland; chris.lynch@ucc.ie
5. Center for Stem Cell Biology & Regenerative Medicine, University of Maryland School of Medicine, Baltimore, MD 21201, USA
6. Marlene and Stewart Greenebaum Cancer Center, University of Maryland School of Medicine, Baltimore, MD 21201, USA
* Correspondence: mmelo@umaryland.edu (M.A.S.M.); lidianykarla@ufc.br (L.K.A.R.); HXu@umaryland.edu (H.H.K.X.)

Received: 8 October 2018; Accepted: 27 October 2018; Published: 2 November 2018

Abstract: Cariogenic oral biofilms cause recurrent dental caries around composite restorations, resulting in unprosperous oral health and expensive restorative treatment. Quaternary ammonium monomers that can be copolymerized with dental resin systems have been explored for the modulation of dental plaque biofilm growth over dental composite surfaces. Here, for the first time, we investigated the effect of bis(2-methacryloyloxyethyl) dimethylammonium bromide (QADM) on human overlying mature oral biofilms grown intra-orally in human participants for 7–14 days. Seventeen volunteers wore palatal devices containing composite specimens containing 10% by mass of QADM or a control composite without QADM. After 7 and 14 days, the adherent biofilms were collected to determine bacterial counts via colony-forming unit (CFU) counts. Biofilm viability, chronological changes, and percentage coverage were also determined through live/dead staining. QADM composites caused a significant inhibition of *Streptococcus mutans* biofilm formation for up to seven days. No difference in the CFU values were found for the 14-day period. Our findings suggest that: (1) QADM composites were successful in inhibiting 1–3-day biofilms in the oral environment in vivo; (2) QADM significantly reduced the portion of the *S. mutans* group; and (3) stronger antibiofilm activity is required for the control of mature long-term cariogenic biofilms. Contact-killing strategies using dental materials aimed at preventing or at least reducing high numbers of cariogenic bacteria seem to be a promising approach in patients at high risk of the recurrence of dental caries around composites.

Keywords: antibacterial; biofilm; caries; dental composite; quaternary ammonium monomers; human in situ study

1. Introduction

In the past decade, dental materials research has intensified attempts at reducing or modulating dental plaque biofilm growth over dental composite surfaces [1] because recurrent caries around restorations (CARS) are identified as some of the major reasons for the failure of composite restorations [2,3]. Replacement rates of failed restorations have been reported to be 37–70%, with consequences that can seriously compromise oral health status [4,5]. CARS are frequently located at the gingival margins of the proximal restorations, which are common areas for food impaction [6]. Patients consider the cleaning of dental biofilm at the proximal space challenging, and very often fail to control biofilm build-up over time.

Dental caries are marked by continuous mineral loss promoted by organic acids released by bacteria after sugar metabolization [7], mainly streptococci from the *mutans* group. Cariogenic bacteria are characterized as pathologically shifted species, with the ability to generate large amounts of acid and to survive in acidic microenvironments [8]. The introduction of novel treatment approaches, supplementary to conventional therapeutic strategies, is thus considered as crucial for the efficient control of CARS. Cariogenic oral biofilms influence the initiation and progression of carious lesions not just in their primary development but also in their recurrence [9]. Reports in the literature have stated a temporal relationship between changes in biofilm composition and enamel demineralization following exposure to sucrose [10,11]. An undisturbed dental biofilm exposed to frequent sucrose leads to enamel demineralization after seven days of biofilm accumulation [10]. As the cariogenic biofilm becomes more mature, some acidogenic and aciduric bacteria become dominant in the biofilm [12].

Resin composites facilitate cariogenic biofilm growth [13]. Dental monomers such as bisphenol A-glycidyl dimethacrylate (BisGMA) and triethylene glycol dimethacrylate (TEGMA) may alter the metabolism and promote the proliferation of *Streptococcus mutans* [14]. Therefore, the synthesis of free radical monomers that have quaternary ammonium groups in their chemical structures paved the way for a noninvasive, biofilm-targeted method that can be used against oral biofilms [15]. Reactive and easily miscible quaternary ammonium monomers have the advantage of copolymerizing with the current dental resin systems through covalent bonding with the polymer network. These polymers are referred to as nonleaching antimicrobial or contact-killing agents. The antibacterial action results from the direct contact of the polymer with the microorganisms, with no release of active molecules. Although the exact antimicrobial mechanism of action has not been fully elucidated, the predominant mode of action is disruption of the cell membrane [16]. This imparts a durable and permanent antibacterial capability to dental composites.

Studies have presented different positions of the functional groups and alkyl chain length for improved balance between mechanical properties, antibacterial effects, and biocompatibility [17]. The majority of the synthetic quaternary ammonium monomers have only one methacrylate group, as monomethacrylates. Incorporating a high content of monomethacrylates could compromise the overall cross-linked polymer matrix [18].

Several in vitro studies have investigated the antibacterial performance of bis(2-methacryloyloxyethyl) dimethylammonium bromide (QADM), a quaternary ammonium monomer containing two methacrylate groups [18]. QADM was loaded at 10 wt % into different parental formulations, such as commercial and experimental adhesive systems [19,20] and nanocomposites [21], rendering reductions in *S. mutans* and total micro-organisms. Overall, these studies achieved a significant reduction of biofilm viability, metabolic activity, lactic acid, and bacterial counts using a 48-h human saliva microcosm biofilm model [22]. The incorporation of QADM also did not compromise the mechanical or bonding performance of the parental materials, and its antibacterial and mechanical properties were long-term and maintained after a one-year follow-up [23].

Although encouraging results were found in vitro [19–22], only a few studies have used native in situ dental plaque to study the effects of quaternary ammonium methacrylate [24,25]. In these studies, bacterial colonization over a short period (from hours to three days) was assessed. Antibacterial dental composites using QADM on an overlying mature cariogenic biofilm formed over seven days have not been studied to date. A longer-term in situ study would give insight into the in vivo antibacterial performance of this material in challenging conditions that mimic the clinical scenario of retentive proximal areas where the biofilm could not be removed in patients with a high risk of caries. Moreover, over a seven-day period, dysbiosis was present due to the proliferation or overgrowth of cariogenic bacteria in a low-pH econiche, and enamel was prone to demineralization.

In light of the evidence available to support quaternary ammonium monomers on initial oral biofilm, the present study for the first time evaluated the antibacterial performance of QADM in a relatively long-term study (beyond three days) by challenging the effectiveness of QADM-containing materials against mature oral biofilms formed in situ. Intact oral biofilms were grown under a cariogenic challenge in situ on composites within the oral cavity for 7 and 14 days. In addition to the determination of bacterial counts, the chronological changes in the biofilm were also visualized by live/dead staining, and the percentages were measured.

2. Results

All 17 volunteers completed the study, and no protocol deviation was identified. Treatment compliance was satisfactory. The mean and standard deviation values of colony-forming unit (CFU) counts of biofilms collected at 7 and 14 days are plotted in Figure 1A,B. The QADM composite had a significant effect on the viability of S. *mutans* at the 7 days ($p = 0.0303$). This effect corresponded to a 43% reduction of both solutions compared with the control. The QADM composite's effect on the viability of total streptococci, lactobacilli, and total microorganisms on the in situ biofilms was measured. However, no statistical significance was observed between the groups ($p > 0.05$). During the 14-day period, the microbiological composition of the biofilms formed on restoration was statistically similar for all evaluated conditions ($p > 0.5$).

Figure 2A shows the statistically significant difference between the tested groups ($p = 0.0385$) at the 7-day period, expressed by the variable percentage of mutans streptococci related to total streptococci (MS/TS). However, the percentage of mutans streptococci related to total micro-organisms was similar during the same period. These variables showed no difference during the 14-day period (Figure 3B).

Figure 3A–D,F–I shows live/dead staining images of biofilms grown on the QADM and control composites on the 1st, 3rd, 7th, and 14th days. Biofilms grown on the control composite on the 1st and 3rd days were primarily alive, which was indicated by continuous green staining (Figure 3A,B). Widespread bacterial cell killing was more pronounced on the 1st and 3rd day of biofilm accumulation on the QADM composite (Figure 3). At the 7th and 14th day, the overly mature biofilm structure was compact, with numerous layers showing a mushroom-like configuration with channels in the outer layer (Figure 3C,D for the control composites and Figure 3G,H for the QADM composites). Complete coverage of the composite surface was observed after seven days. No significant difference between the control and the QADM was observed.

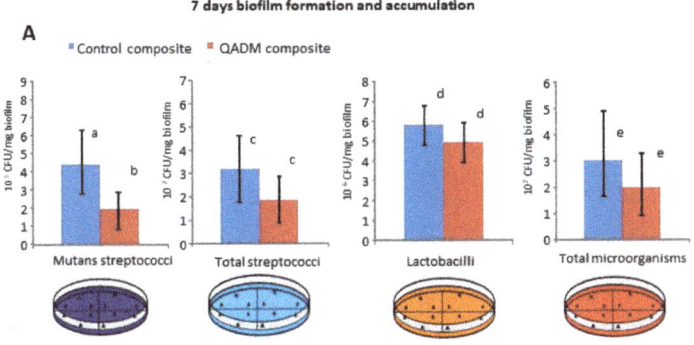

Figure 1. *Cont.*

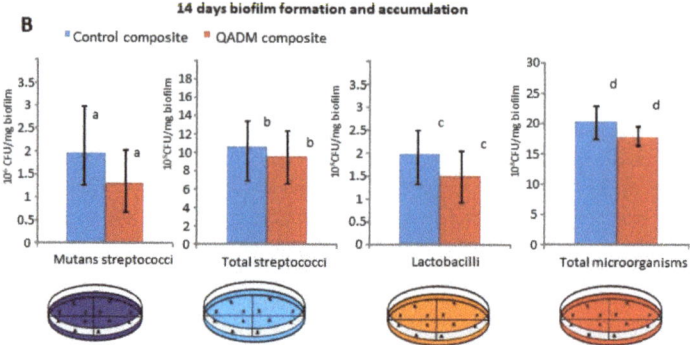

Figure 1. (**A**) Colony-forming unit (CFU) counts for the viability of mutans streptococci (MS), total streptococci (TS), lactobacilli, and total microorganisms (TM) present in the biofilms formed in situ after a 7-day and (**B**) 14-day period. Error bars represent the standard deviation of the mean, and data followed by different letters differed statistically ($p < 0.05$). The reduction in CFU counts from biofilms adherent to the QADM (bis(2-methacryloyloxyethyl) dimethylammonium bromide) composites was significantly different from the control for *Streptococcus mutans* after 7 days of growth. After 14 days, no further reduction was observed for *S. mutans*.

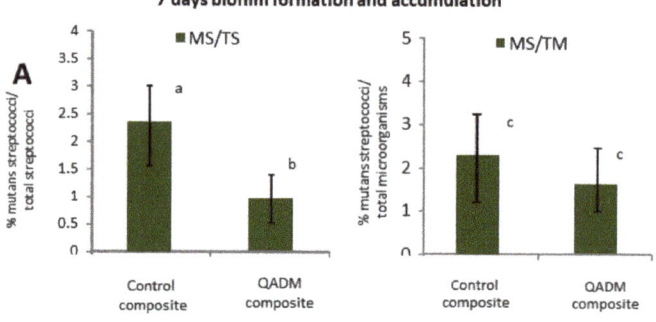

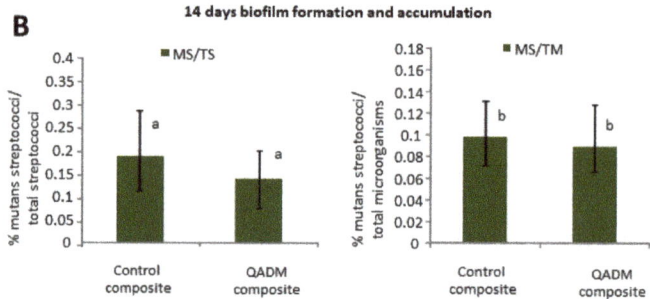

Figure 2. (**A**) The percentage of mutans streptococci related to total streptococci (MS/TS) and the percentage of mutans streptococci related to total micro-organisms (MS/TM) present in biofilms formed in situ after a 7-day and (**B**) 14-day period. The MS/TS was greatly reduced for biofilms adherent to the QADM composite in relation to the control at the 7-day period. Error bars represent SD, and data followed by different letters differed statistically ($p < 0.05$).

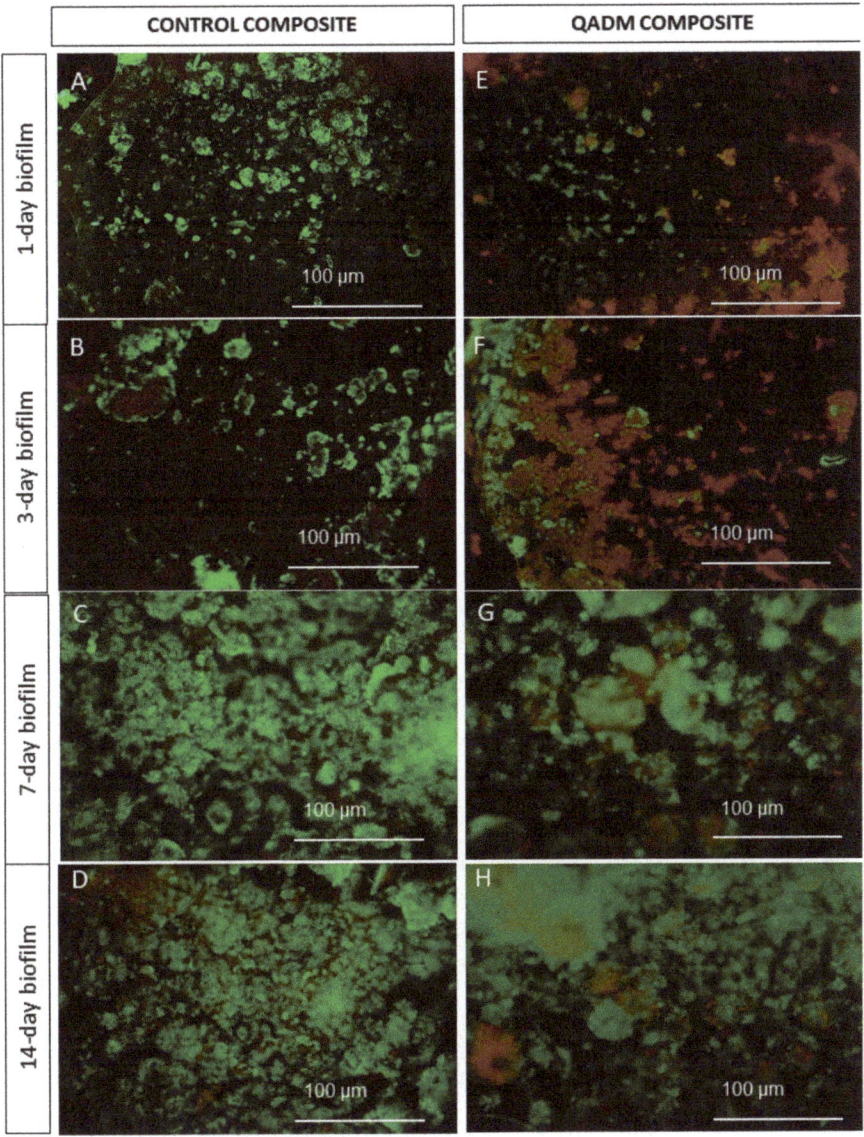

Figure 3. Live/dead staining images of biofilms grown on the QADM and control composites during the 1-, 3-, 7-, and 14-day periods. (**A,B**) Biofilms grown on the control composite on the 1st and 3rd days were primarily alive with continuous green staining. (**E,F**) Widespread cell-killing of bacteria was more pronounced on the 1- and 3-day biofilm accumulation on the QADM composite. On the 7th and 14th day, the overly mature biofilm structure was compact, with numerous layers presenting a mushroom-like configuration with channels in the outer layer: (**C,D**) for the control and (**G,H**) for the QADM.

Figure 4 image analyses show that the living cells grown over control composites accounted for 93% ± 3% (±SD) and 93% ± 7% (±SD) of the total biofilm cells for the 1- and 3-day biofilms.

Percentages of living cells (46% ± 8% for the first day and 37% ± 3% for the third day) were determined on the QADM-containing composites for the same time. They were inactive and dead.

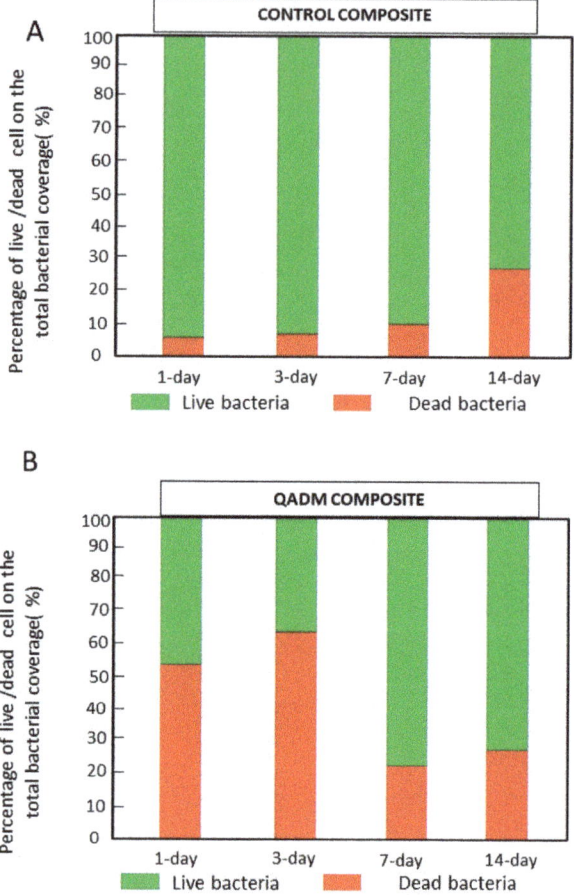

Figure 4. The percentage of live/dead bacterial cells found for the total biofilm coverage over the (**A**) control and (**B**) QADM composites. An increase in the dead percentage was observed for the biofilm grown over the QADM composite during the first and third day.

3. Discussion

Bacterial attachment to dental composite surfaces and subsequent cariogenic biofilm formation is a complex process [26]. There is an interplay between biological factors (i.e., the bacterial ability to rapidly convert dietary sugars to acids lowers the pH and demineralizes the tooth structure), patient-related factors, and physicochemical factors such as surface topography, surface charge, and surface energy of the dental materials [27,28]. Typical treatment for biofilm-mediated recurrence of caries lesions around the composite restorations involves operative replacement of composites, which incurs additional healthcare costs and additional loss of tooth structure.

To avoid biofilms on dental materials, an attractive alternative and complementary method to dental caries management is the use of biomaterials that possess antibacterial surfaces [27]. The contact-active antibacterial material is effective in preventing biofilm formation by killing bacteria [29], reducing bacteria amounts in the surrounding microenvironment, and extending the material's service life.

Antibiofilm effects of new dental materials with releasing fillers or ions were also investigated in the literature [30]. The incorporation of silver nanoparticles was among the bioactive compounds investigated that could provide antibacterial effects on oral bacteria. However, long-term release kinetics have challenged this approach's use in dental materials [31,32].

As contact-killing agents, the quaternary ammonium monomers for dental applications have been intensely investigated in vitro in the past years, with a positive outcome overall. Investigations have shown immediate and robust antibacterial effects (more than 3 log reductions) against oral micro-organisms [21,23,25]. QADM di-functional monomers are effective in producing active surfaces with high densities of immobilized antimicrobial agents [33,34]. For the specific quaternary ammonium used in our study, the previous in vitro studies showed reductions of 68% and 79% in biofilm CFU counts and lactic acid production, respectively, on cured primers specimens [20].

For this in situ study, for the first time we challenged the antibacterial performance of a dual methacrylate group QADM incorporated into composites against overly mature oral biofilms formed inside the oral cavity for up to 14 days in an acidogenic biofilm structure capable of causing substantial mineral loss and deep lesions on the enamel surface. This approach has high clinical relevance since anti-caries therapies aimed at controlling the assembly of cariogenic biofilms should contribute to the prevention of the onset of early carious lesions, clinically known as white spots.

The data revealed that QADM compromised the *S. mutans* group's biofilm accumulation during the 7-day period. The two most common species constantly linked to caries formation are *S. mutans* and *S. sobrinus* [35]. The cariogenic potential of *S. mutans* is accentuated when sucrose is available [36]. Sucrose-mediated biofilm formation creates spatial organizations expressed by a complex network of microcolonies, which modulate the development of compartmentalized acidic microenvironments across the 3D biofilm architecture [37]. Furthermore, within 3D biofilms, *S. mutans* display properties that are dramatically distinct from their planktonic counterparts, including much higher resistance to antibacterial approaches, which makes the biofilm much more difficult to kill than planktonic bacteria [35].

Although our research revealed the inhibitory effect of QADM on *S. mutans* biofilms, the exact mechanism of this inhibition is still unclear. The antibacterial efficacy may be related to the contact-killing mechanism of quaternization of the amino groups of QADM available on the bottom layer of the biofilm adjacent to the composite. The negatively charged counter-ions that stabilize the bacterial membrane were displaced by the positively charged cationic N^+ sites in the chemical structure of the quaternary ammonium-based resin. Indeed, the live/dead images obtained at the initial period of biofilm formation showed the presence of a higher proportion of nonviable bacteria (the red-orange color areas). Previous studies have highlighted the similar viability of biofilms growing on resin-based materials containing quaternary ammonium monomers [35,36,38,39]. Beyth and co-workers have suggested an intracellularly mediated death program, in which the bacterial lysis promoted by the presence of quaternary ammonium on the resin surface functions as a stressful condition triggering programmed cell death in the bacteria further away in the biofilm [25,40].

No expressive microbial reduction results on CFU values or micro-organism proportions were observed for *S. mutans*. These results point out the challenge faced by anti-caries approaches against mature biofilms. The bacterial adhesion processes under in vivo and in vitro conditions differ considerably [33]. Bacteria in biofilms are far less sensitive to antibacterial agents because of the exopolymeric matrix, extracellular polysaccharides, specific gene expression, and metabolic activity, all factors that protect antibacterial therapies to reach target bacteria [41]. The live/dead images of biofilms show a well-developed dense and compact extracellular polymeric substance - EPS matrix and the presence of bacterial cell clusters or microcolonies (Figure 4C,D for the control composites and Figure 4G,H for the QADM composites). The relative alteration of the proportion of live/dead found in the 7–14-day images was related to the uneven spatial distribution of vital and dead microorganisms found in matured and thick dental biofilms, with decreased vitality toward the outer layers [41]. Tawakoli et al. [42] also supported the high variability of the live/dead distribution and the CFU counts as challenges found in in situ biofilm models. Recently, new investigations have started to emerge using nanoparticles to improve the penetration of therapeutic agents into the biofilm matrix of oral cariogenic biofilms [43].

Another aspect to consider is the spatial arrangement, charge density, and counter-anion of the quaternary ammonium monomers and their antibacterial activity [32]. Previous studies have designed antibacterial monomers containing an eight-carbon or longer chain, which has correlated with significant antibacterial activity in vitro [44]. Surface charge density has displayed antibacterial performance in vitro. These monomers need to translate their antibacterial performance from in vitro to a human in vivo [45–48]. Future studies are warranted, especially to investigate whether the charge density is a relevant factor in the antibacterial effect of these new quaternary ammonium monomers.

In summary, the findings in this paper demonstrate that QADM composites at 10% promote a substantial bacterial reduction of *S. mutans* biofilm. This was achieved in the initial days of contact and also reached a 7-day period, an interval where patients at high risk of caries would develop initial enamel carious lesions. However, dental caries result from interactions over time. An undisturbed cariogenic biofilm well-established on the composite surface over long periods is extremely difficult to eradicate. Its inhibition should not rely only on contact with an antibacterial surface. Removing or disturbing biofilm from all tooth and composite surfaces and reducing sugar intake within three days is expected to control carious lesions. Concomitant and multitargeting strategies are needed against mature long-term cariogenic biofilms.

4. Materials and Methods

4.1. Study Design and Participants

This study involved a prospective, randomized, single-blind, split-mouth in situ design conducted according to the code of ethics of the World Medical Association (Declaration of Helsinki) for experiments involving humans. The region's ethical committee (protocol #1232012) also approved it. Seventeen healthy volunteers of both genders, aged from 21 to 36 years, accepted participation in this study, fulfilling the required criteria. Inclusion criteria were a normal salivary flow rate, good general and oral health with no active caries lesions or periodontal treatment needs, an ability to comply with the experiment protocol, no use of antibiotics during the three months before the study, and no use of fixed or removable orthodontic devices. Exclusion criteria were failing to use the device according to the established protocol and taking medication interfering with saliva flow rate or containing antimicrobial agents. The sample size was determined by a power analysis and was based on previous data [25]. Seventeen volunteers were recruited: 16 for the study and 1 volunteer to allow temporal visualization of the biofilm formation through live/dead staining. After being screened, the volunteers were verbally informed about the study aims and procedures, and received written information and the informed consent form. At the next appointment, maxillary alginate impressions were made for the fabrication of the palatal devices.

During the experimental period, each volunteer used a removable acrylic custom-made palatal device containing the tested materials, as shown in Figure 5A. Seven days before the experiment began (i.e., the washout period) and during the whole experiment, the volunteers were asked to use a standard toothbrush and nonfluoridated paste. Each acrylic palatal device enclosed four composite specimens ($5 \times 5 \times 2$ mm^3): two specimens for the control composites and two specimens for the QADM composites. To promote plaque accumulation and to protect it from disarrangement, recessions were created by placing the surface of the composite specimens about 1 mm below the covered plastic mesh (as seen in the details of Figure 1A) [45].

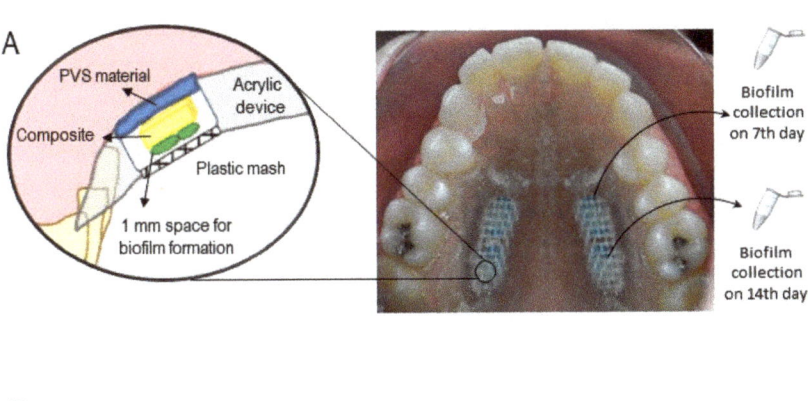

Figure 5. (**A**) In situ palatal devices used by 17 volunteers. Each device contained four slabs, two filled with the QADM composite on one side and two filled with the control composite on the other side. The slabs were held by polyvinyl siloxane (PVS) and covered with a plastic mash to avoid disturbance in the biofilm growth. The biofilm was collected from the surface of the specimens on the 7th and 14th days. The top-left is a magnified description showing details of biofilm formation over the composite specimens inside the device. (**B**) The synthesis route of the bis(2-methacryloyloxyethyl) dimethylammonium bromide monomer via a Menshutkin reaction and the details for dual-polymerizable groups and the bacterial terminal. BEMA: 2-bromoethyl methacrylate; DMAEMA: 2-(*N*,*N*-dimethylamino)ethyl methacrylate.

To divert any possible carry-across effect, the sequence in which the experiment units were assigned in the palatal device took into consideration that antibacterial dental materials should be placed on one side of the palatal appliance and, consequently, that control materials should be placed on the opposite side (Figure 5A). The split-mouth experimental design was a practical approach for testing the effects of various agents on the composition of dental plaque [44]. Within each side of the palatal device, the positions of the specimens were randomly determined according to a computer-generated randomization list [45]. The outcome variables evaluated were colony-forming unit counts for total microorganisms, total streptococci, mutans streptococci, and lactobacilli on the specimens.

4.2. Specimen Preparation

The light-curable composite was made by blending a monomer resin consisting of BisGMA (bisphenol-glycidyl dimethacrylate) and TEGDMA (triethylene glycol dimethacrylate) at a 1:1 ratio (all by mass) with 0.2% camphorquinone and 0.8% ethyl 4-*N*,*N*-dimethylaminobenzoate. As reinforcement co-fillers, barium-boroaluminosilicate glass particles with a median diameter of 1.4 μm (Caulk/Dentsply, Milford, DE, USA) were silanized with 4% 3-methacryloxypropyltrimethoxysilane and 2% *n*-propylamine [21]. The fillers were mixed with the resin at a total filler mass fraction of 60% to form a cohesive paste.

The synthesis of bis(2-methacryloyloxyethyl) dimethylammonium bromide via a modified Menshutkin reaction was previously described [18], and it is summarized in Figure 1B. Briefly, 10 mmol of 2-(*N*,*N*-dimethylamino)ethyl methacrylate (DMAEMA; Sigma-Aldrich, St. Louis, MO, USA) and 10 mmol of 2-bromoethyl methacrylate (BEMA; Monomer-Polymer Labs, Trevose, PA, USA) were combined with 3 g of ethanol in a closed vial. After stirring at 60 °C for 24 h so the reaction could complete, the solvent was removed via evaporation, using a vacuum. This process yielded QADM as a clear and viscous liquid. QADM was mixed with the BisGMA–TEGDMA resin at a QADM mass fraction of 10%. A preliminary study showed that this mass fraction yielded strong antibacterial properties without compromising the resin's mechanical properties [21].

Thirty-six light-curable composite specimens were fabricated for the experimental composition (QADM at 10 wt % each), and a further 36 specimens without antimicrobial monomer served as a control. The composite was inserted and light activated for 20 s using a light-emitting diode (Radii-cal, SDI Limited Victoria, Australia; standard curing mode, irradiance output provided of 689 mW/cm^2). The specimens were mounted in standardized sample chambers inside the device, with an anterior–posterior position, using body impression material (Aquasil Ultra, Dentsply DeTray GmbH, Konstanz, Germany), as demonstrated in Figure 1A.

4.3. Clinical Phase

Audiovisual orientation and written instructions of the in situ protocol were given to the volunteers to assure their adhesion and avoid protocol deviation during the study. To provide a cariogenic challenge during the clinical phase, the application of a 20% sucrose solution extra-orally on the restored specimens was performed by the volunteers eight times per day at predetermined times. According to previous studies, the sucrose was gently dried after 5 min, and the device was reinserted into the mouth [42,44]. No restriction was made about the volunteers' diet, but they were instructed to avoid F-rich food containing bioavailable F, such as black tea. They did, however, drink fluoridated water (about 0.7 ppm fluoride).

4.4. Microbiological and Biochemical Analysis

On the 7th and 14th examination days, the subjects refrained from eating, drinking, and tooth cleaning 12 h after the last application of the sucrose solution and dentifrice before presenting at the clinic (46-48). On the 7th day, the device was removed from the mouth, and the biofilm and one enamel slab from each side were respectively carefully removed and collected (Figure 5B). Then, the device with the remaining specimens was reinserted into the mouth. On the 14th day, a similar process was performed to collect the two residual biofilms from the specimens. After the collection, the biofilm was processed for analysis. First, it was weighed (±1 mg) in preweighed microcentrifuge tubes and agitated during a 2 min period in a Disrupter Genie Cell Disruptor (Precision Solutions, Rice Lake, WI, USA). A 50 μL aliquot of the sonicated suspension was diluted in 0.9% NaCl, and serial decimal dilutions were inoculated in triplicate using the drop-counting technique in the following culture media: (1) in mitis salivarius agar containing 20% sucrose to determine total streptococci (TS), and in mitis salivarius agar plus 0.2 bacitracin/mL to determine mutans streptococci (MS); (2) in Rogosa agar supplemented with 0.13% glacial acetic acid to assess the number of CFU of lactobacilli (LB); and (3) in brain–heart infusion enhanced with 5% sterile defibrinated sheep blood agar plates to determine total micro-organisms (TM). The plates were incubated in 10% CO_2 at 37 °C for 48 h. The CFU were counted, and the results were expressed as CFU/mg biofilm wet weight, the percentage of MS in relation to TM, and the percentage of MS in relation to TS.

4.5. Live/Dead Assays

To visualize the micro-organisms during the initial phase of formation as well as during the experimental periods, one volunteer used a palatal device containing eight composite specimens: four specimens for the control composite and four specimens for the QADM composite. One specimen from each group was removed during the 1st, 3rd, 7th, and 14th days. The specimens were immediately washed with phosphate-buffered saline (PBS) and stained using the LIVE/DEAD BacLight Bacterial Viability Kit (Molecular Probes, USA) to qualify bacterial cell viability. This assay employs two nucleic acid stains: the green-fluorescent SYTO 9 stain and the red-fluorescent propidium iodide stain [39]. These stains differ in their ability to penetrate healthy bacterial cells. When used alone, the SYTO 9 stain labels both live and dead bacteria.

In contrast, propidium iodide penetrates only bacteria with damaged membranes, reducing SYTO 9 fluorescence when both dyes are present. Thus, live bacteria with intact membranes fluoresce green, while dead bacteria with damaged membranes fluoresce red. A volume of 100 µL of the previously described fluorescence dyes was pipetted onto the specimens and incubated in a dark chamber for 15 min. The biofilms grown over the specimens were then examined using an epifluorescence microscope (TE2000-U, Nikon, Melville, NY, USA) at a magnification of 100×. Images ($n = 4$) were acquired and analyzed (NIS Elements software, Nikon Instruments Inc, Melville, NY, USA) for the quantification of live (green fluorescence) and dead (red fluorescence) bacteria.

4.6. Statistical Analysis

The assumptions of equality of variances and normal distribution of errors were checked for all the response variables tested, and those that did not satisfy these assumptions were transformed using the Box–Cox power transformation [42]. To determine the differences between test and control values in the in situ experiment, the viable bacteria counts, percent MS/TS, and percent MS/TM were submitted to a two-sample independent Student's t-test. The significance level was set at $\alpha = 0.05$. The statistical appraisal was computed with SPSS for Windows XP 17.0 (SPSS Inc., Chicago, IL, USA).

5. Conclusions

The results of the present in situ study provide a more realistic perspective on the value of integrating bioactive restorative materials with traditional caries management approaches into clinical practice. Contact-killing strategies via dental materials aiming at preventing or at least reducing high numbers of cariogenic bacteria seem to be a promising approach for helping patients at high risk of recurrence of dental caries around composites.

Author Contributions: Conceptualization, M.A.S.M., L.K.A.R. and H.H.K.X.; C.D.L., L.K.A.R. and H.H.K.X; methodology, M.D.W., V.F.P., J.P.M.R.; software, M.A.S.M..; validation, L.K.A.R and H.H.K.X; formal analysis, M.A.S.M.; investigation, V.F.P., J.P.M.R.; resources, L.K.A.R.; data curation, M.A.S.M.; writing—original draft preparation, M.A.S.M.; writing—review and editing, H.H.K.X.; visualization, M.A.S.M.; supervision, M.A.S.M.; project administration, M.A.S.M.; funding acquisition, H.H.K.X., L.K.A.R., please turn to the CRediT taxonomy for the term explanation. Authorship must be limited to those who have contributed substantially to the work reported.

Funding: This study was supported by CNPq/Brazil (141791/2010-1), the CAPES/ Fulbright Doctoral Program (BEX 0574/06-6), the University of Maryland Baltimore Seed grant (HX), and the University of Maryland School of Dentistry bridge fund (HX).

Acknowledgments: The authors thank the volunteers for their valuable participation and Agostinho Soares de Alcântara Neto (Faculdade de Veterinária, Universidade Estadual do Ceará) for his technical assistance and laboratory facilities for the cell viability assay.

Conflicts of Interest: The authors declare no conflicts of interest.

References

1. Hwang, G.; Koltisko, B.; Jin, X.; Koo, H. Nonleachable Imidazolium-Incorporated Composite for Disruption of Bacterial Clustering, Exopolysaccharide-Matrix Assembly, and Enhanced Biofilm Removal. *ACS Appl. Mater. Interfaces* **2017**, *9*, 38270–38280. [CrossRef] [PubMed]
2. Hollanders, A.C.C.; Kuper, N.K.; Maske, T.; Huysmans, M.D.N.J.M. Secondary Caries in situ Models: A Systematic Review. *Caries Res.* **2018**, *52*, 454–462. [CrossRef] [PubMed]
3. Bernardo, M.; Luis, H.; Martin, M.D.; Leroux, B.G.; Rue, T.; Leitao, J. Survival and reasons for failure of amalgam versus composite posterior restorations placed in a randomized clinical trial. *J. Am. Dent. Assoc.* **2007**, *138*, 775–783. [CrossRef] [PubMed]
4. Owen, B.; Guevara, P.H.; Greenwood, W. Placement and replacement rates of amalgam and composite restorations on posterior teeth in a military population. *US Army Med. Dep. J.* **2017**, *2*, 88–94.
5. Roumanas, E.D. The frequency of replacement of dental restorations may vary based on a number of variables, including type of material, size of the restoration, and caries risk of the patient. *J. Evid. Based Dent. Pract.* **2010**, *1*, 23–24. [CrossRef] [PubMed]
6. Mjör, I.A. The location of clinically diagnosed secondary caries. *Quintes Int.* **1998**, *29*, 313–317.
7. Fejerskov, O. Changing paradigms in concepts on dental caries: Consequences for oral health care. *Caries Res.* **2004**, *38*, 182–191. [CrossRef] [PubMed]
8. Marsh, P.D. Are dental diseases examples of ecological catastrophes? *Microbiology* **2003**, *149*, 279–294. [CrossRef] [PubMed]
9. Paes Leme, A.F.; Koo, H.; Bellato, C.M.; Bedi, G.; Cury, J.A. The role of sucrose in cariogenic dental biofilm formation—New insight. *J. Dent. Res.* **2006**, *85*, 878–887. [CrossRef] [PubMed]
10. Vale, G.C.; Tabchoury, C.P.; Arthur, R.A.; Del Bel Cury, A.A.; Paes Leme, A.F.; Cury, J.A. Temporal relationship between sucrose-associated changes in dental biofilm composition and enamel demineralization. *Caries Res.* **2007**, *41*, 406–412. [CrossRef] [PubMed]
11. Cury, J.A.; Tenuta, L.M. Enamel remineralization: Controlling the caries disease or treating early caries lesions? *Braz. Oral Res.* **2009**, *23*, 23–30. [CrossRef] [PubMed]
12. NTakahashi, B. Nyvad. Caries ecology revisited: Microbial dynamics and the caries process. *Caries Res.* **2008**, *42*, 409–418. [CrossRef] [PubMed]
13. Singh, J.; Khalichi, P.; Cvitkovitch, D.G.; Santerre, J.P. Composite resin degradation products from BisGMA monomer modulate the expression of genes associated with biofilm formation and other virulence factors in Streptococcus mutans. *J. Biomed. Mater. Res. A* **2009**, *88*, 551–560. [CrossRef] [PubMed]
14. Sadeghinejad, L.; Cvitkovitch, D.G.; Siqueira, W.L.; Santerre, J.P.; Finer, Y. Triethylene Glycol Up-Regulates Virulence-Associated Genes and Proteins in Streptococcus mutans. *PLoS ONE* **2016**, *11*, 665–760. [CrossRef] [PubMed]
15. Melo, M.A.; Weir, M.D.; Li, F.; Cheng, L.; Zhang, K.; Xu, H.H.K. Control of Biofilm at the Tooth-Restoration Bonding Interface: A Question for Antibacterial Monomers? A Critical Review. *Rev. Adhes. Adhes.* **2017**, *3*, 287–305. [CrossRef]
16. Jiao, Y.; Niu, L.N.; Ma, S.; Li, J.; Tay, F.R.; Chen, J.H. Quaternary ammonium-based biomedical materials: State-of-the-art, toxicological aspects and antimicrobial resistance. *Prog. Pol. Sci.* **2017**, *71*, 53–90. [CrossRef]
17. Zhang, N.; Ma, Y.; Weir, M.; Xu, H.H.K.; Bai, Y.; Melo, M.A. Current Insights into the Modulation of Oral Bacterial Degradation of Dental Polymeric Restorative Materials. *Materials* **2017**, *10*, 507. [CrossRef] [PubMed]
18. Antonucci, J.M.; Zeiger, D.N.; Tang, K.; Lin-Gibson, S.; Fowler, B.O.; Lin, N.J. Synthesis and characterization of dimethacrylates containing quaternary ammonium functionalities for dental applications. *Dent. Mater.* **2012**, *28*, 219–228. [CrossRef] [PubMed]

19. Melo, M.A.S.; Cheng, L.; Weir, M.D.; Hsia, R.; Rodrigues, L.K.A.; Xu, H.H.K. Novel dental bonding agents containing antibacterial agents and calcium phosphate nanoparticles. *J. Biomed. Mater. Res. B Appl. Biomater.* **2013**, *101*, 620–629. [CrossRef] [PubMed]
20. Cheng, L.; Zhang, K.; Melo, M.A.S.; Weir, M.D.; Zhou, X.; Xu, H.H.K. Anti-biofilm dentin primer with quaternary ammonium and silver nanoparticles. *J. Dent. Res.* **2012**, *91*, 598–604. [CrossRef] [PubMed]
21. Cheng, L.; Weir, M.D.; Xu, H.H.; Antonucci, J.M.; Kraigsley, A.M.; Lin, N.J.; Lin-Gibson, S.; Zhou, X. Antibacterial amorphous calcium phosphate nanocomposites with a quaternary ammonium dimethacrylate and silver nanoparticles. *Dent. Mater.* **2012**, *28*, 561–572. [CrossRef] [PubMed]
22. Zhang, K.; Melo, M.A.S.; Cheng, L.; Weir, M.D.; Bai, Y.; Xu, H.H.K. Effect of quaternary ammonium and silver nanoparticle-containing adhesives on dentin bond strength and dental plaque microcosm biofilms. *Dent. Mater.* **2012**, *28*, 842–852. [CrossRef] [PubMed]
23. Cheng, L.; Zhang, K.; Zhou, C.C.; Weir, M.D.; Zhou, X.D.; Xu, H.H. One-year water-ageing of calcium phosphate composite containing nano-silver and quaternary ammonium to inhibit biofilms. *Int. J. Oral Sci.* **2016**, *29*, 172–181. [CrossRef] [PubMed]
24. Feng, J.; Cheng, L.; Zhou, X.; Xu, H.H.; Weir, M.D.; Meyer, M.; Maurer, H.; Li, Q.; Hannig, M.; Rupf, S. In situ antibiofilm effect of glass-ionomer cement containing dimethylaminododecyl methacrylate. *Dent. Mater.* **2015**, *31*, 992–1002. [CrossRef] [PubMed]
25. Beyth, N.; Yudovin-Farber, I.; Perez-Davidi, M.; Domb, A.J.; Weiss, E.I. Polyethyleneimine nanoparticles incorporated into resin composite cause cell death and trigger biofilm stress in vivo. *Proc. Natl. Acad. Sci. USA* **2010**, *107*, 22038–22043. [CrossRef] [PubMed]
26. Marsh, P.D. Microbiology of dental plaque biofilms and their role in oral health and caries. *Dent. Clin. N. Am.* **2010**, *54*, 441–454. [CrossRef] [PubMed]
27. Chen, C.; Weir, M.D.; Wang, L.; Zhou, X.; Xu, H.H.K.; Melo, M.A. Dental Composite Formulation Design with Bioactivity on Protein Adsorption Combined with Crack-Healing Capability. *J. Funct. Biomater.* **2017**, *8*, 40. [CrossRef] [PubMed]
28. Melo, M.A.; Codes, B.M.; Passos, V.F.; Lima, J.P.M.; Rodrigues, L.K. In Situ Response of Nanostructured Hybrid Fluoridated Restorative Composites on Enamel Demineralization, Surface Roughness, and Ion Release. *Eur. J. Prosthodont. Restor. Dent.* **2014**, *22*, 185–190. [PubMed]
29. Chatzistavrou, X.; Lefkelidou, A.; Papadopoulou, L.; Pavlidou, E.; Paraskevopoulos, K.M.; Fenno, J.C.; Flannagan, S.; González-Cabezas, C.; Kotsanos, N.; Papagerakis, P. Bactericidal and Bioactive Dental Composites. *Front. Physiol.* **2018**, *16*, 99–103. [CrossRef] [PubMed]
30. Sjollema, J.; Zaat, S.A.J.; Fontaine, V.; Ramstedt, M.; Luginbuehl, R.; Thevissen, K.; Li, J.; van der Mei, H.C.; Busscher, H.J. In vitro methods for the evaluation of antimicrobial surface designs. *Acta Biomater.* **2018**, *70*, 12–24. [CrossRef] [PubMed]
31. Khvostenko, D.; Hilton, T.J.; Ferracane, J.L.; Mitchell, J.C.; Kruzic, J.J. Bioactive glass fillers reduce bacterial penetration into marginal gaps for composite restorations. *Dent. Mater. Off. Publ. Acad. Dent. Mater.* **2016**, *32*, 73–81. [CrossRef] [PubMed]
32. Stencel, R.; Kasperski, J.; Pakieła, W.; Mertas, A.; Bobela, E.; Barszczewska-Rybarek, I.; Chladek, G. Properties of Experimental Dental Composites Containing Antibacterial Silver-Releasing Filler. *Materials* **2018**, *11*, 1031. [CrossRef] [PubMed]
33. Kasraei, S.; Sami, L.; Hendi, S.; AliKhani, M.-Y.; Rezaei-Soufi, L.; Khamverdi, Z. Antibacterial properties of composite resins incorporating silver and zinc oxide nanoparticles on Streptococcus mutans and Lactobacillus. *Restor. Dent. Endod.* **2014**, *39*, 109–114. [CrossRef] [PubMed]
34. Hu, X.; Lin, X.; Zhao, H.; Chen, Z.; Yang, J.; Li, F.; Liu, C.; Tian, F. Surface Functionalization of Polyethersulfone Membrane with Quaternary Ammonium Salts for Contact-Active Antibacterial and Anti-Biofouling Properties. *Materials* **2016**, *17*, 376. [CrossRef] [PubMed]
35. Bowen, W.H.; Koo, H. Biology of Streptococcus mutans-derived glucosyltransferases: Role in extracellular matrix formation of cariogenic biofilms. *Caries Res.* **2011**, *45*, 69–86. [CrossRef] [PubMed]
36. Hanada, N.; Kuramitsu, H.K. Isolation and characterization of the Streptococcus mutans gtfD gene, coding for primer-dependent soluble glucan synthesis. *Infect. Immun.* **1989**, *57*, 2079–2085. [PubMed]
37. Xiao, J.; Hara, A.T.; Kim, D.; Zero, D.T.; Koo, H.; Hwang, G. Biofilm three-dimensional architecture influences in situ pH distribution pattern on the human enamel surface. *Int. J. Oral Sci.* **2017**, *9*, 74–79. [CrossRef] [PubMed]

38. Zhou, H.; Weir, M.D.; Antonucci, J.M.; Schumacher, G.E.; Zhou, X.D.; Xu, H.H. Evaluation of three-dimensional biofilms on antibacterial bonding agents containing novel quaternary ammonium methacrylates. *Int. J. Oral Sci.* **2014**, *6*, 77–86. [CrossRef] [PubMed]
39. Zhou, H.; Liu, H.; Weir, M.D.; Reynolds, M.A.; Zhang, K.; Xu, H.H. Three-dimensional biofilm properties on dental bonding agent with varying quaternary ammonium charge densities. *J. Dent.* **2016**, *53*, 73–81. [CrossRef] [PubMed]
40. Beyth, N.; Yudovin-Farber, I.; Bahir, R.; Domb, A.J.; Weiss, E.I. Antibacterial activity of dental composites containing quaternary ammonium polyethylenimine nanoparticles against Streptococcus mutans. *Biomate* **2006**, *27*, 995–4002. [CrossRef] [PubMed]
41. Auschill, T.M.; Arweiler, N.B.; Netuschil, L.; Brecx, M.; Reich, E.; Sculean, A. Spatial distribution of vital and dead microorganisms in dental biofilms. *Arch. Oral Biol.* **2001**, *46*, 471–476. [CrossRef]
42. Tawakoli, P.N.; Al-Ahmad, A.; Hoth-Hannig, W.; Hannig, M.; Hannig, C. Comparison of different live/dead stainings for detection and quantification of adherent microorganisms in the initial oral biofilm. *Clin. Oral Investig.* **2013**, *17*, 841–850. [CrossRef] [PubMed]
43. Liu, Y.; Naha, P.C.; Hwang, G.; Kim, D.; Huang, Y.; Simon-Soro, A.; Jung, H.I.; Ren, Z.; Li, Y.; Gubara, S.; et al. Topical ferumoxytol nanoparticles disrupt biofilms and prevent tooth decay in vivo via intrinsic catalytic activity. *Nat. Commun.* **2018**, *9*, 2920. [CrossRef] [PubMed]
44. Li, F.; Weir, M.D.; Xu, H.H. Effects of Quaternary Ammonium Chain Length on Antibacterial Bonding Agents. *J. Dent. Res.* **2013**, *92*, 932–938. [CrossRef] [PubMed]
45. Li, F.; Weir, M.D.; Chen, J.; Xu, H.H. Effect of charge density of bonding agent containing a new quaternary ammonium methacrylate on antibacterial and bonding properties. *Dent. Mater.* **2014**, *30*, 433–441. [CrossRef] [PubMed]
46. Melo, M.A.S.; Weir, M.D.; Rodrigues, L.K.A.; Xu, H.H.K. Novel calcium phosphate nanocomposite with caries-inhibition in a human in situ model. *Dent. Mater.* **2013**, *29*, 231–240. [CrossRef] [PubMed]
47. Hara, A.T.; Turssi, C.P.; Ando, M.; González-Cabezas, C.; Zero, D.T.; Rodrigues, A.L., Jr.; Serra, M.C.; Cury, J.A. Influence of fluoride-releasing restorative material on root dentine secondary caries in situ. *Caries Res.* **2006**, *40*, 435–439. [CrossRef] [PubMed]
48. Melo, M.A.; Morais, W.A.; Passos, V.F.; Lima, J.P.M.; Rodrigues, L.K. Fluoride releasing and enamel demineralization around orthodontic brackets by fluoride-releasing composite containing nanoparticles. *Clin. Oral Investig.* **2013**, *18*, 1343–1350. [CrossRef] [PubMed]

© 2018 by the authors. Licensee MDPI, Basel, Switzerland. This article is an open access article distributed under the terms and conditions of the Creative Commons Attribution (CC BY) license (http://creativecommons.org/licenses/by/4.0/).

Article

Degree of Conversion and BisGMA, TEGDMA, UDMA Elution from Flowable Bulk Fill Composites

Edina Lempel [1,*], Zsuzsanna Czibulya [2,3], Bálint Kovács [2,3], József Szalma [4], Ákos Tóth [5], Sándor Kunsági-Máté [2,3], Zoltán Varga [1] and Katalin Böddi [6]

1. Department of Restorative Dentistry and Periodontology, University of Pécs, 5 Dischka Gy Street, Pécs H-7621, Hungary; varga.zoltan@gmail.com
2. Department of General and Physical Chemistry, University of Pécs, 6 Ifjúság Street, Pécs H-7624, Hungary; czibulya.zsuzsanna@gmail.com (Z.C.); balint621@gmail.com (B.K.); kunsagi@gamma.ttk.pte.hu (S.K.-M.)
3. János Szentágothai Research Center, University of Pécs, 20 Ifjúság Street, Pécs H-7624, Hungary
4. Department of Oral and Maxillofacial Surgery, University of Pécs, 5 Dischka Gy Street, Pécs H-7621, Hungary; szalma.jozsef@pte.hu
5. Faculty of Sciences, University of Pécs, 6 Ifjúság Street, Pécs H-7621, Hungary; totha@gamma.ttk.pte.hu
6. Department of Biochemistry and Medical Chemistry, University of Pécs, 12 Szigeti Street, Pécs H-7624, Hungary; katalin.boddi@aok.pte.hu
* Correspondence: lempel.edina@pte.hu; Tel.: +36-72-535-926; Fax: +36-72-535-905

Academic Editor: Ihtesham ur Rehman
Received: 20 March 2016; Accepted: 9 May 2016; Published: 20 May 2016

Abstract: The degree of conversion (DC) and the released bisphenol A diglycidyl ether dimethacrylate (BisGMA), triethylene glycol dimethacrylate (TEGDMA) and urethane dimethacrylate (UDMA) monomers of bulk-fill composites compared to that of conventional flowable ones were assessed using micro-Raman spectroscopy and high performance liquid chromatography (HPLC). Four millimeter-thick samples were prepared from SureFil SDR Flow (SDR), X-tra Base (XB), Filtek Bulk Fill (FBF) and two and four millimeter samples from Filtek Ultimate Flow (FUF). They were measured with micro-Raman spectroscopy to determine the DC% of the top and the bottom surfaces. The amount of released monomers in 75% ethanol extraction media was measured with HPLC. The differences between the top and bottom DC% were significant for each material. The mean DC values were in the following order for the bottom surfaces: SDR_4mm_20s > FUF_2mm_20s > XB_4mm_20s > FBF_4mm_20s > XB_4mm_10s > FBF_4mm_10s > FUF_4mm_20s. The highest rate in the amount of released BisGMA and TEGDMA was found from the 4 mm-thick conventional flowable FUF. Among bulk-fills, FBF showed a twenty times higher amount of eluted UDMA and twice more BisGMA; meanwhile, SDR released a significantly higher amount of TEGDMA. SDR bulk-fill showed significantly higher DC%; meanwhile XB, FBF did not reach the same level DC, as that of the 2 mm-thick conventional composite at the bottom surface. Conventional flowable composites showed a higher rate of monomer elution compared to the bulk-fills, except FBF, which showed a high amount of UDMA release.

Keywords: bulk-fill composite; degree of conversion; monomer elution; micro-Raman spectroscopy; HPLC

1. Introduction

The evolution led to contemporary resin-based composites (RBCs) showing high clinical success and survival rates [1–4]. However, despite the continuous development of RBCs, there are some shortcomings, such as polymerization volume shrinkage, incomplete degree of conversion (DC) of the matrix monomers and their release into the oral cavity and pulp space. The degree of conversion of an RBC is an important factor in determining the mechanical properties of the material and its biocompatibility [5,6]. A lower conversion rate will influence the physical performance of the RBC, and increased elution of monomers have been reported [7,8]. However, the DC% for adequate clinical performance has not yet been determined; only a negative correlation of *in vivo* abrasive wear depth with DC has been established in the range of 55–65 DC% [9,10]. To prevent clinical failures and decrease the elution of unreacted monomers, some practical strategies are recommended, including alternative light curing protocols [11], the use of flowable cavity liners [12] and incremental filling techniques [13]. The generally-accepted maximal layer thickness that provides adequate light penetration and photo-polymerization is 2 mm [7,14–16]. However, restoring cavities with RBC increments of 2 mm in thickness is time consuming; voids may be included; and this implies a risk of contaminations between the increments [17]. Thus, bulk-fill RBCs were developed to avoid the aforementioned disadvantages [18]. Literature data suppose better light transmittance of these materials to allow for a reported depth of cure in excess of 4 mm [19–21]. According to the manufacturer's recommendation, bulk-fill materials are indicated as basing materials or permanent restorative materials. *In vitro* studies showed that the application of one thicker increment of bulk-fill composite could be equally successful in marginal adaptation, cavity-bottom adhesion and in depth of cure as the conventional layering technique [20,22]. Bulk-fill materials showed lower shrinkage stress and exhibited an acceptable creep deformation and reduced cuspal deflection when compared to conventional RBCs [17,23,24]. In a recent investigation, mechanical properties, including degree of cure, were shown to be constant within the 4-mm increment [25]. In the dental literature, studies investigated the amount of eluted monomers; thus, the biocompatibility and clinical performance of bulk-fill flowable RBC base materials are limited [20–22,24,26]. Biocompatibility depends on the quality and quantity of released monomers and their derivates, which can irritate the pulp, the soft tissues of the oral cavity and may lead eventually to a toxic reaction [27,28]. Several factors, such as the DC, the specimen thickness, the chemical composition, the filler particle type and content, the porosity and the solvent can influence the amount of released monomers [29].

The aim of this study was to assess the DC and the amount of released BisGMA, TEGDMA and UDMA monomers of some low-viscosity bulk-fill composite materials in a 4-mm layer thickness compared to that of conventional flowable one in a 2-(positive control) and a 4-mm (negative control) thickness, using micro-Raman spectroscopy and HPLC.

2. Results

2.1. Degree of Conversion-Micro-Raman Spectroscopy

The top and bottom surface DC values of the materials are presented in Figures 1 and 2.

The mean DC values on the top surface of the materials were in the following order: SDR_4mm_20s > FBF_4mm_20s > FUF_4mm_2mm > FUF_4mm_20s > FBF_4mm_10s > XB_4mm_20s > XB_4mm_10s; however, the order of the mean DC values on the bottom surface was significantly different from the top surface: SDR_4mm_20s > FUF_2mm_20s > XB_4mm_20s > FBF_4mm_20s > XB_4mm_10s > FBF_4mm_10s > FUF_4mm_20s. Dunnett's *t*-test showed that all of the investigated materials had statistically significant differences in DC% at the bottom surface when compared to each other, except between FBF_4mm_20s and XB_4mm_20s ($p = 0.221$). On the contrary, between FUF_2mm_20s and FUF_4mm_20s ($p = 0.306$), FUF_2mm_20s and FBF_4mm_20s ($p = 1.000$), FUF_4mm_20s and FBF_4mm_20s ($p = 0.241$), FUF_4mm_20s and FBF_4mm_10s ($p = 0.059$) and between XB_4mm_20s and FBF_4mm_10s ($p = 0.063$), Dunnett's *t*-test did not show significant difference in DC% at the top surface. The conventional flowable composite FUF_4mm_20s had the lowest DC value at the bottom (16.53%), while SDR_4mm_20s (50.05%) had the highest DC value not only at the bottom, but at the top surface, as well. The extended curing time of FBF and XB from 10–20 s significantly increased the DC%, especially at the bottom surface.

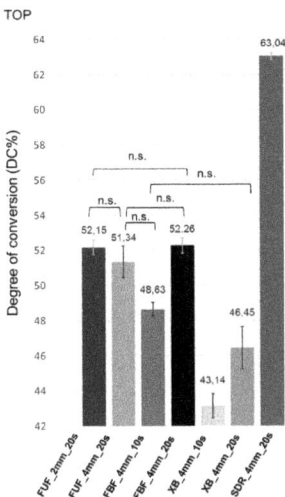

Figure 1. Mean DC% and 95% confidence intervals of the top surface of the samples (abbreviations: DC, degree of conversion; n.s., not significant difference; FUF_2mm_20s, Filtek Ultimate Flow in a 2-mm layer thickness cured for 20 s; FUF_4mm_20s, Filtek Ultimate Flow in a 4-mm layer thickness cured for 20 s; FBF_4mm_10s, 4 mm-thick Filtek Bulk Fill light cured for 10 s; FBF_4mm_20s, 4 mm-thick Filtek Bulk Fill light cured for 20 s; XB_4mm_10s, 4 mm-thick X-tra Base light cured for 10 s; XB_4mm_20s, 4 mm-thick X-tra Base light cured for 20 s; SDR_4mm_20s, SureFil SDR Flow in a 4-mm layer thickness cured for 20 s).

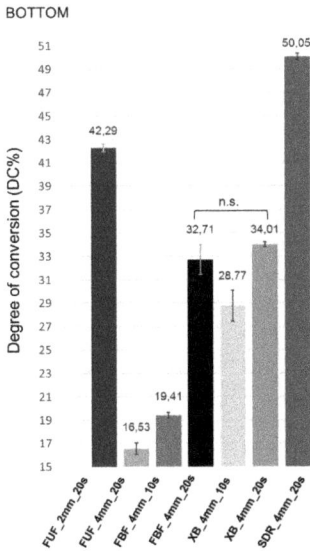

Figure 2. Mean DC% and 95% confidence intervals of the bottom surface of the samples (abbreviations: DC, degree of conversion; n.s., not significant difference; FUF_2mm_20s, Filtek Ultimate Flow in a 2-mm layer thickness cured for 20 s; FUF_4mm_20s, Filtek Ultimate Flow in a 4-mm layer thickness cured for 20 s; FBF_4mm_10s, 4 mm-thick Filtek Bulk Fill light cured for 10 s; FBF_4mm_20s, 4 mm-thick Filtek Bulk Fill light cured for 20 s; XB_4mm_10s, 4 mm-thick X-tra Base light cured for 10 s; XB_4mm_20s, 4 mm-thick X-tra Base light cured for 20 s; SDR_4mm_20s, SureFil SDR Flow in a 4-mm layer thickness cured for 20 s).

2.2. Monomer Elution: HPLC

Figure 3 shows the amount of eluted BisGMA, UDMA and TEGDMA from the 2 mm- and 4 mm-thick conventional flowable and the bulk-fill flowable RBC materials.

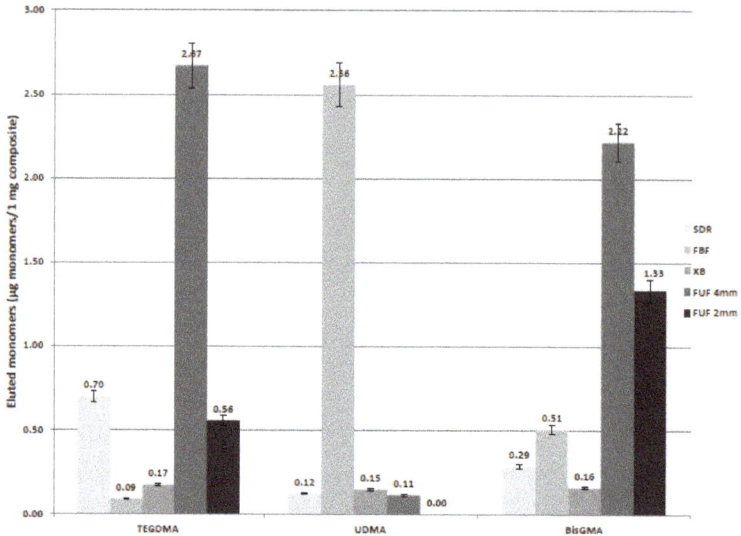

Figure 3. Amount of eluted monomers from bulk-fill and conventional flowable composites. (Abbreviations: SDR, SureFil SDR Flow in a 4-mm layer thickness light cured for 20 s; FBF, Filtek Bulk Fill in a 4-mm layer thickness light cured for 10 s; XB, X-tra Base in a 4-mm layer thickness light cured for 10 s; FUF 4 mm, Filtek Ultimate Flow in a 4-mm layer thickness light cured for 20 s; FUF 2 mm, Filtek Ultimate Flow in a 2-mm layer thickness light cured for 20 s).

In spite of the fact that there is no BisGMA in SDR and XB according to the manufacturer's information, there was a detectable amount of this monomer from these bulk-fill materials, such as TEGDMA from XB. In comparison, with Dunnett's t-test, there were significant differences between the groups. More than a five-times higher elution rate was found in the amount of released BisGMA in the case of the 4 mm-thick conventional flowable FUF when this material was compared to the bulk-fill materials, and the difference in BisGMA release between the 2 mm- and 4 mm-thick FUF was almost twice ($p < 0.001$). The leached BisGMA from bulk-fill RBCs showed the following order: FBF > SDR > XB; the differences were statistically significant ($p < 0.001$). The FUF_4mm_20s had a statistically-significant higher rate of TEGDMA elution, as well, compared to the other materials ($p < 0.001$). Among bulk-fills, SDR showed a seven-times higher amount of TEGDMA elution. More than a twenty-times higher amount of eluted UDMA was observed in the case of FBF compared to SDR, XB and FUF; meanwhile, UDMA was not listed in the FUF's technical product profile.

In the case of FUF and FBF, a high intensity peak was detected on the HPLC chromatogram at a different retention time than that of the above-mentioned three monomers, which was not identified and quantified in the lack of a standard monomer. However, according to the manufacturer's information, it is probably the peak of procrylate monomer.

3. Discussion

In this study, the DC and the elution of unreacted monomers of different bulk fill and commercial flowable dental composites were assessed using micro-Raman spectroscopy and HPLC.

The setting process has a major influence on the mechanical and biological properties of RBCs [30]. Resin polymerization depends on the chemical structure of the monomer, filler characteristics, the photoinitiator concentration and the polymerization conditions [31]. Since polymerization conditions, such as layer thickness, intensity of the curing unit and exposure time, were standardized in this study, differences in the DC value of conventional and bulk-fill RBCs can be attributed to the different composition of the materials, mostly to variations in the chemistry of their resin matrix and the filler loading. In general, the manufacturers of bulk-fill RBCs were able to improve polymerization depth by the use of potent photoinitiator systems along with an increased translucency [18,32]. As light transmission is strongly dependent on material opacity [33], the observed higher DC% at a 4-mm specimen thickness for the investigated bulk-fills compared to the 4 mm-thick conventional flowable RBC might be a result of their reduced opacity. However, when the DC% of bulk fills was compared to the 2 mm-thick conventional flow, only SDR_4mm_20s produced a higher conversion rate. Higher translucency can also be achieved by the reduction in filler content [34]. It has been demonstrated by Halvorson et al. that increasing the filler-matrix ratio progressively decreases conversion, because an increased amount of filler particles is an obstacle for polymeric chain propagation [35]. According to Nomoto and Hirasawa [36], the depth of cure and, thus, the DC are affected by the filler's light permeability, the monomer composition, the type and concentration of the initiator and the inhibitor/accelerator systems in the RBCs. In the present study, a significantly higher DC value was observed with the SDR_4mm_20s, and a tendency for a higher DC value was detected in FBF in comparison to XB cured for both 10 and 20 s. This finding is supported by other investigators [19,37–39] and might be explained by the higher filler content of XB, which may increase light scattering, causing a concurrent decrease of translucency for blue light [18].

Due to the presence of a photo-active modulator in the matrix system and the increased translucency, the manufacturer's recommendation for the exposure time is 10 s for the universal shades, with the intensity of the curing unit ranging from 550–1000 mW·cm^{-2}, while 20 s for SDR and FUF irrespective of the output intensity of the curing unit. Based on this fact, 5.5–10 J·cm^2 of delivered energy should be enough for the adequate polymerization of bulk-fills. For a conventional RBC, the recommendation of a 21-J·cm^2 and a 24-J·cm^2 energy density has been made for the satisfactory conversion of a 2 mm-thick composite specimen [40]. Thus, as was expected, a 20-J·cm^2 energy density was not enough for the acceptable polymerization rate of conventional flowable restorative samples in a 4-mm layer thickness; meanwhile, the decrease of layer thickness to 2 mm increased the polymerization rate with 61.5%. In the case of bulk fill XB and FBF, the extended curing time resulted in a 15.5% and a 40.7% increase in the rate of polymerization. In contrast with our findings, Finan et al. [19] observed higher DC% for SDR (59%) and XB (48%) using a quartz tungsten halogen (QTH) light curing unit (LCU) for polymerization, which operates at an output intensity of 650 mW/cm^2 for 20 s; while Zorzin et al. [41] measured a 52% DC value for SDR, 63% for XB, 66% for FBF and 66% for FUF at a 4-mm layer thickness cured with an LED unit (1200 mW/cm^2) for 20 s in the case of A2 shades. The possible explanation for this difference may be that wider samples in diameter were used to allow a higher degree of light penetration for polymerization or that Fourier transform infrared spectrophotometer was used to analyze the DC of RBC samples. However, Zorzin et al. concluded that extended curing time (30 s) had a positive effect on polymerization properties, so enhanced light curing of bulk-fills in deep cavities is recommended [41]. Similarly to our study, Li et al. [39] used micro-Raman spectroscopy to map the DC along a cross-section of bulk-fill and a conventional composite block. They measured 80% mean maximum DC for FBF and 77.3% for SDR; however, their study design was different from our design, as the investigators tested the curing profile of a thicker (16 mm × 6 mm × 12 mm) rectangular bulk-fill RBC block.

Besides the filler-matrix ratio, the DC is affected by the viscosity and reactivity of the polymerizable monomer, as well [42]. The DC of different monomer systems increases in the following order: BisGMA < BisEMA < UDMA < TEGDMA [43]. BisGMA is considered the most viscous monomer due to the strong intramolecular hydrogen bonding, which can decrease the reactivity and mobility of the monomer during the polymerization process. This might be one of the explanations for the significantly lower DC of the conventional flowable FUF_4mm_20s than that of other materials, as FUF contains the highest amount of BisGMA monomer in the resin matrix. In our study, the investigated bulk-fill RBCs are UDMA-based materials in combination with different types of monomers. Sideridou *et al.* found that UDMA, combining relatively high molecular weight with a high concentration of double bonds and low viscosity, was shown to reach higher final DC% values than BisGMA [43]. Low viscosity has a high impact on free radical migration. It was proven that the DC of RBC monomers is strongly influenced by the nature of the polymerizing monomers, for example more flexible monomers increase the rate of conversion [44]. Although the viscosity of UDMA is much lower than that of BisGMA, when it is mixed with the high molecular weight BisEMA or EBPADMA, it can significantly restrict the mobility of UDMA monomers and decrease their reactivity and conversion value [37,45]. This may explain the significantly lower DC value of FBF and XB cured for both 10 and 20 s than that of SDR_4mm_20s, as was reported by Alshali *et al.* [37]. SDR is similarly an UDMA/EBPADMA-based bulk-fill flowable composite; however, it contains TEGDMA, which has a synergistic effect on the rate of polymerization, and thus, the DC value of this monomer is significantly higher than that of the other investigated bulk-fill materials.

In addition, according to the manufacturer's technical information, a photo-active modulator in SDR may cooperate with camphorquinone (CQ), thereby facilitating polymerization.

In the present study, monomer elution from RBC was quantified using HPLC. This is a standard method used for the determination of monomer elution from RBCs [46]. The release of components has a potential effect on the structural stability and wear rate, as well as the biocompatibility of the material. The analysis of the elution of selected unreacted BisGMA, TEGDMA and UDMA will not provide an absolute measure of the quality of released components; thus, it is a limitation of this study.

Several factors may influence the monomer elution, such as the rate of polymerization, the chemical features of the solvent and the chemical nature of the leached components [5,47]. In the present study, 75% ethanol was used to extract most of the examined unreacted monomers from the polymerized composite samples in order to identify monomer quantity. The elution pattern of unreacted monomers is higher in ethanol than in water storage medium, because of their hydrophobic character, which can significantly reduce and rationalize examination periods. Water storage may simulate oral conditions better than ethanol; however, changeful oral parameters (pH, temperature, enzyme activities) are hardly simulated in water medium. According to our results there was a strong correlation between DC% and the amount of eluted monomers in the case of the conventional flowable RBC (FUF). This is a BisGMA/TEGDMA/procrylate-based material, and the highest amount of leached TEGDMA and BisGMA was observed with the lowest DC%. Meanwhile, the reduced layer thickness decreased the amount of released TEGDMA and BisGMA by five- and two-times, respectively. The possible reasons for the different values for these monomers could be the chemical nature, the different molecular weight and the reactivity of the molecules. Tanaka *et al.* [48] found that TEGDMA has higher mobility caused by its low molecular weight, resulting in a higher and faster rate of elution than the larger BisGMA and UDMA [47,48]. As a TEGDMA/UDMA-based material, SDR showed a high amount of TEGDMA release following FUF_4mm_20s. Similar to our findings, Cebe *et al.* also found a higher amount of eluted TEGDMA monomers than from the other bulk-fills, and the cumulative amount of eluted TEGDMA increased with time [49]. In their study, Łagocka *et al.* detected lower (8.4 µg/g) TEGDMA elution from SDR during the first 24 h of storage in 75% ethanol solvent; however, the elution rate rapidly decreased with time [50]. The reason for the increased elution of the high molecular weight BisGMA could be the low rate of polymerization explained by the hampered light penetration and the decreased photoinitiator activation in the conventional flowable composite at a 4-mm layer thickness. Among bulk-fill composites, FBF showed the highest rate of released UDMA and BisGMA. Cebe *et al.* also detected a higher rate of eluted BisGMA from FBF, especially at the 30-day time interval [49].

Considering the filler content, FBF has the lowest filler value among the investigated bulk-fill materials, which may influence the release of unreacted monomers. Comparing the molecular weight of BisGMA and UDMA, BisGMA has a higher weight; thus, there is quicker and more substantial

UDMA release in a certain time interval. Among bulk fills, XB showed a significantly lower rate of eluted monomers. The lower solubility might be based on the higher (75%) filler content in contrast with the other bulk-fills. There are reports that illustrate a lower absorption rate in composite materials with high filler contents compared to materials with lower filler content [51–53]. The results of the present study, which are in line with these reports, showed a lower monomer elution rate from XB with higher filler content compared to the other investigated bulk-fills. The possible explanation might be the lower solvent absorption in XB samples, which resulted in less leachable component elution. This leads to the conclusion that the elution mechanism is complex and cannot be explained only by the degree of conversion. On the other hand, Sideridou *et al.* concluded that higher silane content has a positive effect on interfacial adhesion between filler and matrix [54]. As the filler particles are chemically bonded to the matrix monomers and oligomers by the silane coupling agent, their higher volume fraction can provide a more stable ligation for the unreacted, leachable monomers, decreasing their release into the solvent. The structure of the silane coupling agent and its bonding to the filler particle has a high impact on the solubility of the RBC [55].

The detected BisGMA from the XB samples might be impurities of the monomer matrix complex.

Considering the BisEMA content in the investigated bulk-fill materials, there was no information from the degree of ethoxylation of the bisphenol A molecule; therefore, it was not possible to identify in the lack of standard monomer. Compared to other composite resin monomers, BisEMA is not a single monomer molecule, rather belonging to a large series of ethoxylated bisphenol A-based dimethacrylates with an ethoxylation reaction of a very reactive ethylene oxide [56,57]. Therefore, the ethoxylation reaction is unselective and difficult to control, leading to different ethoxylated products and byproducts, which must be separated analytically [58].

4. Materials and Methods

BisGMA (98%), UDMA ($\geqslant$97%) and TEGDMA (95%) (Sigma-Aldrich, Steinheim, Germany) were used as standard materials for the identification of the monomer peaks in the chromatograms. Filtek Ultimate Flow flowable nanocomposite samples were prepared as references. The investigated materials were the following: SureFil SDR Flow, X-tra Base and Filtek Bulk Fill. Table 1 shows the composition of the materials. All samples were stored in a 75% ethanol/water solution (Spektrum-3D, Debrecen, Hungary). Acetonitrile (ACN) (VWR International, Leuven, Belgium) was used for the preparation of the mobile phase for the HPLC separation.

Table 1. Materials, manufactures and composition.

Group	Material	Code	Manufacturer	Shade	Organic Matrix	Filler	Filler Loading	LOT Number
Bulk-fill composite	SureFil SDR Flow	SDR	Dentsply Caulk, Milford, DE, USA	U	Modified UDMA, EBPADMA, TEGDMA	Ba-Al-F-B silicate glass, Sr-Al-F silicate glass	68 wt %	1202174
	x-tra base	XB	Voco, Cuxhaven, Germany	U	UDMA, BisEMA	no information	75 wt %	1305261
	Filtek Bulk Fill	FBF	3M ESPE, St Paul, MN, USA	U	BisGMA, UDMA, BisEMA(6), TEGDMA, substituted dimethacrylate, Procrylat resin	silane treated zirconia/silica, ytterbium trifluoride	64.5 wt %	N414680
Conventional flowable composite	Filtek Ultimate Flow	FUF	3M ESPE, St Paul, MN, USA	A2	BisGMA, TEGDMA, substituted dimethacrylate, Procrylat resin	silane treated zirconia/silica, ytterbium trifluoride	65 wt %	N652740

Abbreviations: U, universal; UDMA, urethane dimethacrylate; EBPADMA, ethoxylated Bisphenol A dimethacrylate; TEGDMA, triethylene glycol dimethacrylate; BisEMA, Bisphenol A polyethylene glycol diether dimethacrylate; BisGMA, Bisphenol A diglycidil ether dimethacrylate; Procrylate, reacted polycaprolactone polymer. Missing entries are not specified by the manufacturer.

4.1. Preparation of the Composite Resin Specimens

The flowable bulk-fill RBCs (SDR, XB, FBF) were poured into a stainless steel mold with a size of 3 mm in diameter × 4 mm in thickness ($n = 3 \times 5$) and positioned on a glass slide. As the negative control, conventional flowable composite (FUF) samples were used also with a size of 3 mm in diameter × 4 mm in thickness ($n = 5$) to be comparable with bulk-fill RBCs and with a size of 3 mm in diameter × 2 mm in thickness ($n = 5$) as a positive control. During preparation, each sample was measured to obtain samples of similar weight and volume. The top and the bottom of the RBC were covered with a polyester (Mylar, Dentamerica Inc., San Jose Ave, CA, USA) strip in order to avoid contact with oxygen, which is an inhibitor of the polymerization. The specimens were irradiated with a light-emitting diode (LED) curing unit (λ = 420–480 nm; LED.C, Woodpecker, Guilin, China) with the recommended exposure time at a light intensity of 1100 mW·cm^{-2} with an irradiated diameter of 10 mm. The manufacturer's instruction for curing time at a 4-mm thickness and universal shades is 10 s in FBF and XB bulk-fill RBCs. In the case of these two materials, the effect of extended curing time (20 s) was also investigated. For SDR, the recommended exposure time is 20 s without giving a suggested value for the light intensity. In the case of A2 shade conventional flowable RBC, a 20-s exposure time at a 2-mm thickness is recommended; however, to investigate and to compare the DC value and the amount of released monomers, this product was also used in a 4-mm thickness and was irradiated only for 20 s. Table 2 summarizes the abbreviations of the prepared samples.

Table 2. Sample preparation with layer thickness and exposure time.

Abbreviation	Material	Layer Thickness (mm)	Exposure Time (s)
FUF_2mm_20s	Filtek Ultimate Flow	2	20
FUF_4mm_20s	Filtek Ultimate Flow	4	20
FBF_4mm_10s	Filtek Bulk Fill	4	10
FBF_4mm_20s	Filtek Bulk Fill	4	20
XB_4mm_10s	X-tra Base	4	10
XB_4mm_20s	X-tra Base	4	20
SDR_4mm_20s	SureFil SDR Flow	4	20

A radiometer (SDS, Kerr, Danbury, CT, USA) was used to control the intensity of the curing unit before and after the light exposition. The tip of the curing light guide was positioned parallel and 1 mm above the composite sample. One day after polymerization, the specimens were measured with micro-Raman spectroscopy. For the dissolution of the unreacted monomers the specimens were stored in 1 mL of 75% ethanol/water solution for 72 h in darkness at room temperature. After 3 days, the amount of dissolved unpolymerized monomers was analyzed with reverse-phase high-performance liquid chromatography (RP-HPLC) from the ethanol solutions.

4.2. Micro-Raman Spectroscopy Measurement

The polymerized composite samples were examined using a Labram HR 800 Confocal Raman spectrometer (HORIBA JobinYvon S.A.S., Longjumeau Cedex, France) 24 h after polymerization. During the micro-Raman measurements, a 20-mW He-Ne laser with a 632.817-nm wavelength was applied; the spatial resolution was ~1.5 µm; the spectral resolution was ~2.5 cm^{-1}; with magnification of 100× (Olympus UK Ltd., London, UK), applying a D 0.3 filter (~1.98 mW on the sample). The spectra were taken on the top and also on the bottom surface of the composite specimens at three random locations. The integration time was 10 s, and ten acquisitions were averaged for each geometrical point. Uncured composite spectra were measured as a reference. These samples were placed between two non-fluorescent glass slides. Post-processing of spectra was performed using the dedicated software LabSpec 5.0 (HORIBA JobinYvon S.A.S., Longjumeau Cedex, France), and the Levenberg–Marquardt non-linear peak fitting method was applied for the best fit [59,60]. The following equation was used to calculate the ratio of the double-bond content of monomer to polymer in the composite:

$$\text{DC\%} = \left(1 - \left(\frac{R_{\text{cured}}}{R_{\text{uncured}}}\right)\right) \times 100 \qquad (1)$$

where R is the ratio of aromatic and aliphatic C=C bonds at peak intensities of 1639 cm^{-1} and 1609 cm^{-1} in cured and uncured composite samples, respectively [43,61].

4.3. RP-HPLC Measurements

The RP-UHPLC system (Dionex Ultimate 3000, Thermo Fisher Scientific Inc, Sunnyvale, CA, USA) consists of a Dionex LPG 3400 SD gradient pump, Dionex ACC 3000 autosampler and a Dionex UWD 3400 RS UV–VIS detector (Dionex GmbH, Germering, Germany). Data acquisition was completed using Chromeleon software integrated in Hystar (version: 3.2). The separations were performed on a Synergi HYDRO-RP (particle size: 4 µm; pore size: 8) (Phenomenex, Gen-Lab, Budapest, Hungary) column (150 mm × 2.00 mm) with gradient elution. The composition of Eluent "A" was 40% v/v ACN in bidistilled water, whereas Mobile Phase "B" was composed of 95% v/v ACN and 5% bidistilled water. During the 30-min chromatographic separation, the "B" eluent content increased from 20%–100%. The low rate was 0.3 mL·min^{-1}. As the regeneration of the stationary phase, Mobile Phase B content was decreased from 100% down to 20% in 1 min, and after 31–46 min, the system was washed with 20% "A".

The detection of the eluted monomers was at the following wavelengths: 205, 215, 227 and 254 nm. Two hundred five nanometers was found to be optimal; therefore, the evaluation relied on the data collected at this wavelength. Each separation was implemented at room temperature.

The amount of the eluted monomers was calculated by the calibration curve with the areas under the curve of peaks produced by the monomers, respectively. The TEGDMA, UDMA and BisGMA standard solutions had retention times of 2.95, 5.08 and 6.88 min, respectively, whereas the peaks were well separated from each other.

4.4. Validation of the Monomer Determination the Limit of Detection and the Limit of Quantification

The detection limit (determined as the amount of monomers giving a peak height 5-times higher than the noise level) of TEGDMA is 0.018 pmol (5.19 pg), UDMA, 0.015 pmol (7.196 pg), and BisGMA, 0.007 pmol (3.556 pg). The quantification limit of the method (the peak height of the monomers 10-times higher than the noise level) was low for TEGDMA is 0.036 pmol (10.382 pg), for UDMA, 0.031 pmol (14.392 pg), and for BisGMA, 0.014 pmol (7.1120 pg). Calibration was carried out in the range of 1–50.0 µg·mL^{-1} monomers, respectively. A calibration curve was plotted by the measurement of standard solutions at 205 nm (R^2) 0.9924 for TEGDMA, 0.9966 for UDMA and 0.9996 for BisGMA, respectively. All injections were repeated three times.

4.5. Statistical Analysis

The statistical analysis was performed using SPSS (Statistical Package for Social Science, SPSS Inc., Chicago, IL, USA) software for Windows. The values for the degree of conversion and for residual monomers between the studied test groups were compared by a one-way analysis of variance (ANOVA) test followed by Dunnett's t-test at the α = 0.05 level.

5. Conclusions

Within the limitation of the present study, the following can be concluded:

(1) Among the investigated low viscosity bulk fill and conventional flowable RBCs, SDR showed the highest DC value at the top and bottom surface of the samples.
(2) The DC values of the 4 mm-thick bulk-fill composites SDR, FBF, XB were significantly higher than that of the 4 mm-thick conventional composite (negative control) studied; meanwhile, only

SDR bulk-fill resulted in a higher DC value compared to that of the 2 mm-thick conventional flowable RBC (positive control).
(3) Although the recommended exposure time by the manufacturers for the universal shade FBF and XB is 10 s (with a 1000-mW/cm^2 curing unit), extended (20 s) curing time significantly increased the DC% value.
(4) The amount of released BisGMA and TEGDMA monomers from the bulk-fill composite materials was generally lower than from the conventional composite.
(5) Among bulk fills, in spite of the highest DC%, SDR showed the highest rate of TEGDMA elution; meanwhile, the highest amount of UDMA was eluted from FBF.

Acknowledgments: This work was supported by Pécsi Tudományegyetem Általános Orvostudományi Kar-Kutatási Alap 2013/5 and Pécsi Tudományegyetem Általános Orvostudományi Kar-Kutatási Alap-2016/1 Research Grants. The present scientific contribution is dedicated to the 650th anniversary of the foundation of the University of Pécs, Hungary.

Author Contributions: Edina Lempel conceived of and designed the experiments and wrote the paper. Zsuzsanna Czibulya and Bálint Kovács performed the micro-Raman spectroscopy measurements. József Szalma and Ákos Tóth analyzed the data. Sándor Kunsági-Máté contributed analysis tools. Zoltán Varga prepared the samples. Katalin Böddi designed experiments and performed the HPLC measurements.

Conflicts of Interest: The authors declare no conflict of interest.

Abbreviations

DC	degree of conversion
HPLC	high performance liquid chromatography
RP-HPLC	reverse-phase high-performance liquid chromatography
SDR	SureFil SDR Flow
XB	X-tra Base
FBF	Filtek Bulk Fill
FUF	Filtek Ultimate Flow
FUF_2mm_20s	Filtek Ultimate Flow in a 2-mm layer thickness cured for 20 s
FUF_4mm_20s	Filtek Ultimate Flow in a 4-mm layer thickness cured for 20 s
FBF_4mm_10s	4 mm-thick Filtek Bulk Fill light cured for 10 s
FBF_4mm_20s	4 mm-thick Filtek Bulk Fill light cured for 20 s
XB_4mm_10s	4 mm-thick X-tra Base light cured for 10 s
XB_4mm_20s	4 mm-thick X-tra Base light cured for 20 s
SDR_4mm_20s	SureFil SDR Flow in a 4-mm layer thickness cured for 20 s
UDMA	urethane dimethacrylate
BisGMA	bisphenol A diglycidyl ether dimethacrylate
TEGDMA	triethylene glycol dimethacrylate
BisEMA	bisphenol A polyethylene glycol diether dimethacrylate
EBPADMA	ethoxylated bisphenol A dimethacrylate
RBC	resin-based composite
SD	standard deviation
QTH	quartz tungsten halogen
LCU	light curing unit
CQ	camphorquinone
CAN	acetonitrile
LED	light emitting diode
LOD	limit of detection
LOQ	limit of quantification

1. Da Rosa Rodolpho, P.A.; Donassollo, T.A.; Cenci, M.S.; Loguércio, A.D.; Moraes, R.R.; Bronkhorst, E.M.; Opdam, N.J.M.; Demarco, F.F. 22-year clinical evaluation of the performance of two posterior composites with different filler characteristics. *Dent. Mater.* **2011**, *27*, 955–963. [CrossRef] [PubMed]
2. Lempel, E.; Tóth, Á.; Fábián, T.; Krajczár, K.; Szalma, J. Retrospective evaluation of posterior direct composite restorations: 10-Year findings. *Dent. Mater.* **2015**, *31*, 115–122. [CrossRef] [PubMed]
3. Opdam, N.J.M.; Bronkhorst, E.M.; Loomans, B.A.; Hujsmans, M.C. 12-year survival of composite *vs.* amalgam restorations. *J. Dent. Res.* **2010**, *89*, 1063–1067. [CrossRef] [PubMed]
4. Pallesen, U.; van Dijken, J.W.V.; Halken, J.; Hallosten, A.L.; Höigaard, R. Longevity of posterior resin composite restorations in permanent teeth in Public Dental Health Service: A prospective 8 years follow up. *J. Dent.* **2013**, *41*, 297–306. [CrossRef] [PubMed]
5. Ferracane, J.L. Elution of leachable components from composites. *J. Oral Rehabil.* **1994**, *21*, 441–452. [CrossRef] [PubMed]
6. Ferracane, J.L.; Greener, E.H. The effect of resin formulation on the degree of conversion and mechanical properties of dental restorative resins. *J. Biomed. Mater. Res.* **1986**, *20*, 121–131. [CrossRef] [PubMed]
7. Lempel, E.; Czibulya, Z.; Kunsági-Máté, S.; Szalma, J.; Sümegi, B.; Böddi, K. Quantification of conversion degree and monomer elution from dental composite using HPLC and micro-raman spectroscopy. *Chromatographia* **2014**, *77*, 1137–1144. [CrossRef]
8. Poggio, C.; Lombardini, M.; Gaviati, S.; Chiesa, M. Evaluation of Vickers hardness and depth of cure of six composite resins photo-activated with different polymerization modes. *J. Conserv. Dent.* **2012**, *15*, 237–241. [CrossRef] [PubMed]
9. Silikas, N.; Eliades, G.; Watts, D.C. Light intensity effects on resin-composite degree of conversion and shrinkage strain. *Dent. Mater.* **2000**, *16*, 292–296. [CrossRef]
10. Turssi, C.P.; Ferracane, J.L.; Vogel, K. Filler features and their effects on wear and degree of conversion of particulate dental resin composites. *Biomaterials* **2005**, *26*, 4932–4937. [CrossRef] [PubMed]
11. Ilie, N.; Jelen, E.; Hickel, R. Is the soft-start polymerization concept still relevant for modern curing units? *Clin. Oral Investig.* **2011**, *15*, 21–29. [CrossRef] [PubMed]
12. Alomari, Q.D.; Reinhardt, J.W.; Boyer, D.B. Effect of liners on cusp deflection and gap formation in composite restorations. *Oper. Dent.* **2001**, *26*, 406–411. [PubMed]
13. Park, J.; Chang, J.; Ferracane, J.; Lee, I.B. How should composite be layered to reduce shrinkage stress: Incremental or bulk filling. *Dent. Mater.* **2008**, *24*, 1501–1505. [CrossRef] [PubMed]
14. Obici, A.C.; Sinhoreti, M.A.C.; Frollini, E.; Correr-Sobrinho, L.; Fernando de Goes, M.; Henriques, G.E.P. Monomer conversion at different dental composite depths using six light-curing methods. *Polym. Test.* **2006**, *25*, 282–288. [CrossRef]
15. Watts, D.C.; Cash, A.J. Analysis of optical transmission by 400–500 nm visible light into aesthetic dental biomaterials. *J. Dent.* **1994**, *22*, 112–117. [CrossRef]
16. Flury, S.; Hayoz, S.; Peutzfeldt, A.; Hüsler, J.; Lussi, A. Depth of cure of resin composites: Is the ISO 4049 method suitable for bulk fill materials? *Dent. Mater.* **2012**, *28*, 521–528. [CrossRef] [PubMed]
17. El-Safty, S.; Silikas, N.; Watts, D.C. Creep deformation of restorative resin-composites intended for bulk-fill placement. *Dent. Mater.* **2012**, *28*, 928–935. [CrossRef] [PubMed]
18. Bucuta, S.; Ilie, N. Light transmittance and micro-mechanical properties of bulk fill vs. conventional resin based composites. *Clin. Oral Investig.* **2014**, *18*, 1991–2000. [CrossRef] [PubMed]
19. Finan, L.; Palin, W.M.; Moskwa, N.; McGinley, E.L.; Fleming, G.J.P. The influence of irradiation potential on the degree of conversion and mechanical properties of two bulk-fill flowable RBC base materials. *Dent. Mater.* **2013**, *29*, 906–912. [CrossRef] [PubMed]
20. Van Ende, A.; De Munch, J.; Van Landuyt, K.L.; Poitevin, A.; Peumans, M.; Van Meerbeek, B. Bulk-filling of high C-factor posterior cavities, effect on adhesion to cavity bottom dentin. *Dent. Mater.* **2013**, *29*, 269–277. [CrossRef] [PubMed]
21. Campodonico, C.E.; Tantbirojn, D.; Olin, P.S.; Versluis, A. Cuspal deflection and depth of cure in resin-based composite restorations filled by using bulk; incremental and transtooth illumination techniques. *J. Am. Dent. Assoc.* **2011**, *142*, 1176–1182. [CrossRef] [PubMed]

22. Roggendorf, M.J.; Kramer, N.; Appelt, A.; Naumann, M.; Frankenberger, R. Marginal quality of flowable 4-mm base vs conventionally layered resin composite. *J. Dent.* **2011**, *39*, 643–647. [CrossRef] [PubMed]
23. Ilie, N.; Hickel, R. Investigations on a methacrylate-based flowable composite based on the SDR™ technology. *Dent. Mater.* **2011**, *27*, 348–355. [CrossRef] [PubMed]
24. Moorthy, A.; Hogg, C.H.; Dowling, A.H.; Grufferty, B.F.; Benetti, A.R.; Fleming, G.J.P. Cuspal deflection and microleakage in premolar teeth restored with bulk-fill flowable resin-based composite base materials. *J. Dent.* **2012**, *40*, 500–505. [CrossRef] [PubMed]
25. Czasch, P.; Ilie, N. *In vitro comparison* of mechanical properties and degree of cure of *bulk fill* composites. *Clin. Oral Investig.* **2013**, *17*, 227–235. [CrossRef] [PubMed]
26. Van Dijken, J.W.V.; Pallesen, U. A randomized controlled three year evaluation of "bulk-filled" posterior resin restorations based on stress decreased resin technology. *Dent. Mater.* **2014**, *30*, 245–251. [CrossRef] [PubMed]
27. Goldberg, M. *In vitro* and *in vivo* studies on the toxicity of dental resin components, a review. *Clin. Oral Investig.* **2008**, *12*, 1–8. [CrossRef] [PubMed]
28. Bakopoulou, A.; Mourelatos, D.; Tsiftsoglou, A.S.; Giassin, N.P.; Mioglou, E.; Garefis, P. Genotoxic and cytotoxic effects of different types of dental cement on normal cultured human lymphocytes. *Mutat. Res.* **2009**, *672*, 103–112. [CrossRef] [PubMed]
29. Silva, G.S.; Almeida, G.S.; Poskus, L.T.; Guimarães, J.G. Relationship between the degree of conversion; solubility and salivary sorption of a hybrid and nanofilled resin composite: Influence of the light activation mode. *Appl. Oral Sci.* **2008**, *16*, 161–166. [CrossRef]
30. Cramer, N.B.; Stansbury, J.W.; Bowman, C.N. Recent advances and developments in composite dental restorative materials. *J. Dent. Res.* **2011**, *90*, 402–416. [CrossRef] [PubMed]
31. Leprince, J.G.; Palin, W.M.; Hadis, M.A.; Devaux, J.; Leloup, G. Progress in dimethacrylate-based dental composite technology and curing efficiency. *Dent. Mater.* **2013**, *29*, 139–156. [CrossRef] [PubMed]
32. Alrahlah, A.; Silikas, N.; Watts, D.C. Post-cure depth of cure of bulk fill dental resin-composites. *Dent. Mater.* **2014**, *30*, 149–154. [CrossRef] [PubMed]
33. Shortall, A.C. How light source and product shade influence cure depth for a contemporary composite. *J. Oral Rehabil.* **2005**, *32*, 906–911. [CrossRef] [PubMed]
34. Lee, Y.K. Influence of filler on difference between the transmitted and reflected colors of experimental resin composites. *Dent. Mater.* **2008**, *24*, 1243–1247. [CrossRef] [PubMed]
35. Halvorson, R.H.; Erickson, R.L.; Davidson, C.L. The effect of filler and silane content on conversion of resin-based composite. *Dent. Mater.* **2003**, *19*, 327–333. [CrossRef]
36. Nomoto, R.; Hirasawa, T. Residual monomer and pendant methacryloyl group in light-cured composite resins. *Dent. Mater. J.* **1992**, *11*, 177–188. [CrossRef] [PubMed]
37. Alshali, R.Z.; Silikas, N.; Satterthwaite, J.D. Degree of conversion of bulk-fill compared to conventional resin-composites at two time intervals. *Dent. Mater.* **2013**, *29*, 213–217. [CrossRef] [PubMed]
38. Tarle, Z.; Attin, T.; Marovic, D.; Andermatt, L.; Ristic, M.; Tauböck, T.T. Influence of irradiation time on subsurface degree of conversion and microhardness of high-viscosity bulk-fill resin composites. *Clin. Oral Investig.* **2015**, *19*, 831–840. [CrossRef] [PubMed]
39. Li, X.; Pongprueksa, P.; van Meerbeek, B.; de Munk, J. Curing profile of bulk-fill resin-based composites. *J. Dent.* **2015**, *43*, 664–672. [CrossRef] [PubMed]
40. Emami, N.; Söderholm, K.J.M. How light irradiance and curing time affect monomer conversion in light-cured resin composites. *Eur. J. Oral Sci.* **2003**, *111*, 536–542. [CrossRef] [PubMed]
41. Zorzin, J.; Maier, E.; Harre, S.; Fey, T.; Belli, R.; Lohbauer, U.; Petschelt, A.; Taschner, M. Bulk-fill resin composites: Polymerization properties and extended light curing. *Dent. Mater.* **2015**, *31*, 293–301. [CrossRef] [PubMed]
42. Baroudi, K.; Saleh, A.M.; Silikas, N.; Watts, D.C. Shrinkage behaviour of flowable resin-composites related to conversion and filler-fraction. *J. Dent.* **2007**, *35*, 651–655. [CrossRef] [PubMed]
43. Sideridou, I.D.; Karabela, M.M. Effect of the amount of 3-methacryloxypropyltrimethoxysilane coupling agent on physical properties of dental resin nanocomposites. *Dent. Mater.* **2009**, *25*, 1315–1324. [CrossRef] [PubMed]
44. Asmussen, E.; Peutzfeldt, A. Influence of UEDMA, BisGMA and TEGDMA on selected mechanical properties of experimental resin composites. *Dent. Mater.* **1998**, *14*, 51–56. [CrossRef]

45. Khatri, C.A.; Stansbury, J.W.; Schultheisz, C.R.; Antonucci, J.M. Synthesis characterization and evaluation of urethane derivates of Bis-GMA. *Dent. Mater.* **2003**, *19*, 584–588. [CrossRef]
46. Uzunova, Y.; Lukanov, L.; Filipov, I.; Vladimirov, S. High-performance liquid chromatographic determination of unreacted monomers and other residues contained in dental composites. *J. Biomech. Biophys. Methods* **2008**, *70*, 883–888. [CrossRef] [PubMed]
47. Caughmann, W.F.; Caughmann, G.B.; Shiflett, R.; Rueggeberg, F.; Schuster, G. Correlation of cytotoxicity, filler loading and curing time of dental composites. *Biomaterials* **1991**, *12*, 737–740. [CrossRef]
48. Tanaka, K.; Taira, M.; Shintani, H.; Wakasa, K.; Yamaki, M. Residual monomers (TEGDMA and Bis-GMA) of a set visible-light-cured dental composite resin when immersed in water. *J. Oral Rehabil.* **1991**, *18*, 353–362. [CrossRef] [PubMed]
49. Cebe, M.A.; Cebe, F.; Cengiz, M.F.; Cetin, A.R.; Arpag, O.F.; Ozturk, B. Elution of monomer from different bulk fill dental composite resin. *Dent. Mater.* **2015**, *31*, 141–149. [CrossRef] [PubMed]
50. Łagocka, R.; Jakubowska, K.; Chlubek, D.; Buczkowska-Radlińska, J. Elution study of unreacted TEGDMA from bulk-fill composite (SDR™ Dentsply) using HPLC. *Adv. Med. Sci.* **2015**, *60*, 191–198. [CrossRef] [PubMed]
51. Ortengren, U.; Wellendorf, H.; Karlsson, S.; Ruyter, IE. Water sorption and solubility of dental composites and identification of monomers released in an aqueous environment. *J. Oral Rehabil.* **2001**, *28*, 1106–1115. [CrossRef] [PubMed]
52. Braden, M.; Clarke, R.L. Water absorption characteristics of dental microfine composite filling materials. I. Proprietary materials. *Biomaterials* **1984**, *5*, 369–372. [CrossRef]
53. Braden, M.; Davy, K.W.M. Water absorption characteristics of some unfilled resins. *Biomaterials* **1986**, *7*, 474–475. [CrossRef]
54. Sideridou, I.; Tserki, V.; Papanastasiou, G. Effect of chemical structure on degree of conversion in light-cured dimethacrylate-based dental resins. *Biomaterials* **2002**, *23*, 1819–1829. [CrossRef]
55. Karabela, M.M.; Sideridou, I.D. Effect of the structure of silane coupling agent on sorption characteristics of solvents by dental resin-nanocomposites. *Dent. Mater.* **2008**, *24*, 1631–1639. [CrossRef] [PubMed]
56. Alshali, R.Z.; Salim, N.A.; Sung, R.; Satterthwaite, J.D.; Silikas, N. Qualitative and quantitative characterization of monomers of uncured bulk-fill and conventional resin-composites using liquid chromatography/mass spectrometry. *Dent. Mater.* **2015**, *31*, 711–720. [CrossRef] [PubMed]
57. Smith, M.B.; March, J. *March's Advanced Organic Chemistry, Reactions, Mechanisms and Structure*, 6th ed.; John Wiley & Sons: Hoboken, NJ, USA, 2007.
58. Durner, J.; Schrickel, K.; Watts, D.C.; Ilie, N. Determination of homologous distributions of BisEMAdimethacrylates in bulk-fill resin-composites by GC-MS. *Dent. Mater.* **2015**, *31*, 473–480. [CrossRef] [PubMed]
59. Marquardt, D.W. An algorithm for least-squares estimation of nonlinear parameters. *J. Soc. Ind. Appl. Math.* **1963**, *11*, 431–441. [CrossRef]
60. Santini, A.; Miletic, V.; Koutsaki, D. Degree of conversion of three fissure sealants cured by different light curing units using micro-Raman spectroscopy. *J. Dent. Sci.* **2012**, *7*, 26–32. [CrossRef]
61. Ogliari, F.A.; Ely, C.; Zanchi, C.H.; Fortes, C.B.B.; Samuel, S.M.W.; Demarco, F.F.; Petzhold, C.L.; Piva, E. Influence of chain extender length of aromatic dimethacrylates on polymer network development. *Dent. Mater.* **2008**, *20*, 165–171. [CrossRef] [PubMed]

© 2016 by the authors. Licensee MDPI, Basel, Switzerland. This article is an open access article distributed under the terms and conditions of the Creative Commons Attribution (CC BY) license (http://creativecommons.org/licenses/by/4.0/).

Review

The Role of Transient Receptor Potential (TRP) Channels in the Transduction of Dental Pain

Mohammad Zakir Hossain [1,*], Marina Mohd Bakri [2], Farhana Yahya [2], Hiroshi Ando [3], Shumpei Unno [1] and Junichi Kitagawa [1]

1. Department of Oral Physiology, School of Dentistry, Matsumoto Dental University, 1780 Gobara Hirooka, Shiojiri, Nagano 399-0781, Japan; shumpei.unno@mdu.ac.jp (S.U.); junichi.kitagawa@mdu.ac.jp (J.K.)
2. Department of Oral and Craniofacial Sciences, Faculty of Dentistry, University of Malaya, Kuala Lumpur 50603, Malaysia; marinab@um.edu.my (M.M.B.); farhanayahya@um.edu.my (F.Y.)
3. Department of Biology, School of Dentistry, Matsumoto Dental University, 1780 Gobara, Hirooka, Shiojiri, Nagano 399-0781, Japan; hiroshi.ando@mdu.ac.jp
* Correspondence: mohammad.zakir.hossain@mdu.ac.jp; Tel./Fax: +81-263-51-2053

Received: 26 December 2018; Accepted: 24 January 2019; Published: 27 January 2019

Abstract: Dental pain is a common health problem that negatively impacts the activities of daily living. Dentine hypersensitivity and pulpitis-associated pain are among the most common types of dental pain. Patients with these conditions feel pain upon exposure of the affected tooth to various external stimuli. However, the molecular mechanisms underlying dental pain, especially the transduction of external stimuli to electrical signals in the nerve, remain unclear. Numerous ion channels and receptors localized in the dental primary afferent neurons (DPAs) and odontoblasts have been implicated in the transduction of dental pain, and functional expression of various polymodal transient receptor potential (TRP) channels has been detected in DPAs and odontoblasts. External stimuli-induced dentinal tubular fluid movement can activate TRP channels on DPAs and odontoblasts. The odontoblasts can in turn activate the DPAs by paracrine signaling through ATP and glutamate release. In pulpitis, inflammatory mediators may sensitize the DPAs. They could also induce post-translational modifications of TRP channels, increase trafficking of these channels to nerve terminals, and increase the sensitivity of these channels to stimuli. Additionally, in caries-induced pulpitis, bacterial products can directly activate TRP channels on DPAs. In this review, we provide an overview of the TRP channels expressed in the various tooth structures, and we discuss their involvement in the development of dental pain.

Keywords: dental pain; dentine hypersensitivity; pulpitis; TRP channels; dental primary afferent neurons; odontoblasts; transduction mechanism

1. Introduction

Dental pain or odontogenic pain is the pain that initiates from the teeth or their supporting structures. The most common cause of dental pain is dental caries or tooth decay, the worldwide prevalence of which is very high. It was reported that in 2010, dental caries in permanent teeth remained the most prevalent global health problem, affecting 2.4 billion people, and dental caries in deciduous teeth constituted the tenth most prevalent health condition, affecting 621 million children worldwide [1]. Untreated dental caries lead to pulpitis (inflammation of the dental pulp) [2–6]. Typically, pulpitis is caused by invasion of the commensal oral microorganisms into the pulp due to caries [2,3]. Irritation of the dental pulp by mechanical, chemical, thermal or electrical stimuli may also cause pulpal inflammation [2–6]. Other causes of pulpitis include trauma, cracks on the tooth and periodontal infections [4,6]. Symptomatic pulpitis can be an extremely painful condition and is one of the main reasons that patients seek dental treatment [4,6,7]. It is often associated with intense lingering pain to thermal stimuli. The pain can be spontaneous, diffuse or referred [4,6,7].

Dentine hypersensitivity is another common odontogenic pain condition, the prevalence of which varies widely, ranging from 3–98% [8–14]. It is characterized by short, sharp pain arising from exposed dentine in response to stimuli—typically, thermal, evaporative, tactile, osmotic or chemical—and which cannot be ascribed to any other form of dental defect or disease [13–15]. The dentine can be exposed by chemical erosion, mechanical abrasion/attrition of enamel, and by loss of cementum following gingival recession [6,13–15]. The modern lifestyle increases the consumption of acidic foods and drinks that can result in significant tooth wear and exposure of dentine on any aspect of the tooth surface [13–17]. Dentine hypersensitivity is a special condition where dental pain arises in response to non-noxious stimuli on the exposed dentine that normally does not elicit pain in healthy teeth [6,13–15]. Even light tactile stimuli (weak air puff or water spray directed to the exposed dentine), which may only produce light touch sensation on the oral mucosa or skin, provokes abrupt intense pain [6,13–15]. There are three widely-held theories on the pathogenesis of this type of pain: (1) Dentinal fluid hydrodynamic theory, in which it has been proposed that external stimuli cause movement of the dentinal fluid that ultimately excites the nerve fibers in the pulp and initiates pain; (2) Neural theory, in which it has been suggested that the nerve fibers that project into the dentinal tubules directly respond to external stimuli; (3) Odontoblast transducer theory, in which odontoblasts themselves have been suggested as pain transducers [13–15,18–20]. Among these, the dentinal fluid hydrodynamic theory is the most widely accepted, although not without controversy [16,18–27]. In one study, water application onto human dentine did not evoke pain; however, it caused dentinal tubular fluid movement in vitro [24]. Another study demonstrated a lack of correlation between dentinal fluid flow and pain in patients after cold stimulation of the exposed dentine, suggesting that cold-sensitive receptors might also be involved in pain transduction [26]. Recently, based on their findings, Shibukawa et al. proposed the "odontoblast receptor hydrodynamic theory", in which they posit that the movement of the dentinal fluid mechanically stimulates mechanoreceptors in both odontoblasts and the nerve fibers in the pulp [27]. Odontoblasts, movement of dentinal tubular fluid and nerves in the dental pulp may all be involved in dentine hypersensitivity; however, the underlying mechanisms are not yet fully understood [15].

In addition to painful pulpitis and dentine hypersensitivity, pain may also occur when intense thermal stimuli are applied on the surface of a normal intact tooth [28–31]. In the clinic, thermal pulp testing (applying heat or cold onto the tooth surface) is routinely used to test the vitality of the dental pulp of a tooth [28,29]. Thermal pulp testing induces a localized sharp pain in the tooth being tested if the tooth is vital [28–31]. Drinking/eating of very cold or hot drink/food may also induce dental pain [28,30,31].

Although dental pain is a common health problem, its molecular and cellular pathophysiology, particularly, how the external stimuli (e.g., physical, chemical or thermal) are transduced into electrical signals in the nerve that are ultimately perceived as pain, remain unclear. Various ion channels (e.g., voltage-sensitive Na^+, mechanosensitive K^+, L-type voltage-dependent Ca^{2+} channels) have been reported to be expressed in the dental pulpal afferents and in the odontoblasts, and may play an important role in the transduction process [32–35]. In recent decades, TRP channels have also been detected in odontoblasts and dental primary afferent neurons (DPAs), and have been implicated in the transduction of external stimuli into pain signals in the tooth [15,27,35,36]. This review focuses on the involvement of TRP channels in the transduction of dental pain.

2. Dental Innervation

Teeth are highly vascularized and richly innervated structures [6,37,38]. The nerves of a tooth are mainly confined to the dental pulp [37,38]. The sensory stimuli-detecting nerve networks in the dental pulp differ in many ways from those in the skin or oral mucosa [34,39]. The various types of stimuli (e.g., mechanical, chemical or thermal) to the dental pulp or exposed dentine generally elicit only pain sensation, while these stimuli applied to the skin or oral mucosa produce distinct types of sensation [20,34,35,39,40]. The majority of the axons of the nerves that innervate a tooth enter the dental pulp through the apex [37,38,41,42]. Electron microscopic analysis of the nerve fibers within the dental pulp show that around 70–90% of the axons are unmyelinated (C-fibers) [40,41]. The remaining axons are mostly thinly myelinated (Aδ-fibers), and a very small amount are thickly myelinated (Aβ-fibers) [37,38,40,41]. However, animal studies in which retrograde labeling techniques were used to evaluate the size and histochemical composition of pulpal sensory neurons within the trigeminal ganglion (TG) suggest that the parent axons of the dental pulpal nerves are largely myelinated (Figure 1) [42–44].

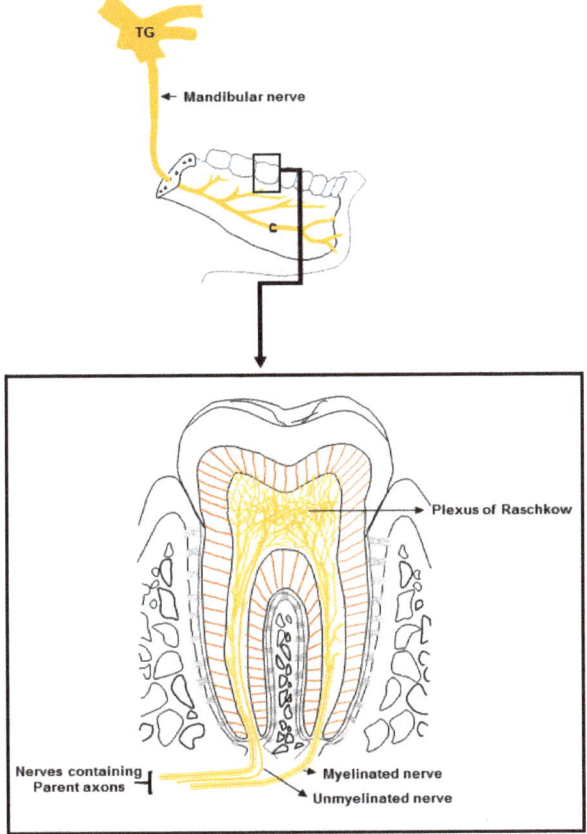

Figure 1. Innervation of a tooth. The cell bodies of the dental primary afferent neurons are located in the trigeminal ganglion (TG). The axons of the afferent neurons project to the dental pulp through the two major branches of the trigeminal nerve, namely, the mandibular (shown in the figure) and maxillary nerves. A large number of the parent axons of the afferent neurons before entering into the dental pulp are myelinated. After entering the dental pulp, they extend branches and gradually lose their myelin sheath. In the crown area, the axons branch extensively to form the plexus of Raschkow. Many axons terminate very close to the odontoblasts and sub-odontoblastic cells, and some enter the dentinal tubules for a short distance into the inner part of the dentine.

It has been observed that the ratio of myelinated axons to unmyelinated axons is reduced in nerves closer to teeth, compared with more distant sites, indicating progressive loss of the myelin sheath as axons course toward the tooth [44,45]. Using electron microscopic analysis, a study reported that the percentage of unmyelinated axons was higher in the crown area compared with the root area in rat molars [46]. In that study, the unmyelinated axons of the dental pulp showed immunoreactivity to a marker of myelinated nerves, neurofilament (NF)-200, and the percentage of NF-200-positive unmyelinated axons was greater in the crown area than in the root area [46]. The conduction velocity in the nerve fibers outside the dental pulp was higher than in the nerve fibers located inside the pulp [47,48]. A study on human teeth also found that unmyelinated nerve fibers within the dental pulp show immunoreactivity to markers for myelinated nerves (neurofilament-200, neurofilament-52) [49]. Furthermore, studies have reported that the majority of parent axons before entering the dental pulp are thinly myelinated (Aδ), with a higher conduction velocity compared with nerve fibers inside the dental pulp [44,47–49]. These findings suggest that many parent axons in the dental pulp afferents are myelinated. Consistently, studies have reported that the cell bodies of the majority of the afferent fibers from the dental pulp that are located in the TG are medium-sized (Aδ-neurons), while a minority are large (Aβ-neurons) or small (C-neurons) [42,43]. After entering the dental pulp through the apex of the tooth, the parent axons of the afferent nerves extend branches and gradually lose their myelin sheath [43,46,49,50] (Figure 1). In the radicular pulp, the axons give off only a few branches, but in the coronal pulp, the axons branch extensively to form the plexus of Raschkow (Figure 1) [6,37,38,41]. The axons lose their myelin sheath mostly in the coronal pulp and emerge as free nerve endings (brush or fan-shaped) [49–51]. Many axons terminate very close to the odontoblasts, as well as sub- and peri-odontoblastic cells, and some enter the dentinal tubules and continue along odontoblast processes for a short distance, no further than 0.2 mm into the inner part of the dentine [51]. Some axons make endings in both the pulp and the dentinal tubule [51]. It has been reported that around 30–70% of odontoblast processes of a tooth are in close association with nerve endings in the inner part of the dentine [50]. The dental pulp also contains sparsely distributed sympathetic postganglionic efferent nerve fibers (unmyelinated), which mostly innervate the blood vessels, while parasympathetic nerve fibers have not been observed [40,41,52,53].

Cytochemically, the DPAs are distinct from the afferent neurons of the skin [34,39,43]. Generally, in the skin, the afferent neurons can be clearly divided into two major groups—peptidergic and non-peptidergic [54,55]. The peptidergic neurons express a variety of neuropeptides and signaling proteins, including calcitonin gene-related peptide (CGRP), substance P, nerve growth factor (NGF) receptor and tyrosine kinase A (TrkA) receptor, and these neurons are responsive to NGF. In comparison, the non-peptidergic neurons express isolectin B4 (IB4), glial cell line-derived neurotrophic factor (GDNF) family receptor alpha-1 (GFRα1) and receptors for other GDNF family members, and these neurons are responsive to GDNF [32,55–57]. The non-peptidergic neurons also express purinergic receptors (P2X3, a receptor for ATP) [58–60]. The pulpal afferents cannot be categorized into these two broad groups. They display the cytochemical features of both peptidergic and non-peptidergic neurons, which is rare for skin somatosensory afferents [61,62]. Although most pulpal afferents do not express IB4, they express other markers of non-peptidergic neurons such as GFRα1 and P2X3 [61,62], and a large group of pulpal afferents also express markers for peptidergic neurons such as CGRP, NGF receptor, substance P and TrkA receptors [63–65]. Another unique feature of dental pulp afferents is that they express markers for both mechanoreceptors and nociceptors [39,43,49,61,62]. It has been observed that pulpal nerve fibers express various cytochemical markers of mechanosensitive nerve fibers, such as neurofilament (NF) markers (e.g., NF-200, NF-52), parvalbumin, calbindin, epithelial sodium channels (ENaCs), acid-sensing ion channel 3 (ASIC3) and mechano-gated potassium channels [32,34,39,43,61,62,66,67]. They also express various cytochemical markers of nociceptive nerve fibers in the skin, such as CGRP, GFRα1, TrkA and substance P [39,43,61,68,69].

3. TRP Channels and Their Presence in Dental Tissues

TRPs are integral pore-forming membrane proteins that function primarily as non-selective ion channels [70–73]. They have a putative six-transmembrane-spanning protein domain with a pore region localized between transmembrane segments 5 and 6 [70–73]. They were discovered in the fruit fly (Drosophila) in studies of their phototransduction (light detection) mechanism [74–76]. Later, they were found in vertebrates, and to date, seven TRP subfamilies have been identified: TRPA (ankyrin), TRPC (canonical), TRPM (melastatin), TRPML (mucolipin), TRPP (polycystin), TRPN (Drosophila no mechanoreceptor potential C or NOMPC) and TRPV (vanilloid) [70–73,77]. Recently, the TRPY subfamily was reported in yeast [78]. TRPN and TRPY are absent in mammals [70–73,77]. The TRPV subfamily has six members (TRPV1–6), TRPA has one member (TRPA1), TRPC has seven members (TRPC1–7), TRPM has eight members (TRPM1–8), TRPML has three members (TRPML1–3), TRPP has three members (TRPP1–3), and TRPN has one member (TRPN1, found in fish) [70–73,77]. Among the 28 TRP channel genes that have been identified in mammals, 17 have been detected in the mouse TG at the mRNA level [79]. Many TRP channel genes have also been identified in the human TG [80].

To date, to our knowledge four members of the TRPV subfamily (TRPV1, TRPV2, TRPV3, TRPV4), four members of the TRPM subfamily (TRPM2, TRPM3, TRPM7, TRPM8), two members of the TRPC subfamily (TRPC1, TRPC6), and one member of the TRPA subfamily (TRPA1) have been detected in dental tissues (e.g., DPAs, odontoblasts).

3.1. TRPV

Among the six members of the TRPV subfamily, TRPV1–4 are weakly Ca^{2+}-selective cation channels [70–73]. They can be activated by thermal stimulation, and are therefore referred to as thermo–TRPs [70–73,81]. In vitro studies show that TRPV1–4 can be activated by temperatures ranging from ~34 (TRPV4) to ~52 °C (TRPV2) [81]. The temperature range for channels activation is not strict, and a thermal threshold is not the optimal parameter to describe thermo-TRPs, because, the sensitivity for thermal activation of these channels is substantially modified by cellular and environmental factors [82–84]. For example, TRPV1 (normal activation temperature ~43 °C) can be activated at much lower temperatures when the membrane depolarized than when it hyperpolarized [82,83]. TRPV5–6 are highly Ca^{2+}-selective and are not activated by heat. They play an important role in Ca^{2+} homeostasis [70–73].

TRPV1, which is also known as the capsaicin receptor, was the first member of the TRPV subfamily to be isolated [78]. It is activated by various exogenous and endogenous stimuli, such as capsaicin (found in hot chili peppers), acids (pH < 5.9), heat, inflammatory mediators (e.g., bradykinin, prostaglandins), NGF, anandamide (arachidonoyl ethanol amide), arachidonic acid metabolites (e.g., N-arachidonoyl-dopamine), lipoxygenase products (e.g., 12-hydroperoxyeicosatetraenoic acid), adenosine and ATP [85–91]. The threshold for activation of TRPV1 is dynamic [87,90,91]. For example, the threshold is decreased by inflammatory mediators, but after activation by capsaicin, there is a sustained refractory state (desensitization) [70–73,87,90,91]. Activation of TRPV1 has been shown to promote the release of neuropeptides such as substance P and CGRP [70–73,87,91]. TRPV1 is expressed predominantly in C-fibers and, to a lesser extent, in Aδ fibers [70–73,87,91]. In addition to sensory neurons, TRPV1 expression has been detected in various other tissues, including keratinocytes of the skin and oral mucosa, epithelium of the respiratory system, digestive tract, urinary bladder, cardiac muscle, vascular smooth muscle, and the endothelium of blood vessels [87,92]. TRPV1 has been reported to play an important role in thermal nociception [70–73,87,91]. The importance of TRPV1 in pain sensation has been demonstrated by studies in TRPV1 knockout mice [93]. It plays a role in many physiological functions, including satiety, olfaction, gastrointestinal motility, and energy homeostasis [72,87,92–95].

TRPV2 channels share 50% sequence identity with TRPV1 [70,73,81,96]. This channel has been reported to function as a thermoreceptor, mechanoreceptor and osmoreceptor [70,73,81,96,97]. TRPV2 is activated by noxious temperature (~52 °C) in vitro [97–99]. However, in vivo, TRPV2 knockout mice exhibit a thermal response similar to wild-type mice [100]. TRPV2 is expressed in the peripheral and central nervous systems [70,73,81,97,99,101–104]. In the spinal dorsal root ganglia (DRG) and in the trigeminal ganglia (TG), expression of TRPV2 was observed in the medium to large-diameter primary afferent neurons [97,99,102–104]. In culture, one-third of TRPV2-expressing DRG neurons are CGRP-positive, and activation of these neurons by a TRPV2 agonist results in CGRP release [102]. In the central nervous system, TRPV2 expression has been observed in a number of regions that are involved in osmoregulation and other autonomic functions [97,99,101].

TRPV3 can be activated by innocuous temperature (31–39 °C), various natural compounds (such as camphor and carvacrol), synthetic agents (such as 2-aminoethoxydiphenyl borate), and the endogenous ligand farnesyl pyrophosphate (FPP) [105–108]. TRPV3 expression in sensory neurons of the peripheral nervous system is not consistent across species, and it is minimally expressed in sensory neurons in the rodent; however, the channel is relatively abundantly expressed in skin keratinocytes [109–112]. TRPV3 is also expressed in the epithelial cells of the oral and nasal cavities, as well as the cornea, where it is reported to be involved in wound healing and thermo-sensation [111,112]. TRPV3 in keratinocytes is reported to be involved in nociception mediated by ATP [110]. Indeed, increased pain sensitivity is observed in transgenic mice overexpressing TRPV3 in skin keratinocytes [113].

TRPV4 was identified as an osmoreceptor [97,114–116]. Later, it was shown to be a polymodal receptor that can be activated by various stimuli, including innocuous warm temperature (27–35 °C) and mechanical stimuli (membrane stretch and shear stress) [97,117–121]. This channel can be activated by arachidonic acid metabolites, synthetic chemical agents such as 4α-phorbol 12,13-didecanoate, GSK1016790A, and the plant extract bisandrographolide A [70,72,73,122]. TRPV4 has also been implicated in nociception, itch, inflammatory pain, neuropathic pain and visceral pain [70,72,73,123–127]. Expression of TRPV4 is observed in DRG neurons [123,126,127] and TG [121,128–130] neurons. It is also expressed in satellite glial cells (cells surrounding the neurons) in the DRG that regulate neuronal excitability in pain and inflammatory conditions [131]. Inflammatory mediators (e.g., prostaglandins and proteases) may sensitize TRPV4, resulting in hyperalgesia [123,132]. Furthermore, escape latency in the hot plate test is increased following tissue injury and inflammation in TRPV4 knockout mice, suggesting involvement of TRPV4 in thermal hyperalgesia [124].

3.1.1. TRPV in DPAs

TRPV channels have been detected in the cell bodies of DPAs located in the TG [133–138]. Among the six members of the TRPV subfamily, TRPV1 has been the most extensively studied [133–141]. In animals, the percentage of TRPV1-expressing dental pulp TG afferent neurons varies among the published papers from 8% to 85%, and TRPV1 expression is observed in small, medium and large neurons [133–138]. In 2001, Ichikawa & Sugimoto reported that approximately 8% of rat DPAs express TRPV1, and that 20% of these co-express the neuropeptide CGRP [133]. They also compared TRPV1-immunoreactive afferent neurons in the dental pulp and the facial skin [133]. The percentage of TRPV1-immunoreactive neurons was lower in the dental pulp (17%) than in the facial skin (26%) [133]. The same year, Chaudhary et al. reported TRPV1 mRNA expression in the cell bodies of DPAs in the TG, and 65% of these cells were excited by capsaicin. The capsaicin-evoked excitation was attenuated by a TRPV1 antagonist (capsazepine) [134]. Similar to Ichikawa & Sugimoto [133], Chung et al. reported that approximately 10% of the dental pulpal afferent neurons expressed TRPV1, and that the majority of these were small to medium neurons [135]. They also observed that TRPV1 expression was upregulated by application of lipopolysaccharides (a product of Gram-negative bacteria) into the dentine [135]. Stenholm et al. reported that 21–34% of dental pulpal afferent neurons were TRPV1-immunoreactive, slightly higher than among gingival primary afferent neurons (21–26%) [136]. Gibbs et al. found that approximately 17% of the afferent neurons in the TG from the dental pulp expressed TRPV1, lower than the percentage of TRPV1-immunoreactive afferent neurons from the periodontal ligament (26%) in the rat [104]. They also found that 70% of the TRPV1-immunoreactive neurons were myelinated, and that a majority (82%) of these were medium to large-diameter neurons [104]. Furthermore, 60% of the TRPV1-immunoreactive neurons co-expressed CGRP [104]. A high percentage of TRPV1-immunoreactive DPAs was observed by Kim et al. [137] and Park et al. [138]. Kim et al. showed that 85% of the retrograde-labeled rat DPAs were immunoreactive for TRPV1, and that 71% of them produced inward cationic currents in response to application of a TRPV1 agonist (capsaicin) [137]. Park et al. observed that 45% of the DPAs expressed TRPV1 [138]. They also showed that application of a TRPV1 agonist increased the intracellular Ca^{2+} concentrations and produced inward currents in these neurons. Temperature changes (>42 °C) also increased the intracellular Ca^{2+} concentrations and produced inward currents [138].

Expression of TRPV1 has been reported in the nerve fibers within the dental pulp in animal and human studies [137,139–141]. Kim et al. [137] found TRPV1-immunoreactive nerve fibers within the rat dental pulp; however, Gibbs et al. did not observe TRPV1-immunoreactive nerve fibers within the dental pulp of rat molars [104]. TRPV1 immunoreactivity has also been detected in the dental pulp nerve fibers of the human permanent teeth [139–141], and immunoreactivity is significantly increased in carious teeth compared with non-carious teeth [141]. Interestingly, TRPV1 expression tended to be increased in painful carious teeth compared with non-painful carious teeth, although the difference was not significant [141].

TRPV2 is expressed in the nerves within the dental pulp and in the cell bodies of the DPAs located in the TG [103,104,136]. In 2000, Ichikawa & Sugimoto reported TRPV2-immunoreactive nerve fibers within the rat dental pulp [103]. In the root area, TRPV2 immunoreactivity was observed in the nerve bundles, and in the crown area, the TRPV2-immunoreactive nerve fibers were ramified and extended to the base of the odontoblastic cell layer [103]. These investigators also observed that approximately 37% of the dental primary afferent neurons retrogradely traced in the TG expressed TRPV2, and that most of these were medium to large neurons [103]. In addition, 45% of these neurons co-expressed CGRP, and 41% co-expressed parvalbumin (a marker of proprioceptors) [103]. Furthermore, the percentage of TRPV2-immunoreactive afferent neurons (37%) was greater than the percentage of TRPV2-immunoreactive facial skin afferent neurons (9%) [103]. Stenholm et al. reported that 32–51% of DPAs expressed TRPV2, somewhat higher than the percentage of TRPV2-expressing gingival afferent neurons (41%) [136]. Gibbs et al. reported that a higher percentage of rat DPAs expressed TRPV2 compared with TRPV1; 50% were immunoreactive for TRPV2, while 17% were immunoreactive for TRPV1 [104]. They also observed that TRPV2 expression was higher among DPAs (50%) compared with primary afferent neurons from the periodontal ligament (41%) [104]. Approximately 83% of the TRPV2-immunoreactive neurons were myelinated, and the majority of these were medium to large neurons [104]. Around 33% of the TRPV2-immunoreactive neurons co-expressed CGRP [104].

TRPV3 and TRPV4 are expressed in the TG on afferent neurons from the facial skin and temporomandibular joint [105,129,130]. TRPV4 expression is also observed on the nerves of human dental pulp, and expression is upregulated by chronic inflammation of the pulp [142].

3.1.2. TRPV in Odontoblasts

In 2005, Okumura et al. reported expression of TRPV1 on odontoblasts in neonatal rat teeth [143]. They also showed a capsaicin-induced inward current in the odontoblasts using the patch clamp technique that was inhibited by capsaizepine [143]. However, in acutely isolated adult rat odontoblasts, Yeon et al. did not observe expression of TRPV1 by immunohistochemistry, and intracellular Ca^{2+} concentration in the odontoblasts was not increased by application of heat (42 °C) or a TRPV1 agonist (capsaicin) [144]. They also did not detect TRPV1 or TRPV2 mRNA in the odontoblasts [144]. In contrast, Tsumura et al. reported expression of TRPV1 in adult rat odontoblasts, on the cell membrane and on their processes [145]. Application of an agonist (capsaicin/resiniferatoxin/low pH solution) increased the intracellular Ca^{2+} level, which was inhibited by application of antagonists [145]. They also reported that the TRPV1 channels in odontoblasts are functionally coupled with cannabinoid receptor 1 and Na^+–Ca^{2+} exchangers, and that this coupling is mediated by cyclic adenosine monophosphate (cAMP) [145]. TRPV1 has also been detected on the odontoblasts of extracted healthy caries-free human premolar teeth [139]. mRNAs for TRPV1, TRPV2, TRPV3 and TRPV4 were detected in acutely isolated cultured odontoblasts by Son et al. [146].

They also showed that when odontoblasts were stimulated by heat above 32 °C, the intracellular Ca^{2+} concentration increased, suggesting functional involvement of TRPV1, TRPV2 and TRPV3 in the transduction of heat stimuli [146]. In addition, the intracellular Ca^{2+} concentration was increased when hypotonic solution was applied, suggesting involvement of TRPV4 channels [146]. Sato et al. reported expression of TRPV2 and TRPV4 proteins in rat odontoblasts [147]. They also showed that extracellular hypotonic solution-induced membrane stretching in cultured mouse odontoblast lineage cells produces inward currents and increases intracellular Ca^{2+} concentration, which can be inhibited by application of TRPV1, TRPV2 and TRPV4 channel antagonists, suggesting that these channels function as mechanoreceptors in odontoblasts [147]. TRPV1 and TRPV4 expression has been reported by immunohistochemistry, PCR and western blot analysis in human odontoblast-like cells derived from the dental pulp of a permanent tooth [148,149]. Application of agonists to these channels increases the intra-odontoblast Ca^{2+} concentration. Hypotonic solution-induced membrane stretch also increases the intra-odontoblastic Ca^{2+} concentration, which can be reduced by channel antagonists [148,149]. Brief (10 min) application of the pro-inflammatory cytokine tumor necrosis factor (TNF)-α enhances the response to chemical agonists and hypotonic solution-induced membrane stretch, suggesting that inflammation increases TRPV4 channel activation [148,149]. TRPV4 immunoreactivity has been detected in odontoblasts and their process in extracted healthy caries-free immature human third molar teeth [149]. Wen et al. reported expression of TRPV1, TRPV2 and TRPV3 in native human odontoblasts as well as cultured odontoblast-like cells from healthy human third molars by immunohistochemistry, quantitative real-time polymerase chain reaction (qRT-PCR), western blotting and immunoelectron microscopy [150]. Immunoelectron microscopy revealed expression of TRPV1, TRPV2 and TRPV3 in odontoblastic processes, mitochondria and endoplasmic reticulum [150]. Egbuniwe et al. reported expression of TRPV1 and TRPV4 mRNAs in odontoblast-like cells derived from human immortalized dental pulp cells. They observed that the application of a TRPV4 agonist increased the intra-odontoblastic Ca^{2+} concentration. They also observed that the agonist caused release of ATP from the odontoblasts, which was blocked by pretreatment with the antagonist [151]. In odontoblast-like cells obtained from cultured dental pulp cells from newborn rats, mechanical stimulation increases the intracellular Ca^{2+} concentration, and this effect can be attenuated by application of TRPV1, TRPV2 and TRPV4 antagonists, suggesting that these channels function as mechanosensors in these cells [27].

3.2. TRPM

TRPM8 has been detected in DPAs and pulpal fibroblasts. TRPM3 and TRPM7 are expressed in odontoblasts, while TRPM2 is expressed in pulpal fibroblasts. TRPM2 is activated by cellular stress and participates in various cellular functions, including cytokine production, cell motility and cell death [152,153]. It can be activated by cytosolic adenosine diphosphate ribose (ADPR), oxidative stress and moderate heat in various cell types [152–155]. TRPM2 expression has also been detected in afferent sensory neurons in the DRG and TG [79]. This channel is implicated in pathogenic pain [156].

TRPM3 is expressed in non-neural (e.g., epithelium of the kidney, pancreatic β cells) and neural tissues, including sensory neurons in the DRG and TG [157–161]. TRPM3 can be activated by extracellular hypo-osmolarity, noxious heat (>30 °C) and chemical compounds, such as pregnenolone sulfate (endogenous excitatory neurosteroid) and nifedipine (a drug used for hypertension) [157,160,161]. TRPM3 knockout mice exhibit significant attenuation of thermal hyperalgesia under inflammatory conditions [160].

Similar to other TRP channels, TRPM7 is expressed in a wide variety of tissues, including the brain, heart and hematopoietic tissues [162–166]. TRPM7 has been implicated in multiple cellular and physiological functions, including embryonic development, Mg^{2+} homeostasis, cell growth and viability, synaptic transmission, and neuronal degeneration [163–166].

TRPM8 is activated by innocuous cooling (~26−15 °C) as well as by noxious cooling (<15 °C) and by a number of cooling agents, such as menthol and icilin [167,168]. TRPM8 is expressed in small-diameter sensory neurons in the TG and DRG [169–172]. TRPM8 mRNA is more abundantly expressed in the sensory afferents of the TG (especially those that innervate the tongue) compared with the DRG [169–172]. While TRPM8 knockout mice do not lack sensation to noxious cold, they exhibit an attenuation of avoidance behavior to moderately cold temperatures [173,174]. These mice also display reduced cold hypersensitivity following nerve injury or complete Freund's adjuvant (CFA)-induced inflammation [175].

3.2.1. TRPM in DPAs

TRPM8 is expressed in rat DPAs in the TG [137,138,176]. By immunohistochemistry, Park et al. observed that 13% of rat DPAs expressed TRPM8 [138]. Application of menthol (a TRPM8 agonist) or exposure to temperatures less than 25 °C increase intracellular Ca^{2+} concentrations and evoke inward currents in the TRPM8-expressing neurons [138]. In one study, TRPM8 mRNA expression was detected in 58% of rat DPAs [137]. However, Michot et al. detected TRPM8 expression in only 5.7% of mouse DPAs, similar to that in afferent neurons innervating the facial skin and the buccal mucosa [176].

3.2.2. TRPM in Odontoblasts

Expression of TRPM3, TRPM8 and TRPM7 in odontoblasts has been reported in numerous studies [146,148,177–182]. Son et al. detected mRNA expression of osmo-sensitive TRPM3 channels in acutely isolated cultured odontoblasts from neonatal mice [146]. They also suggested that TRPM3 in these cells are involved in the transduction of osmotic stimuli, based on their finding of increased intra-odontoblastic Ca^{2+} concentration following exposure to hypotonic solution [146]. However, these investigators did not observe functional expression of TRPM8 [146]. TRPM8 expression was also not observed by Yeon et al. in acutely isolated adult rat odontoblasts at the mRNA level [144]. They showed that application of cold stimuli (12 °C) or menthol (TRPM8 agonist) did not increase the intra-odontoblastic Ca^{2+} concentration [144]. In contrast, Tsumura et al. reported functional expression of TRPM8 in acutely isolated adult rat odontoblasts [177]. TRPM8 immunoreactivity was observed in the odontoblasts and their processes [177]. Chemical agonists (e.g., menthol, WS3, WS12) and temperature changes (22 ± 1 °C) also increased the intra-odontoblastic Ca^{2+} levels, and these changes could be reduced by an antagonist [177]. TRPM8 expression was also reported in odontoblast-like cells from the dental pulp of a human permanent tooth at both the mRNA (using PCR) and protein levels (using western blotting and immunohistochemistry) [148]. Chemical agonism also increased the intra-odontoblastic Ca^{2+} concentration in an antagonist-sensitive manner [148]. TRPM8 has also been detected in freshly isolated human odontoblasts at the mRNA and protein levels as well as in odontoblasts and their processes in extracted healthy caries-free human permanent teeth [178].

By single-cell RT-PCR and immunocytochemical analysis, TRPM7 was detected in acutely isolated rat odontoblasts [179,180]. Functionality of TRPM7 as a mechanoreceptor in acutely isolated rat odontoblasts was reported by Won et al. [180]. In their study, hypotonic solution-induced membrane stretch or chemical agonism for TRPM7 caused a transient increase in intra-odontoblastic Ca^{2+} concentration, which was blocked by a non-selective mechanosensitive channel blocker or a TRPM7 blocker [180]. TRPM7 in odontoblasts has also been reported to play an important role in dentine mineralization by regulating intracellular Mg^{2+} and alkaline phosphatase activity [181,182].

3.3. TRPA

The only member of the TRPA subfamily, TRPA1 is generally involved in pain, thermal and chemical sensation [70,73,91,97]. It is expressed in many tissues, including the brain, heart, small intestine, lung, bladder and joints [183–185]. It is highly expressed in small and medium neurons in the DRG and TG [91,183,186–188]. TRPA1 channels are co-localized with TRPV1, CGRP, substance P and bradykinin receptors [186,189–191]. TRPA1 is activated by noxious cold (<8 °C) and by exogenous chemical agents, such as allyl isothiocyanate (AITC, present in mustard oil and wasabi), cinnamaldehyde (found in cinnamon), allicin (found in garlic) and acrolein (present in diesel exhaust) [191–193].

3.3.1. TRPA in DPAs

In rat DPAs (labeled using retrograde dye applied to the dental pulp), TRPA1 was observed in 11% of neurons [138]. Chemical stimulation by icilin (a TRPA1 and TRPM8 agonist) and cold stimulation (<17 °C) increase intracellular Ca^{2+} concentrations and produce inward currents in the TRPA1-expressing neurons [138]. Kim et al. [137] reported that 34% of rat DPAs expressed TRPA1 mRNA, and Hermanstyne et al. [34] observed TRPA1 mRNA expression in 64% of retrogradely-labeled DPAs in the rat. TRPA1 protein in the rat TG was increased following experimental exposure of the dental pulp, implicating the protein in hyperalgesia and allodynia following tooth injury [194]. In the mouse, TRPA1 was detected in 18.9% of DPAs, lower than in neurons innervating the buccal mucosa (43%) or the facial skin (24.6%) [176]. In the same study, noxious cold stimulation of the tooth increased the expression of c-Fos (a marker of neuronal excitation) in the brainstem trigeminal nucleus; however, this increase was not blocked by systemic administration of a TRPA1 antagonist [176].

TRPA1 is also expressed in a large number of axons that branch extensively in the peripheral pulp [195]. By electron microscopy, TRPA1 immunoreactivity was observed near the plasma membrane of unmyelinated axons. TRPA1 was also co-localized with a sodium channel, Nav1.8. Furthermore, TRPA1 expression on myelinated nerve fibers was upregulated in teeth with signs of pulpitis [195].

3.3.2. TRPA in Odontoblasts

Tsumura et al. reported expression of TRPA1 on the cell membrane of acutely isolated adult rat odontoblasts and their processes [177]. The agonist AITC increased intracellular Ca^{2+} levels in an antagonist-sensitive manner [177]. Temperature changes (13 ± 1 °C) also similarly increased the intracellular Ca^{2+} level in a TRPA1 antagonist-sensitive manner. In addition, TRPA1 antagonists reduce the hypotonic solution-induced increase in intracellular Ca^{2+} level, suggesting that TRPA1 functions as a mechanoreceptor in odontoblasts [177]. TRPA1 was found to be expressed at both the mRNA and protein levels in human odontoblast-like cells derived from the dental pulp of a permanent tooth [149,196]. A TRPA1 agonist also increased the intra-odontoblast Ca^{2+} concentration in an antagonist-sensitive manner [149,196]. Gene silencing experiments showed that TRPA1 in odontoblasts functions as a receptor for noxious cold temperature. Hypotonic solution-induced membrane stretch and chemical agonism also increased the intra-odontoblastic Ca^{2+} concentration [149,196]. A brief (10 min) application of TNF-α enhanced these responses, suggesting that inflammatory conditions increase TRPA1 activity [149,196]. Long-term (24 h) application of TNF-α also enhanced TRPA1 activity and upregulated its mRNA and protein levels in odontoblast-like cells [149]. Furthermore, TRPA1 immunoreactivity in odontoblasts and their process is increased in carious teeth compared with caries-free human third molar teeth [149,196]. Egbuniwe et al. reported expression of TRPA1 mRNA in odontoblast-like cells from human immortalized dental pulp cells [151]. A TRPA1 agonist increased the intra-odontoblastic Ca^{2+} concentration and caused the release of ATP from these cells, and this effect was blocked by antagonist pretreatment [151]. TRPA1 is also detected on odontoblasts and their processes in extracted healthy caries-free human premolar teeth [195]. However, Tazawa et al. did not observe TRPA1 immunoreactivity on odontoblasts or their processes in extracted healthy caries-free human permanent teeth [178]. In acutely isolated adult rat odontoblasts, Yeon et al. also did not observe TRPA1 mRNA expression [144]. A study using odontoblast-like cells derived from cultured dental pulp cells of newborn rats found that mechanical stimulation increased the intra-odontoblastic Ca^{2+} concentration, which was attenuated by a TRPA1 antagonist, suggesting involvement of this channel in the transduction of mechanical stimuli [27].

3.4. TRPC

TRPC1 and TRPC6 have been reported to be expressed in odontoblasts [179,197,198]. TRPC1 plays an important role in store-operated Ca^{2+} entry (SOCE) in a variety of tissues and cell types [199,200]. TRPC1 is highly expressed in the hippocampus, amygdala, cerebellum, substantia nigra and inferior colliculus [201–203], and participates in important neuronal processes related to synaptic transmission and plasticity [203]. TRPC1 expression is observed on large myelinated sensory neurons in the DRG [204], and it and TRPC6 play critical roles in mechanosensation, hearing [205–208] and cell membrane stretch [206,207]. TRPC1 and TRPC6 are co-expressed with TRPV4 in the DRG, where they are implicated in mechanical hypersensitivity in inflammatory and neuropathic pain conditions [208].

TRPC in Odontoblasts

TRPC1 and TRPC6 mRNA expression has been reported in acutely isolated rat odontoblasts, where they have been suggested to play a role as mechanoreceptors [179]. TRPC1 and TRPC6 immunoreactivity is also detected on odontoblasts in human permanent teeth. TRPC6 protein expression increases with time during odontoblast differentiation from pulp tissue. Furthermore, differentiation is inhibited by downregulation of TRPC1 and TRPC6 expression, indicating involvement of these channels in the odontogenic differentiation of human dental pulp cells [197,198].

3.5. TRP Channels in Pulpal Fibroblasts and Blood Vessels

Expression of TRPV1 mRNA and protein has been reported in human dental pulp fibroblasts cultured from the pulp of third molar teeth [209], and application of the agonist capsaicin induces production of IL-6 (an inflammatory cytokines) in these cells, which is dose-dependently inhibited by the TRPV1 antagonist capsazepine, indicating involvement of this channel in fibroblast-mediated pulpal inflammation [209]. TRPV1 expression was also observed in pulpal fibroblasts of healthy caries-free human permanent teeth [139]. Karim et al. reported mRNA and protein expression of TRPA1 and TRPM8 in human dental pulp fibroblasts [210]. Application of the TRPM8 agonist menthol or cool temperature ($22 \pm 3\ °C$) in combination with the TRPA1 agonist cinnamaldehyde or noxious cold temperature ($12 \pm 2\ °C$) increased the intra-fibroblastic Ca^{2+} concentration, suggesting that TRPA1 and TRPM8 in pulp fibroblasts are involved in sensing environmental stimuli [210]. TRPM2 has also been detected in the fibroblasts of human dental pulp, and its expression is upregulated in teeth with signs of irreversible pulpitis [211].

4. Involvement of TRP Channels in the Transduction of Dental Pain

TRP channels in odontoblasts and DPAs are suggested to function as polymodal receptors involved in the transduction of various external stimuli. TRPV1, TRPV2, TRPV4, TRPM7 and TRPA1 in odontoblasts have been implicated as mechanoreceptors [27,146–149,151,180]. In an odontoblast/TG co-culture preparation, mechanical stimulation of the odontoblast-like cells increased intracellular Ca^{2+} concentration in the mechanically-stimulated odontoblasts as well as in neighboring neuron-like cells [27]. The increase in intracellular Ca^{2+} in these cells was almost completely abolished by application of a cocktail of TRPV1, TRPV2, TRPV4 and TRPA1 channel antagonists [27]. In addition, hypotonic solution-induced mechanical stretching of the cell membrane increases intracellular Ca^{2+} in cultured odontoblast-like cells [146–149,177,180], which is inhibited by antagonists of TRPV1, TRPV2, TRPV4 [146,148], TRPA1 [177] and TRPM7 [180], further suggesting that these channels function as mechanoreceptors and/or osmoreceptors.

Mechanosensitive TRP channels along with other mechanoreceptors expressed in odontoblasts and DPAs can be activated by application of thermal stimuli to an intact tooth. A study on human subjects showed that when a thermal stimulus was applied to a tooth surface, the subjects rapidly sensed pain (1.28 s for hot stimuli and 1.49 s for cold stimuli), even before a noticeable change in temperature at the wall of the dental pulp (3.68 s) [28,30]. Thermal stimulation on the tooth surface can induce fluid movement in the dentinal tubules because of thermal expansion or contraction of the fluid [18,212]. Interestingly, fluid movement was observed before a change in temperature in the dentine, where the dentinal tubules are located [21,22]. A temperature gradient between the enamel and dentine is observed when a thermal stimulus is applied on the surface of a tooth [21–23,25]. Tooth structures expand or contract because of this thermal gradient, which produces mechanical stresses in these structures [21–23,25,28]. Mechanical deformation of the dentine (including pulpal wall dentine) precedes the temperature changes in the dentine/dentine–enamel junction following thermal stimulation of the surface a tooth [22,25]. These observations suggest that intense thermal stimulation-induced mechanical deformation of the dentine may exert mechanical stresses on the odontoblasts as well as on the pulpal tissues. In turn, the mechanical stresses may directly activate the mechanosensitive TRP channels and other mechanoreceptors present in the odontoblasts and pulpal nerve fibers (Figure 2).

The sharp pain perceived soon after thermal stimulation of the tooth surface may be at least partly attributed to the mechanical stress-induced activation of mechanoreceptors on odontoblasts and DPAs. Temperature changes in the dentine caused by thermal stimulation of the surface of an intact tooth or a tooth with exposed dentine may also cause expansion/contraction and movement of dentinal tubular fluid, which can activate mechanosensitive TRP channels (along with other mechanoreceptors) on odontoblasts and nerve fibers within/near the dentinal tubules (Figure 2).

Intense thermal stimulation of the tooth surface may cause vasodilation and changes in pulpal blood flow [213–215]. A study using dogs found that pulpal blood flow increased slightly and gradually when the tooth surface temperature was raised from 35 to 40 °C, and that it increased sharply when the temperature was raised from 35 to 55 °C [213,214]. The increase of pulpal blood flow by thermal stimulation can increase the pressure within the pulpal tissue [213,214], which can excite mechanoreceptors, including TRP channels, in the pulpal nerves.

Patients with dentine hypersensitivity may feel pain caused by osmotic changes. Osmotic stimulation of the exposed dentine may induce movement of dentinal tubular fluid [18,19], which can excite mechanosensitive TRP channels on nearby odontoblasts and nerve fibers (Figure 2). It has been observed that dentine covered with a smear layer is much less responsive to hypertonic solutions than dentine devoid of a smear layer, suggesting that fluid movement is greater when the smear layer is absent [216].

Patients with dentine hypersensitivity may also feel pain after very light mechanical stimulation, such as air puffs or water spray, on the exposed dentine, which may induce very little movement of the dentinal tubular fluid. This suggests the presence of low-threshold mechanoreceptors on odontoblasts and DPAs [39]. Fried et al. proposed that many DPAs may contain low-threshold mechanoreceptors [39]. They termed these "algoneurons" [39]. The researchers suggested that these algoneurons may transduce the nociceptive signal in teeth [39]. Indeed, activation of these neurons causes activation of trigeminal brainstem neurons that deliver a pain message to higher brain centers, resulting in the sensation of pain [39]. Similar low-threshold mechanoreceptors in the skin or oral mucosa generally signal tactile sensation (e.g., light touch) [217]. Air puff on the exposed dentine may also dehydrate the dentine surface and cause outward flow of the dentinal tubular fluid, which can activate mechanosensitive receptors on odontoblasts and nerve fibers within/near the dentinal tubules [18,19]. Mechanosensitive TRP channels may function as mechanoreceptors [218] to transduce these types of stimuli.

When intense thermal stimuli are applied on the tooth surface, they can increase the temperature at the dentine–pulp border (albeit slowly), which may excite the thermosensitive TRP channels on odontoblasts and DPAs (Figure 2). Indeed, a slow increase in temperature to >43 °C on the tooth surface activates intra-pulpal C-fibers in the cat [219,220]. Another study showed that intra-dental A-δ and C-fibers respond to intense cooling of the tooth surface [31]. Thermosensitive TRP channels expressed on the A-δ and C-fibers in the dental pulp can be activated by high-temperature stimulation of the tooth surface. Heat and cool/cold stimuli increase intracellular Ca^{2+} concentrations and elicit inward currents in cultured DPAs [138], indicating that these channels function as thermoreceptors in these cells.

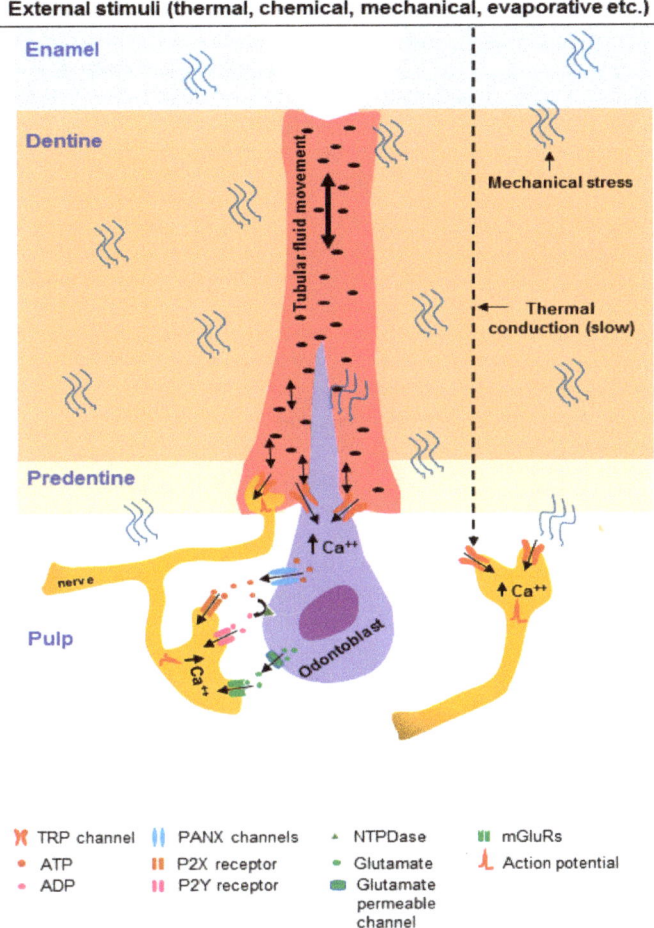

Figure 2. Mechanisms by which TRP channels may transduce dental pain when external stimuli are applied on the exposed dentine or on the surface of an intact tooth. External stimuli on the exposed dentine may create movement (indicated by double way arrows) in the dentinal tubular fluid which can activate the mechanosensitive TRP channels on odontoblasts and pulpal nerves. Intense thermal stimulation on the surface of an intact tooth may induce mechanical stresses within the tooth structures that ultimately excite the mechanosensitive TRP channels. In addition, temperature may conduct through the dental structures (relatively slow) to activate the thermosensitive TRP channels. Odontoblasts may communicate with the pulpal nerves through paracrine signaling mechanisms using ATP and glutamate. Ca^{2+} enters (indicated by single way arrows) odontoblasts through the activated TRP channels. ATP may be released (indicated by a single way arrow) from the odontoblasts through pannexin (PANX) channels and can activate P2X receptors expressed on the pulpal nerves. ATP can be converted (indicated by a curve arrow) by NTPDases to ADP, which can activate P2Y receptors expressed on the pulpal nerves. Furthermore, glutamate released (indicated by a single way arrow) from odontoblasts through glutamate-permeable channels can excite the pulpal nerves via metabotropic glutamate receptors (mGluRs).

The findings described above provide ample evidence that external stimuli can activate the TRP channels expressed in odontoblasts. The odontoblasts may in turn activate the sensory nerves of the dental pulp. Recent studies have begun to provide insight into the mechanisms by which pulpal nerves are activated by stimulated odontoblasts. Activation of TRP channels and other receptors on the odontoblasts by external stimuli increases the intra-odontoblastic Ca^{2+} concentration [27,145–148]. These odontoblasts may then release ATP [27,140,151] and glutamate, which act on their receptors on adjacent nerve fibers of DPAs (Figure 2). ATP plays an important role in pain signaling through activation of purinergic receptors expressed on peripheral sensory nerve fibers [221,222]. Indeed, in an odontoblast/TG neuron co-culture preparation, mechanical stimulation of odontoblast-like cells increases the intracellular Ca^{2+} concentration in these cells as well as in the neighboring neuron-like cells [27]. Application of an inhibitor of ATP-permeable pannexin-1 (PANX-1) channels and ATP-degrading enzyme abolished the increases in intracellular Ca^{2+} in the neuron-like cells, but not in the mechanically stimulated odontoblast-like cells [27]. This suggests that ATP released through PANX-1 in stimulated odontoblasts excites neurons [27]. PANX-1 immunoreactivity in the cell bodies and the processes of odontoblasts has been observed [27]. In addition, purinergic receptor (P2X3/P2Y1/P2Y12) antagonists attenuate the increase in intracellular Ca^{2+} in neurons, but not in the mechanically-stimulated odontoblasts, suggesting that neuronal purinergic receptors are activated by the ATP released by the odontoblasts and its metabolite ADP [27]. Using the same odontoblast/TG neuron co-culture preparation, Sato et al. showed that mechanical stimulation of odontoblast-like cells evokes inward currents in medium-sized neuron-like cells that express NF-200 immunoreactivity (a marker of myelinated neurons), but not IB4 [223]. Medium-sized TG neurons (A-δ) have been implicated in sharp pain associated with dentine hypersensitivity [20,34,38]. Sato et al. also observed that the mechanical stimulation-induced currents were attenuated by application of a cocktail of TRPV1, TRPV2, TRPV4 and TRPA1 channel antagonists, suggesting involvement of these channels in transduction of mechanical stimuli applied to the odontoblast-like cells [223]. Furthermore, application of a P2X3 antagonist attenuated the induced currents in the neuron-like cells, suggesting activation of P2X3 on the neurons by ATP released from mechanically-stimulated odontoblasts [223]. PANX channels have also been detected in the dental pulp [224]. Expression of PANX-1 [27,224] and 2 [224] was observed in odontoblasts and their processes. ATP is hydrolyzed to ADP and other metabolites by ectonucleoside triphosphate diphosphohydrolases (NTPDases) [225]. NTPDase expression has been detected in the odontoblasts and Schwann cells that surround the myelinated pulpal nerves [226]. Functional NTPDase enzymatic activity is observed in odontoblasts and their processes, the sub-odontoblast layer, blood vessels and Schwann cells that surround the myelinated pulpal nerves, suggesting that ATP and its metabolites are present in the dental pulp [224]. Using an in vitro human tooth perfusion model, it has been demonstrated that mechanical or cold stimulation of the exposed dentine releases ATP from the dentine pulp complex, and that application of a PANX channel blocker reduces ATP release [224]. ATP can activate the ionotropic purinergic receptor P2X [227]. A variety of P2X receptor subtypes have been detected in the neurons of the TG [228,229]. Among these, P2X3 expression is comparatively high [229], and it has been detected in the nerves of the dental pulp [229–232]. ADP can activate the metabotropic purinergic receptor P2Y [227]. Expression of P2Y12 receptors is detected in the neurons of the TG [233].

Glutamate may also function as a signaling molecule in odontoblast–TG neuron communication [234]. In odontoblast/TG neuron co-culture preparations, the odontoblast mechanical stimulation-induced increase in Ca^{2+} in the neighboring neuron-like cells is attenuated by the application of a cocktail of antagonists of metabotropic glutamate receptors (mGluRs) [234]. When an ATP-degrading enzyme was incorporated into this cocktail, the neuronal Ca^{2+} increases were further suppressed, suggesting that both ATP and glutamate released by the odontoblasts act in a paracrine manner to signal neighboring neurons [234]. The Ca^{2+} increase in the neighboring neurons, but not in the stimulated odontoblasts, was also reduced by application of antagonists of glutamate-permeable anion channels, suggesting that glutamate may be released from the stimulated odontoblasts through these channels [234]. Furthermore, in the same study, odontoblast-like cells were observed to express group I, II and III mGluRs [234]. Glutamate is detected in the odontoblasts and nerve fibers of the rat dental pulp [235]. The nerve fibers also express mGluR5, which is upregulated following dentine injury [235]. The expression of mGluR5 has also been reported on TRPV1-immunoreactive nerve fibers in the human dental pulp [139]. These observations suggest that glutamate released from stimulated odontoblasts signal through mGluRs expressed on nearby nerve fibers (Figure 2).

TRP channels in the DPAs may play an important role in pain transduction under inflammatory conditions of the dental pulp (pulpitis) (Figure 3).

In symptomatic pulpitis, the tooth is hypersensitive to external stimuli and pain persists after removal of the stimuli (lingering pain). The tooth can be spontaneously painful [6,7]. The pain may be caused by the sensitization of DPAs. Upregulation of various channels, including TRPs, in the odontoblasts and DPAs may lead to hyperexcitability of the nerves. Indeed, TRPV1 in the DPAs is up-regulated by LPS, a product of Gram-negative bacteria [135]. Upregulation of TRPV1 [141] and TRPA1 [149,196] in the nerve fibers of the dental pulp is observed in carious human teeth. In addition, upregulation of TRPA1 [195] and TRPV4 [142] in the pulpal nerve fibers is observed in teeth with signs of pulpitis. TRPA1 expression is also increased in the TG following experimental exposure of the dental pulp [194]. In caries-induced pulpitis, the various structures of the dentine–pulp complex (e.g., odontoblasts, fibroblasts, dendritic cells, resident mast cells, endothelial cells in blood vessels, nerve fibers) sense the invading pathogen-associated molecular patterns shared by microorganisms through specialized pattern recognition receptors, such as toll-like receptors (TLRs) and nucleotide-oligomerization binding domain (NOD)-like receptors [236–239], leading to the initiation of an immune response. Vasodilation and extravasation ensue, leading to the infiltration of blood-borne immune cells, such as neutrophils, monocytes and T-lymphocytes, into the pulp [240,241]. Various inflammatory and immune mediators (e.g., prostaglandins, bradykinin, histamine, cytokines, chemokines) are released from these cells (Figure 3). Activated odontoblasts, fibroblasts and mast cells also release inflammatory mediators [240,241]. These inflammatory mediators act on their receptors expressed on the nerve fibers, leading to sensitization of peripheral nerves [32,33,240–243]. The inflammatory mediators can also modulate the sensitivity of TRP channels to external stimuli [88,244–246]. They signal through their receptors on the sensory nerves, leading to the activation of intra-neuronal signaling pathways (e.g., protein kinases A and C, Src kinase, phospholipase C (PLC), extracellular signal-regulated kinase (ERK)) that induce post-translational modifications on multiple TRP channel proteins, thereby affecting their trafficking to the membrane, channel gating and sensitivity to various stimuli [88,244,246]. Additionally, several growth factors produced during inflammation (e.g., NGF) increase the production and transport of TRP channels to peripheral nerve terminals [88,244,246]. Growth factors may also directly increase the sensitivity of the nerves to stimuli [88,244,246]. The sensitized nerves in turn release various neuropeptides, such as substance P, CGRP and vasoactive intestinal peptide [240–243,246]. Neuropeptides, such as substance P, are also released from fibroblasts [247–250]. Expression of the mRNAs for substance P and its receptor neurokinin-1 has been reported in pulpal fibroblasts, suggesting that substance P may be released from and signal in an autocrine manner in these cells [247]. Substance P can also be released from various immune cells [251]. Local elevation of CGRP and substance P enhances vasodilation and immune cell invasion, further increasing the release of inflammatory mediators, thereby perpetuating and exacerbating the neurogenic inflammation (Figure 3) [32,242,247–250]. Under this inflammatory state, the threshold for activation of TRP channels may decrease, causing hypersensitivity of the tooth to external stimuli.

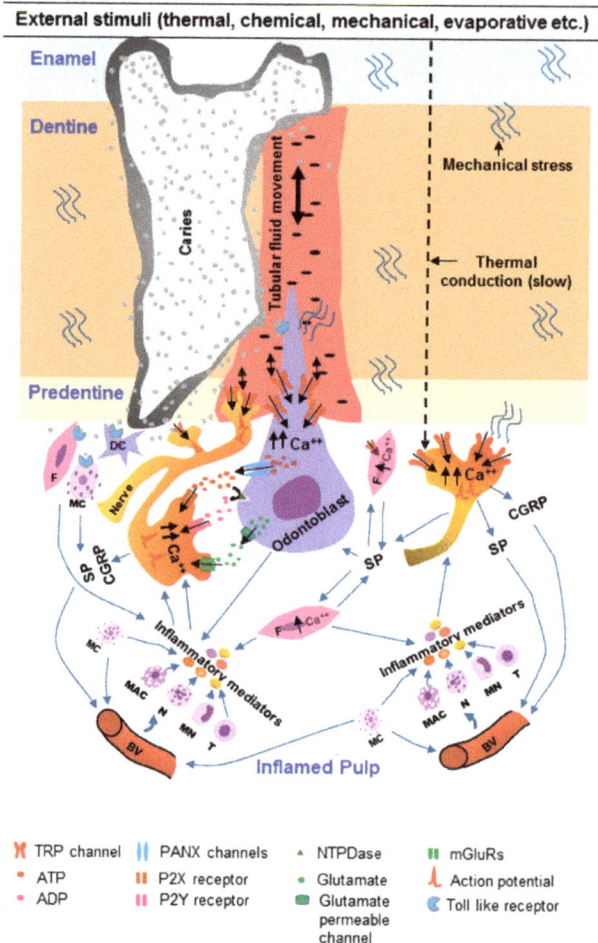

Figure 3. Involvement of TRP channels in the transduction of dental pain under inflammatory conditions. In caries-induced pulpitis, the various structures of the dentine–pulp complex (e.g., odontoblasts, fibroblasts, dendritic cells and resident mast cells etc.) sense the invading pathogens through specialized pattern recognition receptors, such as toll-like receptors (TLRs), leading to the initiation of an immune response. Blood-borne immune cells (e.g., neutrophils, monocytes and T-lymphocytes) infiltrate (indicated by a blue arrow) the pulp from the dilated blood vessels. These immune cells as well as odontoblasts and fibroblasts release (indicated by blue arrows) various inflammatory mediators that activate (indicated by blue arrows) cognate receptors on the nerve fibers, leading to sensitization. Sensitization can involve numerous changes, including enhancement of TRP sensitivity to external stimuli and increased expression on the nerve terminals by mechanisms such as post-translational modification and altered trafficking of these channels. Upregulation of TRP channels is observed in the odontoblasts and the pulpal nerves. The sensitized pulpal nerves release (indicated by blue arrows) various neuropeptides, such as substance P and CGRP. Neuropeptides can also be released (indicated by blue arrows) from fibroblasts and various immune cells. Local elevation of neuropeptides increases the release of inflammatory mediators from blood vessels that further elevate the release of neuropeptides from activated nerve fibers, exacerbating neurogenic inflammation. Besides, bacterial endotoxins can directly activate TRP channels on DPAs or odontoblasts and thereby contributed to the development of pain. SP: substance P; CGRP: calcitonin gene-related peptide; F: fibroblast; DC: dendritic cell; MC: mast cell; MAC: macrophage; T: T-lymphocyte; N: neutrophil; BV: blood vessel.

Interestingly, bacterial components can directly activate neurons before bacterial-induced immune response matured [252–254]. Gram-positive bacterial-derived factors (e.g., N-formylated peptides and α-hemolysin toxin) reported to elicit Ca^{2+} influx directly in mouse nociceptive neurons and contributed to the development of hypersensitivity in vivo [253]. Gram-negative bacterial toxin, lipopolysaccharide (LPS), also reported activating cultured TG neurons and sensitize TRPV1 channels via a toll-like receptor 4 (TLR4) mediated mechanism [254–256]. In a rat model, TLR4 signaling in the TRPV1 expressing TG neurons was implicated for acute dental pulpitis induced pain [257]. Recent studies demonstrate that LPS can activate TRP channels in a TLR4-independent manner [254,258]. It observed in a study that LPS directly activates TRPA1 channels present in the sensory neurons of the nodose and TG in a TLR4-independent mechanism [258]. Pain and neurovascular responses to LPS, including neuropeptide (CGRP) release were dependent on TRPA1 channel activation in the sensory neurons [258]. Another study demonstrates that in addition to TRPA1, LPS can directly activate other TRP channels (TRPV1, TRPM3, TRPM8) present in the sensory neurons [259]. These findings suggest that bacterial products can directly activate sensory nerve fibers before the development of neurogenic inflammation secondary to the immune response to infection [254]. Similarly, in caries-induced pulpitis, bacterial products may directly activate pulpal nerve fibers and contribute to the development of pain (Figure 3). LPS was also observed to activate TRPV4 channels on nonneuronal airway epithelial cells and increased intracellular Ca^{2+} concentration [260]. Since TRPV4 channels are expressed in the odontoblasts, there is a possibility of activation of TRPV4 on the odontoblasts by the bacterial toxin, which increases the intra-odontoblastic Ca^{2+} concentration and in turn, activates the intra-dental sensory neurons.

During inflammation, pulpal nerves undergo sprouting, which may also contribute to tooth hypersensitivity [38,261]. NGF may stimulate nearby pulpal nerves to sprout new branches [38,261], which may in turn increase the number of TRP channels in the dental pulp. Sprouting of the nerves may also lead to innervation of more dentinal tubules, further increasing pain sensitivity [38,261].

Under inflammatory conditions, extravasation of fluid from blood vessels can elevate the pressure within the pulp, because dental pulp is enclosed by hard tissues, creating a low compliance environment [261]. It has been reported that pulpal tissue pressure during inflammation can rise from 15 to 40–50 cm H2O [261,262]. This increase in intra-pulpal pressure can excite the mechanoreceptors (including mechanosensitive TRP channels) on the nerve fibers of the pulp and lead to spontaneous pain.

Fibroblasts in the dental pulp are also reported to be involved in pulpitis. Fibroblasts are abundantly present in the dental pulp and are responsible for the synthesis of extracellular matrix and the maintenance of the structural integrity of the dental pulp [6,263,264]. They are also reported to produce pro-inflammatory cytokines, including interleukin (IL)-1β, IL-6 and IL-8, in response to bacterial stimulation [265,266]. Pro-inflammatory cytokines can be released from fibroblasts by neuropeptides [267–269]. Pulpal fibroblasts also express TRPV1, and activation of this channel leads to the release of IL6, suggesting a role in the development of pulpitis [209]. TRPM2 expression is also increased in the pulpal fibroblasts of teeth with signs of irreversible pulpitis [211].

5. Clinical Significance

Despite the high prevalence of dental pain, effective pain management is lacking because the cellular and molecular mechanisms underlying the pain are unclear, particularly those involved in the transduction of nociceptive signals. Elucidation of these transduction mechanisms is crucial for the development of therapeutic strategies that target the root cause of dental pain. Pharmacologically targeting TRP channels is a novel therapeutic strategy for managing dental pain. Various studies show that pharmacological antagonists of TRP channels attenuate the experimental stimuli-induced increases in intracellular Ca^{2+} concentration in odontoblasts and DPAs. TRP channels in the sensory neurons have been targeted to develop pain-specific local anesthesia in an animal study [270]. Capsaicin (a TRPV1 activator) has been combined with a permanently charged derivative of lidocaine, QX-314, and this combination appears to be effective as a local anesthetic [271–273]. An animal study demonstrated that local application of this drug combination in the gingiva near a tooth before extraction reduced extraction-induced neuronal activation, indicated by reduced expression of c-Fos in the brainstem trigeminal subnucleus caudalis [273]. Eugenol was reported to sensitize and then desensitize TRPV1 channels, which may explain the pain suppressing action of zinc oxide eugenol cements used in temporary restorations of carious teeth [274]. Several small-molecule antagonists of TRPV1 channels have been tested in human clinical trials for dental pain, but unfortunately, outcomes have been poor or the clinical trials were prematurely terminated [275–277] due to side-effects. In one clinical trial, the analgesic effect of the first-generation TRPV1 antagonist AMG 517 (Amgen) was tested following extraction of a third molar tooth. However, the drug increased core body temperature, resulting in termination of the trial [276]. Another TRPV1 antagonist (AZD1386, Astra-Zeneca) was tested in patients with acute pain following removal of the mandibular third molar, but displayed only a short-term analgesic effect [277]. Further studies are required to identify more effective analgesics targeting TRP channels. Incorporation of pharmacological antagonists of TRP channels to dentine desensitizing formulations, dental cements and pulp capping materials is a potential therapeutic strategy for dental pain.

6. Conclusions

The sensory detection system in the tooth is unique. Any type of external stimuli on the exposed dentine or tooth with pulpitis predominantly causes pain sensation. However, the underlying pain transduction mechanisms are not yet fully understood. Various TRP channels that have been detected in the odontoblasts and DPAs may play an important role in the transduction of external stimuli to electrical signals in the nerves, which are conveyed to and perceived as pain by the brain. The TRP channels may serve as potential drug targets for the development of pharmacological strategies to manage dental pain.

Author Contributions: Conceptualization, M.Z.H. and J.K.; Data curation, M.Z.H., M.M.B., F.Y., H.A., S.U. and J.K.; Funding acquisition, J.K.; Resources, M.Z.H., M.M.B., F.Y., H.A., S.U. and J.K.; Supervision, J.K.; Validation, M.Z.H., M.M.B., H.A., F.Y., S.U. and J.K.; Visualization, M.Z.H., and J.K.; Writing—original draft, M.Z.H. and J.K.; Writing—review and editing, M.Z.H. and J.K. All authors read and approved the final version of the manuscript.

Funding: The APC was funded by The Japan Society for the Promotion of Science (JSPS) KAKENHI Grant Number #17K 11656 to Junichi Kitagawa.

Conflicts of Interest: The authors declare no conflict of interest.

References

1. Kassebaum, N.J.; Bernabe, E.; Dahiya, M.; Bhandari, B.; Murray, C.J.; Marcenes, W. Global burden of untreated caries: A systematic review and metaregression. *J. Dent. Res.* **2015**, *94*, 650–658. [CrossRef]
2. Hahn, C.L.; Falkler, W.A., Jr.; Minah, G.E. Microbiological studies of carious dentine from human teeth with irreversible pulpitis. *Arch. Oral Biol.* **1991**, *36*, 147–153. [CrossRef]
3. Bergenholtz, G. Inflammatory response of the dental pulp to bacterial irritation. *J. Endod.* **1981**, *7*, 100–104. [CrossRef]
4. Dummer, P.M.; Hicks, R.; Huws, D. Clinical signs and symptoms in pulp disease. *Int. Endod. J.* **1980**, *13*, 27–35. [CrossRef] [PubMed]
5. Cooper, P.R.; Holder, M.J.; Smith, A.J. Inflammation and regeneration in the dentin-pulp complex: A double-edged sword. *J. Endod.* **2014**, *40*, S46–S51. [CrossRef]
6. Mitchell, D.A.; Mitchell, L. *Oxford Handbook of Clinical Dentistry*, 4th ed.; Oxford University Press: Oxford, UK, 2005; p. 260.
7. McCarthy, P.J.; McClanahan, S.; Hodges, J.; Bowles, W.R. Frequency of localization of the painful tooth by patients presenting for an endodontic emergency. *J. Endod.* **2010**, *36*, 801–805. [CrossRef]
8. Splieth, C.H.; Tachou, A. Epidemiology of dentin hypersensitivity. *Clin. Oral Investig.* **2013**, *17*, S3–S8. [CrossRef]

9. Flynn, J.; Galloway, R.; Orchardson, R. The incidence of 'hypersensitive' teeth in the West of Scotland. *J. Den.* **1985**, *13*, 230–236. [CrossRef]
10. Fischer, C.; Fischer, R.G.; Wennberg, A. Prevalence and distribution of cervical dentine hypersensitivity in a population in Rio de Janeiro, Brazil. *J. Den.* **1992**, *20*, 272–276. [CrossRef]
11. Irwin, C.R.; McCusker, P. Prevalence of dentine hypersensitivity in a general dental population. *J. Ir. Dent. Assoc.* **1997**, *43*, 7–9.
12. Liu, H.C.; Lan, W.H.; Hsieh, C.C. Prevalence and distribution of cervical dentin hypersensitivity in a population in Taipei, Taiwan. *J. Endod.* **1998**, *24*, 45–47. [CrossRef]
13. Dowell, P.; Addy, M.; Dummer, P. Dentine hypersensitivity: Aetiology, differential diagnosis and management. *Br. Dent. J.* **1985**, *158*, 92–96. [CrossRef] [PubMed]
14. Dababneh, R.H.; Khouri, A.T.; Addy, M. Dentine hypersensitivity—An enigma? A review of terminology, mechanisms, aetiology and management. *Br. Dent. J.* **1999**, *187*, 606–611. [PubMed]
15. Sole-Magdalena, A.; Martinez-Alonso, M.; Coronado, C.A.; Junquera, L.M.; Cobo, J.; Vega, J.A. Molecular basis of dental sensitivity: The odontoblasts are multisensory cells and express multifunctional ion channels. *Ann. Anat.* **2018**, *215*, 20–29. [CrossRef] [PubMed]
16. Rahim, Z.H.; Bakri, M.M.; Hm, Z.; Ia, A.; Na, Z. High fluoride and low pH level have been detected in popular flavoured beverages in Malaysia. *Pak. J. Med. Sci.* **2014**, *30*, 404–408. [CrossRef]
17. Bakri, M.M.; Hossain, M.Z.; Razak, F.A.; Saqina, Z.H.; Misroni, A.A.; Ab-Murat, N.; Kitagawa, J.; Saub, R.B. Dentinal tubules occluded by bioactive glass-containing toothpaste exhibit high resistance toward acidic soft drink challenge. *Aust. Dent. J.* **2017**, *62*, 186–191. [CrossRef] [PubMed]
18. Brannstrom, M.; Linden, L.A.; Astrom, A. The hydrodynamics of the dental tubule and of pulp fluid. A discussion of its significance in relation to dentinal sensitivity. *Caries Res.* **1967**, *1*, 310–317. [CrossRef] [PubMed]
19. Brannstrom, M.; Astrom, A. The hydrodynamics of the dentine; its possible relationship to dentinal pain. *Int. Dent. J.* **1972**, *22*, 219–227. [PubMed]
20. Sessle, B.J. The neurobiology of facial and dental pain: Present knowledge, future directions. *J. Dent. Res.* **1987**, *66*, 962–981. [CrossRef]
21. Linsuwanont, P.; Versluis, A.; Palamara, J.E.; Messer, H.H. Thermal stimulation causes tooth deformation: A possible alternative to the hydrodynamic theory? *Arch. Oral Biol.* **2008**, *53*, 261–272. [CrossRef]
22. Jacobs, H.R.; Thompson, R.E.; Brown, W.S. Heat transfer in teeth. *J. Dent. Res.* **1973**, *52*, 248–252. [CrossRef] [PubMed]
23. Lloyd, B.A.; McGinley, M.B.; Brown, W.S. Thermal stress in teeth. *J. Dent. Res.* **1978**, *57*, 571–582. [CrossRef] [PubMed]
24. Horiuchi, H.; Matthews, B. In-vitro observations on fluid flow through human dentine caused by pain-producing stimuli. *Arch. Oral Biol.* **1973**, *18*, 275–294. [CrossRef]
25. Linsuwanont, P.; Palamara, J.E.; Messer, H.H. An investigation of thermal stimulation in intact teeth. *Arch. Oral Biol.* **2007**, *52*, 218–227. [CrossRef] [PubMed]
26. Chidchuangchai, W.; Vongsavan, N.; Matthews, B. Sensory transduction mechanisms responsible for pain caused by cold stimulation of dentine in man. *Arch. Oral Biol.* **2007**, *52*, 154–160. [CrossRef] [PubMed]
27. Shibukawa, Y.; Sato, M.; Kimura, M.; Sobhan, U.; Shimada, M.; Nishiyama, A.; Kawaguchi, A.; Soya, M.; Kuroda, H.; Katakura, A.; et al. Odontoblasts as sensory receptors: Transient receptor potential channels, pannexin-1, and ionotropic ATP receptors mediate intercellular odontoblast-neuron signal transduction. *Pflug. Arch.* **2015**, *467*, 843–863. [CrossRef] [PubMed]
28. Trowbridge, H.O.; Franks, M.; Korostoff, E.; Emling, R. Sensory response to thermal stimulation in human teeth. *J. Endod.* **1980**, *6*, 405–412. [CrossRef]
29. Fuss, Z.; Trowbridge, H.; Bender, I.B.; Rickoff, B.; Sorin, S. Assessment of reliability of electrical and thermal pulp testing agents. *J. Endod.* **1986**, *12*, 301–305. [CrossRef]
30. Närhi, M.; Ngassapa, D.; Shimono, M.M.T.; Suda, H.; Takahashi, K. Function of intradental nociceptors in normal and inflamed teeth. In *Dentin/pulp Complex*; Shimono, M., Maeda, T., Suda, H., Takahashi, K., Eds.; Quintessence: Tokyo, Japan, 1996; pp. 136–140.
31. Jyvasjarvi, E.; Kniffki, K.D. Cold stimulation of teeth: A comparison between the responses of cat intradental A delta and C fibres and human sensation. *J. Physiol.* **1987**, *391*, 193–207. [CrossRef] [PubMed]

32. Magloire, H.; Maurin, J.C.; Couble, M.L.; Shibukawa, Y.; Tsumura, M.; Thivichon-Prince, B.; Bleicher, F. Topical review. Dental pain and odontoblasts: Facts and hypotheses. *J. Orofac. Pain* **2010**, *24*, 335–349.
33. Allard, B.; Magloire, H.; Couble, M.L.; Maurin, J.C.; Bleicher, F. Voltage-gated sodium channels confer excitability to human odontoblasts: Possible role in tooth pain transmission. *J. Biol. Chem.* **2006**, *281*, 29002–29010. [CrossRef] [PubMed]
34. Hermanstyne, T.O.; Markowitz, K.; Fan, L.; Gold, M.S. Mechanotransducers in rat pulpal afferents. *J. Dent. Res.* **2008**, *87*, 834–838. [CrossRef] [PubMed]
35. Chung, G.; Jung, S.J.; Oh, S.B. Cellular and molecular mechanisms of dental nociception. *J. Dent. Res.* **2013**, *92*, 948–955. [CrossRef] [PubMed]
36. Chung, G.; Oh, S.B. TRP channels in dental pain. *Open Pain J.* **2013**, *6*, 31–36. [CrossRef]
37. Narhi, M. The neurophysiology of the teeth. *Dent. Clin. N. Am.* **1990**, *34*, 439–448.
38. Byers, M.R.; Narhi, M.V. Dental injury models: Experimental tools for understanding neuroinflammatory interactions and polymodal nociceptor functions. *Crit. Rev. Oral Biol. Med.* **1999**, *10*, 4–39. [CrossRef] [PubMed]
39. Fried, K.; Sessle, B.J.; Devor, M. The paradox of pain from tooth pulp: Low-threshold "algoneurons"? *Pain* **2011**, *152*, 2685–2689. [CrossRef] [PubMed]
40. Nair, P.N. Neural elements in dental pulp and dentin. *Oral Surg. Oral Med. Oral Pathol. Oral Radiol.* **1995**, *80*, 710–719.
41. Johnsen, D.; Johns, S. Quantitation of nerve fibres in the primary and permanent canine and incisor teeth in man. *Arch. Oral Biol.* **1978**, *23*, 825–829. [CrossRef]
42. Paik, S.K.; Park, K.P.; Lee, S.K.; Ma, S.K.; Cho, Y.S.; Kim, Y.K.; Rhyu, I.J.; Ahn, D.K.; Yoshida, A.; Bae, Y.C. Light and electron microscopic analysis of the somata and parent axons innervating the rat upper molar and lower incisor pulp. *Neuroscience* **2009**, *162*, 1279–1286. [CrossRef]
43. Fried, K.; Arvidsson, J.; Robertson, B.; Brodin, E.; Theodorsson, E. Combined retrograde tracing and enzyme/immunohistochemistry of trigeminal ganglion cell bodies innervating tooth pulps in the rat. *Neuroscience* **1989**, *33*, 101–109. [CrossRef]
44. Fried, K.; Hildebrand, C. Axon number and size distribution in the developing feline inferior alveolar nerve. *J. Neurol. Sci.* **1982**, *53*, 169–180. [CrossRef]
45. Johansson, C.S.; Hildebrand, C.; Povlsen, B. Anatomy and developmental chronology of the rat inferior alveolar nerve. *Anat. Rec.* **1992**, *234*, 144–152. [CrossRef] [PubMed]
46. Paik, S.K.; Lee, D.S.; Kim, J.Y.; Bae, J.Y.; Cho, Y.S.; Ahn, D.K.; Yoshida, A.; Bae, Y.C. Quantitative ultrastructural analysis of the neurofilament 200-positive axons in the rat dental pulp. *J. Endod.* **2010**, *36*, 1638–1642. [CrossRef] [PubMed]
47. Cadden, S.W.; Lisney, S.J.; Matthews, B. Thresholds to electrical stimulation of nerves in cat canine tooth-pulp with A beta-, A delta- and C-fibre conduction velocities. *Brain Res.* **1983**, *261*, 31–41. [CrossRef]
48. Lisney, S.J. Some anatomical and electrophysiological properties of tooth-pulp afferents in the cat. *J. Physiol.* **1978**, *284*, 19–36. [CrossRef] [PubMed]
49. Henry, M.A.; Luo, S.; Levinson, S.R. Unmyelinated nerve fibers in the human dental pulp express markers for myelinated fibers and show sodium channel accumulations. *BMC Neurosci.* **2012**, *13*, 29. [CrossRef]
50. Carda, C.; Peydro, A. Ultrastructural patterns of human dentinal tubules, odontoblasts processes and nerve fibres. *Tissue Cell* **2006**, *38*, 141–150. [CrossRef]
51. Byers, M.R. Terminal arborization of individual sensory axons in dentin and pulp of rat molars. *Brain Res.* **1985**, *345*, 181–185. [CrossRef]
52. Aars, H.; Brodin, P.; Andersen, E. A study of cholinergic and β-adrenergic components in the regulation of blood flow in the tooth pulp and gingiva in man. *Acta Physiol. Scand.* **1993**, *148*, 441–447. [CrossRef]
53. Sasano, T.; Shoji, N.; Kuriwada, S.; Sanjo, D.; Izumi, H.; Karita, K. Absence of parasympathetic vasodilatation in cat dental pulp. *J. Dent. Res.* **1995**, *74*, 1665–1670. [CrossRef] [PubMed]
54. Carr, P.A.; Nagy, J.I. Emerging relationships between cytochemical properties and sensory modality transmission in primary sensory neurons. *Brain Res. Bull.* **1993**, *30*, 209–219. [CrossRef]
55. Basbaum, A.I.; Bautista, D.M.; Scherrer, G.; Julius, D. Cellular and molecular mechanisms of pain. *Cell* **2009**, *139*, 267–284. [CrossRef] [PubMed]

56. Alvarez, F.J.; Morris, H.R.; Priestley, J.V. Sub-populations of smaller diameter trigeminal primary afferent neurons defined by expression of calcitonin gene-related peptide and the cell surface oligosaccharide recognized by monoclonal antibody LA4. *J. Neurocytol.* **1991**, *20*, 716–731. [CrossRef] [PubMed]
57. Silverman, J.D.; Kruger, L. Lectin and neuropeptide labeling of separate populations of dorsal root ganglion neurons and associated "nociceptor" thin axons in rat testis and cornea whole-mount preparations. *Somatosens. Res.* **1988**, *5*, 259–267. [CrossRef] [PubMed]
58. Bradbury, E.J.; Burnstock, G.; McMahon, S.B. The expression of P2X3 purinoreceptors in sensory neurons: Effects of axotomy and glial-derived neurotrophic factor. *Mol. Cell. Neurosci.* **1998**, *12*, 256–268. [CrossRef]
59. Chen, C.C.; Akopian, A.N.; Sivilotti, L.; Colquhoun, D.; Burnstock, G.; Wood, J.N. A P2X purinoceptor expressed by a subset of sensory neurons. *Nature* **1995**, *377*, 428–431. [CrossRef] [PubMed]
60. Vulchanova, L.; Riedl, M.S.; Shuster, S.J.; Stone, L.S.; Hargreaves, K.M.; Buell, G.; Surprenant, A.; North, R.A.; Elde, R. P2X3 is expressed by DRG neurons that terminate in inner lamina II. *Eur. Neurosci.* **1998**, *10*, 3470–3478. [CrossRef]
61. Ichikawa, H.; Sugimoto, T. The co-expression of ASIC3 with calcitonin gene-related peptide and parvalbumin in the rat trigeminal ganglion. *Brain Res.* **2002**, *943*, 287–291. [CrossRef]
62. Ichikawa, H.; Deguchi, T.; Nakago, T.; Jacobowitz, D.M.; Sugimoto, T. Parvalbumin- and calretinin-immunoreactive trigeminal neurons innervating the rat molar tooth pulp. *Brain Res.* **1995**, *679*, 205–211. [CrossRef]
63. Yang, H.; Bernanke, J.M.; Naftel, J.P. Immunocytochemical evidence that most sensory neurons of the rat molar pulp express receptors for both glial cell line-derived neurotrophic factor and nerve growth factor. *Arch. Oral Biol.* **2006**, *51*, 69–78. [CrossRef] [PubMed]
64. Fried, K.; Risling, M. Nerve growth factor receptor-like immunoreactivity in primary and permanent canine tooth pulps of the cat. *Cell Tissue Res.* **1991**, *264*, 321–328. [CrossRef] [PubMed]
65. Pan, M.; Naftel, J.P.; Wheeler, E.F. Effects of deprivation of neonatal nerve growth factor on the expression of neurotrophin receptors and brain-derived neurotrophic factor by dental pulp afferents of the adult rat. *Arch. Oral Biol.* **2000**, *45*, 387–399. [CrossRef]
66. Ichikawa, H.; Deguchi, T.; Fujiyoshi, Y.; Nakago, T.; Jacobowitz, D.M.; Sugimoto, T. Calbindin-D28k-immunoreactivity in the trigeminal ganglion neurons and molar tooth pulp of the rat. *Brain Res.* **1996**, *715*, 71–78. [CrossRef]
67. Ichikawa, H.; Fukuda, T.; Terayama, R.; Yamaai, T.; Kuboki, T.; Sugimoto, T. Immunohistochemical localization of gamma and β subunits of epithelial Na^+ channel in the rat molar tooth pulp. *Brain Res.* **2005**, *1065*, 138–141. [CrossRef] [PubMed]
68. Mori, H.; Ishida-Yamamoto, A.; Senba, E.; Ueda, Y.; Tohyama, M. Calcitonin gene-related peptide containing sensory neurons innervating tooth pulp and buccal mucosa of the rat: An immunohistochemical analysis. *J. Chem. Neuroanat.* **1990**, *3*, 155–163. [PubMed]
69. Jacobsen, E.B.; Fristad, I.; Heyeraas, K.J. Nerve fibers immunoreactive to calcitonin gene-related peptide, substance P, neuropeptide Y, and dopamine β-hydroxylase in innervated and denervated oral tissues in ferrets. *Acta Odontol. Scand.* **1998**, *56*, 220–228. [CrossRef] [PubMed]
70. Nilius, B.; Owsianik, G. The transient receptor potential family of ion channels. *Genome Biol.* **2011**, *12*, 218. [CrossRef] [PubMed]
71. Damann, N.; Voets, T.; Nilius, B. TRPs in our senses. *Curr. Biol.* **2008**, *18*, R880–889. [CrossRef] [PubMed]
72. Nilius, B.; Szallasi, A. Transient receptor potential channels as drug targets: From the science of basic research to the art of medicine. *Pharmacol. Rev.* **2014**, *66*, 676–814. [CrossRef] [PubMed]
73. Jardin, I.; Lopez, J.J.; Diez, R.; Sanchez-Collado, J.; Cantonero, C.; Albarran, L.; Woodard, G.E.; Redondo, P.C.; Salido, G.M.; Smani, T.; et al. TRPs in Pain Sensation. *Front. Physiol.* **2017**, *8*, 392. [CrossRef] [PubMed]
74. Montell, C.; Jones, K.; Hafen, E.; Rubin, G. Rescue of the Drosophila phototransduction mutation trp by germline transformation. *Science* **1985**, *230*, 1040–1043. [CrossRef] [PubMed]
75. Montell, C.; Rubin, G.M. Molecular characterization of the Drosophila trp locus: A putative integral membrane protein required for phototransduction. *Neuron* **1989**, *2*, 1313–1323. [CrossRef]
76. Minke, B.; Selinger, Z. The roles of trp and calcium in regulating photoreceptor function in Drosophila. *Curr. Opin. Neurobiol.* **1996**, *6*, 459–466. [CrossRef]
77. Li, H. TRP Channel Classification. *Adv. Exp. Med. Biol.* **2017**, *976*, 1–8. [PubMed]

78. Caterina, M.J.; Schumacher, M.A.; Tominaga, M.; Rosen, T.A.; Levine, J.D.; Julius, D. The capsaicin receptor: A heat-activated ion channel in the pain pathway. *Nature* **1997**, *389*, 816–824. [PubMed]
79. Vandewauw, I.; Owsianik, G.; Voets, T. Systematic and quantitative mRNA expression analysis of TRP channel genes at the single trigeminal and dorsal root ganglion level in mouse. *BMC Neurosci.* **2013**, *14*, 21. [CrossRef]
80. Flegel, C.; Schobel, N.; Altmuller, J.; Becker, C.; Tannapfel, A.; Hatt, H.; Gisselmann, G. RNA-Seq Analysis of Human Trigeminal and Dorsal Root Ganglia with a Focus on Chemoreceptors. *PLoS ONE* **2015**, *10*, e0128951. [CrossRef]
81. Vay, L.; Gu, C.; McNaughton, P.A. The thermo-TRP ion channel family: Properties and therapeutic implications. *Br. J. Pharm.* **2012**, *165*, 787–801. [CrossRef]
82. Voets, T.; Talavera, K.; Owsianik, G.; Nilius, B. Sensing with TRP channels. *Nat. Chem. Biol.* **2005**, *1*, 85–92. [CrossRef]
83. Voets, T.; Droogmans, G.; Wissenbach, U.; Janssens, A.; Flockerzi, V.; Nilius, B. The principle of temperature-dependent gating in cold- and heat-sensitive TRP channels. *Nature* **2004**, *430*, 748–754. [CrossRef] [PubMed]
84. Vriens, J.; Nilius, B.; Voets, T. Peripheral thermosensation in mammals. *Nat. Rev.* **2014**, *15*, 573–589. [CrossRef] [PubMed]
85. Tominaga, M.; Caterina, M.J.; Malmberg, A.B.; Rosen, T.A.; Gilbert, H.; Skinner, K.; Raumann, B.E.; Basbaum, A.I.; Julius, D. The cloned capsaicin receptor integrates multiple pain-producing stimuli. *Neuron* **1998**, *21*, 531–543. [CrossRef]
86. Jordt, S.E.; Tominaga, M.; Julius, D. Acid potentiation of the capsaicin receptor determined by a key extracellular site. *Proc. Natl. Acad. Sci. USA* **2000**, *97*, 8134–8139. [CrossRef] [PubMed]
87. Szallasi, A.; Cortright, D.N.; Blum, C.A.; Eid, S.R. The vanilloid receptor TRPV1: 10 years from channel cloning to antagonist proof-of-concept. *Nat. Rev.* **2007**, *6*, 357–372.
88. Hwang, S.W.; Cho, H.; Kwak, J.; Lee, S.Y.; Kang, C.J.; Jung, J.; Cho, S.; Min, K.H.; Suh, Y.G.; Kim, D.; et al. Direct activation of capsaicin receptors by products of lipoxygenases: Endogenous capsaicin-like substances. *Proc. Natl. Acad. Sci. USA* **2000**, *97*, 6155–6160. [CrossRef] [PubMed]
89. Moriyama, T.; Higashi, T.; Togashi, K.; Iida, T.; Segi, E.; Sugimoto, Y.; Tominaga, T.; Narumiya, S.; Tominaga, M. Sensitization of TRPV1 by EP1 and IP reveals peripheral nociceptive mechanism of prostaglandins. *Mol. Pain* **2005**, *1*, 3. [CrossRef] [PubMed]
90. Tominaga, M.; Wada, M.; Masu, M. Potentiation of capsaicin receptor activity by metabotropic ATP receptors as a possible mechanism for ATP-evoked pain and hyperalgesia. *Proc. Natl. Acad. Sci. USA* **2001**, *98*, 6951–6956. [CrossRef] [PubMed]
91. Julius, D. TRP channels and pain. *Ann. Rev. Cell Dev. Biol.* **2013**, *29*, 355–384. [CrossRef] [PubMed]
92. Hossain, M.Z.; Ando, H.; Unno, S.; Masuda, Y.; Kitagawa, J. Activation of TRPV1 and TRPM8 Channels in the Larynx and Associated Laryngopharyngeal Regions Facilitates the Swallowing Reflex. *Int. J. Mol. Sci.* **2018**, *19*, 4113. [CrossRef] [PubMed]
93. Caterina, M.J.; Leffler, A.; Malmberg, A.B.; Martin, W.J.; Trafton, J.; Petersen-Zeitz, K.R.; Koltzenburg, M.; Basbaum, A.I.; Julius, D. Impaired nociception and pain sensation in mice lacking the capsaicin receptor. *Science* **2000**, *288*, 306–313. [CrossRef] [PubMed]
94. White, J.P.; Urban, L.; Nagy, I. TRPV1 function in health and disease. *Curr. Pharm. Biotechnol.* **2011**, *12*, 130–144. [CrossRef] [PubMed]
95. Christie, S.; Wittert, G.A.; Li, H.; Page, A.J. Involvement of TRPV1 Channels in Energy Homeostasis. *Front. Endocrinol.* **2018**, *9*, 420. [CrossRef] [PubMed]
96. Caterina, M.J.; Rosen, T.A.; Tominaga, M.; Brake, A.J.; Julius, D. A capsaicin-receptor homologue with a high threshold for noxious heat. *Nature* **1999**, *398*, 436–441. [CrossRef] [PubMed]
97. Liedtke, W. Transient receptor potential vanilloid channels functioning in transduction of osmotic stimuli. *J. Endocrinol.* **2006**, *191*, 515–523. [CrossRef] [PubMed]
98. Ahluwalia, J.; Rang, H.; Nagy, I. The putative role of vanilloid receptor-like protein-1 in mediating high threshold noxious heat-sensitivity in rat cultured primary sensory neurons. *Eur. Neurosci.* **2002**, *16*, 1483–1489. [CrossRef]
99. Story, G.M. The emerging role of TRP channels in mechanisms of temperature and pain sensation. *Curr. Neuropharmacol.* **2006**, *4*, 183–196. [CrossRef] [PubMed]

100. Park, U.; Vastani, N.; Guan, Y.; Raja, S.N.; Koltzenburg, M.; Caterina, M.J. TRP vanilloid 2 knock-out mice are susceptible to perinatal lethality but display normal thermal and mechanical nociception. *J. Neurosci.* **2011**, *31*, 11425–11436. [CrossRef]
101. Nedungadi, T.P.; Dutta, M.; Bathina, C.S.; Caterina, M.J.; Cunningham, J.T. Expression and distribution of TRPV2 in rat brain. *Exp. Neurol.* **2012**, *237*, 223–237. [CrossRef]
102. Qin, N.; Neeper, M.P.; Liu, Y.; Hutchinson, T.L.; Lubin, M.L.; Flores, C.M. TRPV2 is activated by cannabidiol and mediates CGRP release in cultured rat dorsal root ganglion neurons. *J. Neurosci.* **2008**, *28*, 6231–6238. [CrossRef]
103. Ichikawa, H.; Sugimoto, T. Vanilloid receptor 1-like receptor-immunoreactive primary sensory neurons in the rat trigeminal nervous system. *Neuroscience* **2000**, *101*, 719–725. [CrossRef]
104. Gibbs, J.L.; Melnyk, J.L.; Basbaum, A.I. Differential TRPV1 and TRPV2 channel expression in dental pulp. *J. Dent. Res.* **2011**, *90*, 765–770. [CrossRef] [PubMed]
105. Xu, H.; Ramsey, I.S.; Kotecha, S.A.; Moran, M.M.; Chong, J.A.; Lawson, D.; Ge, P.; Lilly, J.; Silos-Santiago, I.; Xie, Y.; et al. TRPV3 is a calcium-permeable temperature-sensitive cation channel. *Nature* **2002**, *418*, 181–186. [CrossRef] [PubMed]
106. Smith, G.D.; Gunthorpe, M.J.; Kelsell, R.E.; Hayes, P.D.; Reilly, P.; Facer, P.; Wright, J.E.; Jerman, J.C.; Walhin, J.P.; Ooi, L.; et al. TRPV3 is a temperature-sensitive vanilloid receptor-like protein. *Nature* **2002**, *418*, 186–190. [CrossRef] [PubMed]
107. Vogt-Eisele, A.K.; Weber, K.; Sherkheli, M.A.; Vielhaber, G.; Panten, J.; Gisselmann, G.; Hatt, H. Monoterpenoid agonists of TRPV3. *Br. J. Pharm.* **2007**, *151*, 530–540. [CrossRef]
108. Xu, H.; Delling, M.; Jun, J.C.; Clapham, D.E. Oregano, thyme and clove-derived flavors and skin sensitizers activate specific TRP channels. *Nat. Neurosci.* **2006**, *9*, 628–635. [CrossRef]
109. Peier, A.M.; Reeve, A.J.; Andersson, D.A.; Moqrich, A.; Earley, T.J.; Hergarden, A.C.; Story, G.M.; Colley, S.; Hogenesch, J.B.; McIntyre, P.; et al. A heat-sensitive TRP channel expressed in keratinocytes. *Science* 2046–2049. [CrossRef]
110. Mandadi, S.; Sokabe, T.; Shibasaki, K.; Katanosaka, K.; Mizuno, A.; Moqrich, A.; Patapoutian, A.; Fukumi-Tominaga, T.; Mizumura, K.; Tominaga, M. TRPV3 in keratinocytes transmits temperature information to sensory neurons via ATP. *Pflug. Arch.* **2009**, *458*, 1093–1102. [CrossRef] [PubMed]
111. Yamada, T.; Ueda, T.; Ugawa, S.; Ishida, Y.; Imayasu, M.; Koyama, S.; Shimada, S. Functional expression of transient receptor potential vanilloid 3 (TRPV3) in corneal epithelial cells: Involvement in thermosensation and wound healing. *Exp. Eye Res.* **2010**, *90*, 121–129. [CrossRef]
112. Aijima, R.; Wang, B.; Takao, T.; Mihara, H.; Kashio, M.; Ohsaki, Y.; Zhang, J.Q.; Mizuno, A.; Suzuki, M.; Yamashita, Y.; et al. The thermosensitive TRPV3 channel contributes to rapid wound healing in oral epithelia. *FASEB J.* **2015**, *29*, 182–192. [CrossRef]
113. Huang, S.M.; Lee, H.; Chung, M.K.; Park, U.; Yu, Y.Y.; Bradshaw, H.B.; Coulombe, P.A.; Walker, J.M.; Caterina, M.J. Overexpressed transient receptor potential vanilloid 3 ion channels in skin keratinocytes modulate pain sensitivity via prostaglandin E2. *J. Neurosci.* **2008**, *28*, 13727–13737. [CrossRef] [PubMed]
114. Liedtke, W.; Choe, Y.; Marti-Renom, M.A.; Bell, A.M.; Denis, C.S.; Sali, A.; Hudspeth, A.J.; Friedman, J.M.; Heller, S. Vanilloid receptor-related osmotically activated channel (VR-OAC), a candidate vertebrate osmoreceptor. *Cell* **2000**, *103*, 525–535. [CrossRef]
115. Strotmann, R.; Harteneck, C.; Nunnenmacher, K.; Schultz, G.; Plant, T.D. OTRPC4, a nonselective cation channel that confers sensitivity to extracellular osmolarity. *Nat. Cell Biol.* **2000**, *2*, 695–702. [CrossRef] [PubMed]
116. Liedtke, W.; Friedman, J.M. Abnormal osmotic regulation in trpv4$^{-/-}$ mice. *Proc. Natl. Acad. Sci. USA* **2003**, *100*, 13698–13703. [CrossRef] [PubMed]
117. Liedtke, W.; Tobin, D.M.; Bargmann, C.I.; Friedman, J.M. Mammalian TRPV4 (VR-OAC) directs behavioral responses to osmotic and mechanical stimuli in Caenorhabditis elegans. *Proc. Natl. Acad. Sci. USA* **2003**, *100*, 14531–14536. [CrossRef] [PubMed]
118. Alessandri-Haber, N.; Yeh, J.J.; Boyd, A.E.; Parada, C.A.; Chen, X.; Reichling, D.B.; Levine, J.D. Hypotonicity induces TRPV4-mediated nociception in rat. *Neuron* **2003**, *39*, 497–511. [CrossRef]
119. Alessandri-Haber, N.; Joseph, E.; Dina, O.A.; Liedtke, W.; Levine, J.D. TRPV4 mediates pain-related behavior induced by mild hypertonic stimuli in the presence of inflammatory mediator. *Pain* **2005**, *118*, 70–79. [CrossRef] [PubMed]

120. Kohler, R.; Heyken, W.T.; Heinau, P.; Schubert, R.; Si, H.; Kacik, M.; Busch, C.; Grgic, I.; Maier, T.; Hoyer, J. Evidence for a functional role of endothelial transient receptor potential V4 in shear stress-induced vasodilatation. *Arterioscler. Thromb. Vasc. Biol.* **2006**, *26*, 1495–1502. [CrossRef]
121. Suzuki, M.; Mizuno, A.; Kodaira, K.; Imai, M. Impaired pressure sensation in mice lacking TRPV4. *J. Biol. Chem.* **2003**, *278*, 22664–22668. [CrossRef]
122. Watanabe, H.; Vriens, J.; Prenen, J.; Droogmans, G.; Voets, T.; Nilius, B. Anandamide and arachidonic acid use epoxyeicosatrienoic acids to activate TRPV4 channels. *Nature* **2003**, *424*, 434–438. [CrossRef]
123. Grant, A.D.; Cottrell, G.S.; Amadesi, S.; Trevisani, M.; Nicoletti, P.; Materazzi, S.; Altier, C.; Cenac, N.; Zamponi, G.W.; Bautista-Cruz, F.; et al. Protease-activated receptor 2 sensitizes the transient receptor potential vanilloid 4 ion channel to cause mechanical hyperalgesia in mice. *J. Physiol.* **2007**, *578*, 715–733. [CrossRef] [PubMed]
124. Todaka, H.; Taniguchi, J.; Satoh, J.; Mizuno, A.; Suzuki, M. Warm temperature-sensitive transient receptor potential vanilloid 4 (TRPV4) plays an essential role in thermal hyperalgesia. *J. Biol. Chem.* **2004**, *279*, 35133–35138. [CrossRef] [PubMed]
125. Moore, C.; Gupta, R.; Jordt, S.E.; Chen, Y.; Liedtke, W.B. Regulation of Pain and Itch by TRP Channels. *Neurosci. Bull.* **2018**, *34*, 120–142. [CrossRef] [PubMed]
126. Cenac, N.; Altier, C.; Chapman, K.; Liedtke, W.; Zamponi, G.; Vergnolle, N. Transient receptor potential vanilloid-4 has a major role in visceral hypersensitivity symptoms. *Gastroenterology* **2008**, *135*, 937–946.e2. [CrossRef] [PubMed]
127. Zhang, Y.; Wang, Y.H.; Ge, H.Y.; Arendt-Nielsen, L.; Wang, R.; Yue, S.W. A transient receptor potential vanilloid 4 contributes to mechanical allodynia following chronic compression of dorsal root ganglion in rats. *Neurosci. Lett.* **2008**, *432*, 222–227. [CrossRef] [PubMed]
128. Chen, L.; Liu, C.; Liu, L. Changes in osmolality modulate voltage-gated calcium channels in trigeminal ganglion neurons. *Brain Res.* **2008**, *1208*, 56–66. [CrossRef] [PubMed]
129. Chen, Y.; Williams, S.H.; McNulty, A.L.; Hong, J.H.; Lee, S.H.; Rothfusz, N.E.; Parekh, P.K.; Moore, C.; Gereau, R.W.; Taylor, A.B.; et al. Temporomandibular joint pain: A critical role for Trpv4 in the trigeminal ganglion. *Pain* **2013**, *154*, 1295–1304. [CrossRef] [PubMed]
130. Chen, Y.; Kanju, P.; Fang, Q.; Lee, S.H.; Parekh, P.K.; Lee, W.; Moore, C.; Brenner, D.; Gereau, R.W.; Wang, F.; et al. TRPV4 is necessary for trigeminal irritant pain and functions as a cellular formalin receptor. *Pain* **2014**, *155*, 2662–2672. [CrossRef]
131. Rajasekhar, P.; Poole, D.P.; Liedtke, W.; Bunnett, N.W.; Veldhuis, N.A. P2Y1 Receptor Activation of the TRPV4 Ion Channel Enhances Purinergic Signaling in Satellite Glial Cells. *J. Biol. Chem.* **2015**, *290*, 29051–29062. [CrossRef]
132. Alessandri-Haber, N.; Dina, O.A.; Joseph, E.K.; Reichling, D.; Levine, J.D. A transient receptor potential vanilloid 4-dependent mechanism of hyperalgesia is engaged by concerted action of inflammatory mediators. *J. Neurosci.* **2006**, *26*, 3864–3874. [CrossRef]
133. Ichikawa, H.; Sugimoto, T. VR1-immunoreactive primary sensory neurons in the rat trigeminal ganglion. *Brain Res.* **2001**, *890*, 184–188. [CrossRef]
134. Chaudhary, P.; Martenson, M.E.; Baumann, T.K. Vanilloid receptor expression and capsaicin excitation of rat dental primary afferent neurons. *J. Dent. Res.* **2001**, *80*, 1518–1523. [CrossRef] [PubMed]
135. Chung, M.K.; Lee, J.; Duraes, G.; Ro, J.Y. Lipopolysaccharide-induced pulpitis up-regulates TRPV1 in trigeminal ganglia. *J. Dent. Res.* **2011**, *90*, 1103–1107. [CrossRef] [PubMed]
136. Stenholm, E.; Bongenhielm, U.; Ahlquist, M.; Fried, K. VR1- and VRL-l-like immunoreactivity in normal and injured trigeminal dental primary sensory neurons of the rat. *Acta Odontol. Scand.* **2002**, *60*, 72–79. [CrossRef] [PubMed]
137. Kim, H.Y.; Chung, G.; Jo, H.J.; Kim, Y.S.; Bae, Y.C.; Jung, S.J.; Kim, J.S.; Oh, S.B. Characterization of dental nociceptive neurons. *J. Dent. Res.* **2011**, *90*, 771–776. [CrossRef] [PubMed]
138. Park, C.K.; Kim, M.S.; Fang, Z.; Li, H.Y.; Jung, S.J.; Choi, S.Y.; Lee, S.J.; Park, K.; Kim, J.S.; Oh, S.B. Functional expression of thermo-transient receptor potential channels in dental primary afferent neurons: Implication for tooth pain. *J. Biol. Chem.* **2006**, *281*, 17304–17311. [CrossRef] [PubMed]
139. Kim, Y.S.; Kim, Y.J.; Paik, S.K.; Cho, Y.S.; Kwon, T.G.; Ahn, D.K.; Kim, S.K.; Yoshida, A.; Bae, Y.C. Expression of metabotropic glutamate receptor mGluR5 in human dental pulp. *J. Endod.* **2009**, *35*, 690–694. [CrossRef]

140. Renton, T.; Yiangou, Y.; Baecker, P.A.; Ford, A.P.; Anand, P. Capsaicin receptor VR1 and ATP purinoceptor P2X3 in painful and nonpainful human tooth pulp. *J. Orofac. Pain* **2003**, *17*, 245–250.
141. Morgan, C.R.; Rodd, H.D.; Clayton, N.; Davis, J.B.; Boissonade, F.M. Vanilloid receptor 1 expression in human tooth pulp in relation to caries and pain. *J. Orofac. Pain* **2005**, *19*, 248–260.
142. Bakri, M.M.; Yahya, F.; Munawar, K.M.M.; Kitagawa, J.; Hossain, M.Z. Transient receptor potential vanilloid 4 (TRPV4) expression on the nerve fibers of human dental pulp is upregulated under inflammatory condition. *Arch. Oral Biol.* **2018**, *89*, 94–98. [CrossRef]
143. Okumura, R.; Shima, K.; Muramatsu, T.; Nakagawa, K.; Shimono, M.; Suzuki, T.; Magloire, H.; Shibukawa, Y. The odontoblast as a sensory receptor cell? The expression of TRPV1 (VR-1) channels. *Arch. Histol. Cytol.* **2005**, *68*, 251–257. [CrossRef] [PubMed]
144. Yeon, K.Y.; Chung, G.; Shin, M.S.; Jung, S.J.; Kim, J.S.; Oh, S.B. Adult rat odontoblasts lack noxious thermal sensitivity. *J. Dent. Res.* **2009**, *88*, 328–332. [CrossRef] [PubMed]
145. Tsumura, M.; Sobhan, U.; Muramatsu, T.; Sato, M.; Ichikawa, H.; Sahara, Y.; Tazaki, M.; Shibukawa, Y. TRPV1-mediated calcium signal couples with cannabinoid receptors and sodium-calcium exchangers in rat odontoblasts. *Cell Calcium* **2012**, *52*, 124–136. [CrossRef] [PubMed]
146. Son, A.R.; Yang, Y.M.; Hong, J.H.; Lee, S.I.; Shibukawa, Y.; Shin, D.M. Odontoblast TRP channels and thermo/mechanical transmission. *J. Dent. Res.* **2009**, *88*, 1014–1019. [CrossRef] [PubMed]
147. Sato, M.; Sobhan, U.; Tsumura, M.; Kuroda, H.; Soya, M.; Masamura, A.; Nishiyama, A.; Katakura, A.; Ichinohe, T.; Tazaki, M.; et al. Hypotonic-induced stretching of plasma membrane activates transient receptor potential vanilloid channels and sodium-calcium exchangers in mouse odontoblasts. *J. Endod.* **2013**, *39*, 779–787. [CrossRef]
148. El Karim, I.A.; Linden, G.J.; Curtis, T.M.; About, I.; McGahon, M.K.; Irwin, C.R.; Lundy, F.T. Human odontoblasts express functional thermo-sensitive TRP channels: Implications for dentin sensitivity. *Pain* **2011**, *152*, 2211–2223. [CrossRef]
149. El Karim, I.; McCrudden, M.T.; Linden, G.J.; Abdullah, H.; Curtis, T.M.; McGahon, M.; About, I.; Irwin, C.; Lundy, F.T. TNF-α-induced p38MAPK activation regulates TRPA1 and TRPV4 activity in odontoblast-like cells. *Am. J. Pathol.* **2015**, *185*, 2994–3002. [CrossRef]
150. Wen, W.; Que, K.; Zang, C.; Wen, J.; Sun, G.; Zhao, Z.; Li, Y. Expression and distribution of three transient receptor potential vanilloid (TRPV) channel proteins in human odontoblast-like cells. *J. Mol. Histol.* **2017**, *48*, 367–377. [CrossRef]
151. Egbuniwe, O.; Grover, S.; Duggal, A.K.; Mavroudis, A.; Yazdi, M.; Renton, T.; Di Silvio, L.; Grant, A.D. TRPA1 and TRPV4 activation in human odontoblasts stimulates ATP release. *J. Dent. Res.* **2014**, *93*, 911–917. [CrossRef]
152. Sumoza-Toledo, A.; Penner, R. TRPM2: A multifunctional ion channel for calcium signalling. *J. Physiol.* **2011**, *589*, 1515–1525. [CrossRef]
153. Aarts, M.M.; Tymianski, M. TRPMs and neuronal cell death. *Pflug. Arch.* **2005**, *451*, 243–249. [CrossRef] [PubMed]
154. Togashi, K.; Hara, Y.; Tominaga, T.; Higashi, T.; Konishi, Y.; Mori, Y.; Tominaga, M. TRPM2 activation by cyclic ADP-ribose at body temperature is involved in insulin secretion. *EMBO J.* **2006**, *25*, 1804–1815. [CrossRef] [PubMed]
155. Perraud, A.L.; Fleig, A.; Dunn, C.A.; Bagley, L.A.; Launay, P.; Schmitz, C.; Stokes, A.J.; Zhu, Q.; Bessman, M.J.; Penner, R.; et al. ADP-ribose gating of the calcium-permeable LTRPC2 channel revealed by Nudix motif homology. *Nature* **2001**, *411*, 595–599. [CrossRef] [PubMed]
156. Jang, Y.; Cho, P.S.; Yang, Y.D.; Hwang, S.W. Nociceptive Roles of TRPM2 Ion Channel in Pathologic Pain. *Mol. Neurobiol.* **2018**, *55*, 6589–6600. [CrossRef] [PubMed]
157. Grimm, C.; Kraft, R.; Sauerbruch, S.; Schultz, G.; Harteneck, C. Molecular and functional characterization of the melastatin-related cation channel TRPM3. *J. Biol. Chem.* **2003**, *278*, 21493–21501. [CrossRef] [PubMed]
158. Lee, N.; Chen, J.; Sun, L.; Wu, S.; Gray, K.R.; Rich, A.; Huang, M.; Lin, J.H.; Feder, J.N.; Janovitz, E.B.; et al. Expression and characterization of human transient receptor potential melastatin 3 (hTRPM3). *J. Biol. Chem.* **2003**, *278*, 20890–20897. [CrossRef] [PubMed]
159. Wagner, T.F.; Loch, S.; Lambert, S.; Straub, I.; Mannebach, S.; Mathar, I.; Dufer, M.; Lis, A.; Flockerzi, V.; Philipp, S.E.; et al. Transient receptor potential M3 channels are ionotropic steroid receptors in pancreatic β cells. *Nat. Cell Biol.* **2008**, *10*, 1421–1430. [CrossRef] [PubMed]

160. Vriens, J.; Owsianik, G.; Hofmann, T.; Philipp, S.E.; Stab, J.; Chen, X.; Benoit, M.; Xue, F.; Janssens, A.; Kerselaers, S. TRPM3 is a nociceptor channel involved in the detection of noxious heat. *Neuron* **2011**, *70*, 482–494. [CrossRef] [PubMed]
161. Held, K.; Voets, T.; Vriens, J. TRPM3 in temperature sensing and beyond. *Temperature* **2015**, *2*, 201–213. [CrossRef] [PubMed]
162. He, Y.; Yao, G.; Savoia, C.; Touyz, R.M. Transient receptor potential melastatin 7 ion channels regulate magnesium homeostasis in vascular smooth muscle cells: Role of angiotensin II. *Circ. Res.* **2005**, *96*, 207–215. [CrossRef] [PubMed]
163. Schmitz, C.; Perraud, A.L.; Johnson, C.O.; Inabe, K.; Smith, M.K.; Penner, R.; Kurosaki, T.; Fleig, A.; Scharenberg, A.M. Regulation of vertebrate cellular Mg^{2+} homeostasis by TRPM7. *Cell* **2003**, *114*, 191–200. [CrossRef]
164. Krapivinsky, G.; Mochida, S.; Krapivinsky, L.; Cibulsky, S.M.; Clapham, D.E. The TRPM7 ion channel functions in cholinergic synaptic vesicles and affects transmitter release. *Neuron* **2006**, *52*, 485–496. [CrossRef] [PubMed]
165. Jin, J.; Desai, B.N.; Navarro, B.; Donovan, A.; Andrews, N.C.; Clapham, D.E. Deletion of Trpm7 disrupts embryonic development and thymopoiesis without altering Mg^{2+} homeostasis. *Science* **2008**, *322*, 756–760. [CrossRef] [PubMed]
166. Turlova, E.; Bae, C.Y.J.; Deurloo, M.; Chen, W.; Barszczyk, A.; Horgen, F.D.; Fleig, A.; Feng, Z.P.; Sun, H.S. TRPM7 Regulates Axonal Outgrowth and Maturation of Primary Hippocampal Neurons. *Mol. Neurobiol.* **2016**, *53*, 595–610. [CrossRef] [PubMed]
167. McKemy, D.D.; Neuhausser, W.M.; Julius, D. Identification of a cold receptor reveals a general role for TRP channels in thermosensation. *Nature* **2002**, *416*, 52–58. [CrossRef]
168. Peier, A.M.; Moqrich, A.; Hergarden, A.C.; Reeve, A.J.; Andersson, D.A.; Story, G.M.; Earley, T.J.; Dragoni, I.; McIntyre, P.; Bevan, S.; et al. A TRP channel that senses cold stimuli and menthol. *Cell* **2002**, *108*, 705–715. [CrossRef]
169. Reid, G.; Flonta, M.L. Ion channels activated by cold and menthol in cultured rat dorsal root ganglion neurones. *Neurosci. Lett.* **2002**, *324*, 164–168. [CrossRef]
170. Thut, P.D.; Wrigley, D.; Gold, M.S. Cold transduction in rat trigeminal ganglia neurons in vitro. *Neuroscience* **2003**, *119*, 1071–1083. [CrossRef]
171. Kobayashi, K.; Fukuoka, T.; Obata, K.; Yamanaka, H.; Dai, Y.; Tokunaga, A.; Noguchi, K. Distinct expression of TRPM8, TRPA1, and TRPV1 mRNAs in rat primary afferent neurons with adelta/c-fibers and colocalization with trk receptors. *J. Comp. Neurol.* **2005**, *493*, 596–606. [CrossRef]
172. Abe, J.; Hosokawa, H.; Okazawa, M.; Kandachi, M.; Sawada, Y.; Yamanaka, K.; Matsumura, K.; Kobayashi, S. TRPM8 protein localization in trigeminal ganglion and taste papillae. *Brain Res.* **2005**, *136*, 91–98. [CrossRef]
173. Dhaka, A.; Murray, A.N.; Mathur, J.; Earley, T.J.; Petrus, M.J.; Patapoutian, A. TRPM8 is required for cold sensation in mice. *Neuron* **2007**, *54*, 371–378. [CrossRef] [PubMed]
174. Bautista, D.M.; Siemens, J.; Glazer, J.M.; Tsuruda, P.R.; Basbaum, A.I.; Stucky, C.L.; Jordt, S.E.; Julius, D. The menthol receptor TRPM8 is the principal detector of environmental cold. *Nature* **2007**, *448*, 204–208. [CrossRef] [PubMed]
175. Colburn, R.W.; Lubin, M.L.; Stone, D.J., Jr.; Wang, Y.; Lawrence, D.; D'Andrea, M.R.; Brandt, M.R.; Liu, Y.; Flores, C.M.; Qin, N. Attenuated cold sensitivity in TRPM8 null mice. *Neuron* **2007**, *54*, 379–386. [CrossRef] [PubMed]
176. Michot, B.; Lee, C.S.; Gibbs, J.L. TRPM8 and TRPA1 do not contribute to dental pulp sensitivity to cold. *Sci. Rep.* **2018**, *8*, 13198. [CrossRef] [PubMed]
177. Tsumura, M.; Sobhan, U.; Sato, M.; Shimada, M.; Nishiyama, A.; Kawaguchi, A.; Soya, M.; Kuroda, H.; Tazaki, M.; Shibukawa, Y. Functional expression of TRPM8 and TRPA1 channels in rat odontoblasts. *PLoS ONE* **2013**, *8*, e82233. [CrossRef] [PubMed]
178. Tazawa, K.; Ikeda, H.; Kawashima, N.; Okiji, T. Transient receptor potential melastatin (TRPM) 8 is expressed in freshly isolated native human odontoblasts. *Arch. Oral Biol.* **2017**, *75*, 55–61. [CrossRef]
179. Kwon, M.; Baek, S.H.; Park, C.K.; Chung, G.; Oh, S.B. Single-cell RT-PCR and immunocytochemical detection of mechanosensitive transient receptor potential channels in acutely isolated rat odontoblasts. *Arch. Oral Biol.* **2014**, *59*, 1266–1271. [CrossRef]

180. Won, J.; Vang, H.; Kim, J.H.; Lee, P.R.; Kang, Y.; Oh, S.B. TRPM7 Mediates Mechanosensitivity in Adult Rat Odontoblasts. *J. Dent. Res.* **2018**, *97*, 1039–1046. [CrossRef]
181. Nakano, Y.; Le, M.H.; Abduweli, D.; Ho, S.P.; Ryazanova, L.V.; Hu, Z.; Ryazanov, A.G.; Den Besten, P.K.; Zhang, Y. A Critical Role of TRPM7 As an Ion Channel Protein in Mediating the Mineralization of the Craniofacial Hard Tissues. *Front. Physiol.* **2016**, *7*, 258. [CrossRef]
182. Won, J.; Kim, J.H.; Oh, S.B. Molecular expression of Mg^{2+} regulator TRPM7 and CNNM4 in rat odontoblasts. *Arch. Oral Biol.* **2018**, *96*, 182–188. [CrossRef]
183. Nilius, B.; Appendino, G.; Owsianik, G. The transient receptor potential channel TRPA1: From gene to pathophysiology. *Pflug. Arch.* **2012**, *464*, 425–458. [CrossRef] [PubMed]
184. Nassenstein, C.; Kwong, K.; Taylor-Clark, T.; Kollarik, M.; Macglashan, D.M.; Braun, A.; Undem, B.J. Expression and function of the ion channel TRPA1 in vagal afferent nerves innervating mouse lungs. *J. Physiol.* **2008**, *586*, 1595–1604. [CrossRef] [PubMed]
185. Nozawa, K.; Kawabata-Shoda, E.; Doihara, H.; Kojima, R.; Okada, H.; Mochizuki, S.; Sano, Y.; Inamura, K.; Matsushime, H.; Koizumi, T.; et al. TRPA1 regulates gastrointestinal motility through serotonin release from enterochromaffin cells. *Proc. Natl. Acad. Sci. USA* **2009**, *106*, 3408–3413. [CrossRef] [PubMed]
186. Ro, J.Y.; Lee, J.S.; Zhang, Y. Activation of TRPV1 and TRPA1 leads to muscle nociception and mechanical hyperalgesia. *Pain* **2009**, *144*, 270–277. [CrossRef] [PubMed]
187. Story, G.M.; Peier, A.M.; Reeve, A.J.; Eid, S.R.; Mosbacher, J.; Hricik, T.R.; Earley, T.J.; Hergarden, A.C.; Andersson, D.A.; Hwang, S.W.; et al. ANKTM1, a TRP-like channel expressed in nociceptive neurons, is activated by cold temperatures. *Cell* **2003**, *112*, 819–829. [CrossRef]
188. Obata, K.; Katsura, H.; Mizushima, T.; Yamanaka, H.; Kobayashi, K.; Dai, Y.; Fukuoka, T.; Tokunaga, A.; Tominaga, M.; Noguchi, K. TRPA1 induced in sensory neurons contributes to cold hyperalgesia after inflammation and nerve injury. *J. Clin. Investig.* **2005**, *115*, 2393–2401. [CrossRef] [PubMed]
189. Bautista, D.M.; Jordt, S.E.; Nikai, T.; Tsuruda, P.R.; Read, A.J.; Poblete, J.; Yamoah, E.N.; Basbaum, A.I.; Julius, D. TRPA1 mediates the inflammatory actions of environmental irritants and proalgesic agents. *Cell* **2006**, *124*, 1269–1282. [CrossRef] [PubMed]
190. Kwan, K.Y.; Allchorne, A.J.; Vollrath, M.A.; Christensen, A.P.; Zhang, D.S.; Woolf, C.J.; Corey, D.P. TRPA1 contributes to cold, mechanical, and chemical nociception but is not essential for hair-cell transduction. *Neuron* **2006**, *50*, 277–289. [CrossRef]
191. Bandell, M.; Story, G.M.; Hwang, S.W.; Viswanath, V.; Eid, S.R.; Petrus, M.J.; Earley, T.J.; Patapoutian, A. Noxious cold ion channel TRPA1 is activated by pungent compounds and bradykinin. *Neuron* **2004**, *41*, 849–857. [CrossRef]
192. Jordt, S.E.; Bautista, D.M.; Chuang, H.H.; McKemy, D.D.; Zygmunt, P.M.; Hogestatt, E.D.; Meng, I.D.; Julius, D. Mustard oils and cannabinoids excite sensory nerve fibres through the TRP channel ANKTM1. *Nature* **2004**, *427*, 260–265. [CrossRef]
193. Bautista, D.M.; Movahed, P.; Hinman, A.; Axelsson, H.E.; Sterner, O.; Hogestatt, E.D.; Julius, D.; Jordt, S.E.; Zygmunt, P.M. Pungent products from garlic activate the sensory ion channel TRPA1. *Proc. Natl. Acad. Sci. USA* **2005**, *102*, 12248–12252. [CrossRef] [PubMed]
194. Haas, E.T.; Rowland, K.; Gautam, M. Tooth injury increases expression of the cold sensitive TRP channel TRPA1 in trigeminal neurons. *Arch. Oral Biol.* **2011**, *56*, 1604–1609. [CrossRef] [PubMed]
195. Kim, Y.S.; Jung, H.K.; Kwon, T.K.; Kim, C.S.; Cho, J.H.; Ahn, D.K.; Bae, Y.C. Expression of transient receptor potential ankyrin 1 in human dental pulp. *J. Endod.* **2012**, *38*, 1087–1092. [CrossRef] [PubMed]
196. El Karim, I.A.; McCrudden, M.T.; McGahon, M.K.; Curtis, T.M.; Jeanneau, C.; Giraud, T.; Irwin, C.R.; Linden, G.J.; Lundy, F.T.; About, I. Biodentine Reduces Tumor Necrosis Factor α-induced TRPA1 Expression in Odontoblastlike Cells. *J. Endod.* **2016**, *42*, 589–595. [CrossRef] [PubMed]
197. Yang, X.; Song, Z.; Chen, L.; Wang, R.; Huang, S.; Qin, W.; Guo, J.; Lin, Z. Role of transient receptor potential channel 6 in the odontogenic differentiation of human dental pulp cells. *Exp. Ther. Med.* **2017**, *14*, 73–78. [CrossRef] [PubMed]
198. Song, Z.; Chen, L.; Guo, J.; Qin, W.; Wang, R.; Huang, S.; Yang, X.; Tian, Y.; Lin, Z. The Role of Transient Receptor Potential Cation Channel, Subfamily C, Member 1 in the Odontoblast-like Differentiation of Human Dental Pulp Cells. *J. Endod.* **2017**, *43*, 315–320. [CrossRef] [PubMed]
199. Dietrich, A.; Fahlbusch, M.; Gudermann, T. Classical Transient Receptor Potential 1 (TRPC1): Channel or Channel Regulator? *Cells* **2014**, *3*, 939–962. [CrossRef]

200. Ambudkar, I.S. TRPC1: A core component of store-operated calcium channels. *Biochem. Soc. Trans.* **2007**, *35*, 96–100. [CrossRef]
201. Strubing, C.; Krapivinsky, G.; Krapivinsky, L.; Clapham, D.E. TRPC1 and TRPC5 form a novel cation channel in mammalian brain. *Neuron* **2001**, *29*, 645–655. [CrossRef]
202. Valero, M.L.; Caminos, E.; Juiz, J.M.; Martinez-Galan, J.R. TRPC1 and metabotropic glutamate receptor expression in rat auditory midbrain neurons. *J. Neurosci. Res.* **2015**, *93*, 964–972. [CrossRef]
203. Broker-Lai, J.; Kollewe, A.; Schindeldecker, B.; Pohle, J.; Nguyen Chi, V.; Mathar, I.; Guzman, R.; Schwarz, Y.; Lai, A.; Weissgerber, P.; et al. Heteromeric channels formed by TRPC1, TRPC4 and TRPC5 define hippocampal synaptic transmission and working memory. *EMBO J.* **2017**, *36*, 2770–2789. [CrossRef] [PubMed]
204. Elg, S.; Marmigere, F.; Mattsson, J.P.; Ernfors, P. Cellular subtype distribution and developmental regulation of TRPC channel members in the mouse dorsal root ganglion. *J. Comp. Neurol.* **2007**, *503*, 35–46. [CrossRef]
205. Garrison, S.R.; Dietrich, A.; Stucky, C.L. TRPC1 contributes to light-touch sensation and mechanical responses in low-threshold cutaneous sensory neurons. *J. Neurophysiol.* **2012**, *107*, 913–922. [CrossRef] [PubMed]
206. Spassova, M.A.; Hewavitharana, T.; Xu, W.; Soboloff, J.; Gill, D.L. A common mechanism underlies stretch activation and receptor activation of TRPC6 channels. *Proc. Natl. Acad. Sci. USA* **2006**, *103*, 16586–16591. [CrossRef] [PubMed]
207. Maroto, R.; Raso, A.; Wood, T.G.; Kurosky, A.; Martinac, B.; Hamill, O.P. TRPC1 forms the stretch-activated cation channel in vertebrate cells. *Nat. Cell Biol.* **2005**, *7*, 179–185. [CrossRef] [PubMed]
208. Alessandri-Haber, N.; Dina, O.A.; Chen, X.; Levine, J.D. TRPC1 and TRPC6 channels cooperate with TRPV4 to mediate mechanical hyperalgesia and nociceptor sensitization. *J. Neurosci.* **2009**, *29*, 6217–6228. [CrossRef] [PubMed]
209. Miyamoto, R.; Tokuda, M.; Sakuta, T.; Nagaoka, S.; Torii, M. Expression and characterization of vanilloid receptor subtype 1 in human dental pulp cell cultures. *J. Endod.* **2005**, *31*, 652–658. [CrossRef]
210. El Karim, I.A.; Linden, G.J.; Curtis, T.M.; About, I.; McGahon, M.K.; Irwin, C.R.; Killough, S.A.; Lundy, F.T. Human dental pulp fibroblasts express the "cold-sensing" transient receptor potential channels TRPA1 and TRPM8. *J. Endod.* **2011**, *37*, 473–478. [CrossRef]
211. Rowland, K.C.; Kanive, C.B.; Wells, J.E.; Hatton, J.F. TRPM2 immunoreactivity is increased in fibroblasts, but not nerves, of symptomatic human dental pulp. *J. Endod.* **2007**, *33*, 245–248. [CrossRef]
212. Hashimoto, M.; Ito, S.; Tay, F.R.; Svizero, N.R.; Sano, H.; Kaga, M.; Pashley, D.H. Fluid movement across the resin-dentin interface during and after bonding. *J. Dent. Res.* **2004**, *83*, 843–848. [CrossRef]
213. Kim, S. Thermal stimuli in dentinal sensitivity. *Endod. Dent. Traumatol.* **1986**, *2*, 138–140. [CrossRef] [PubMed]
214. Kim, S.; Schuessler, G.; Chien, S. Measurement of blood flow in the dental pulp of dogs with the 133xenon washout method. *Arch. Oral Biol.* **1983**, *28*, 501–505. [CrossRef]
215. Narhi, M. Activation of dental pulp nerves of the cat and the dog with hydrostatic pressure. *Proc. Finn. Dent. Soc.* **1978**, *74*, 1–63. [PubMed]
216. Pashley, D.H. Sensitivity of dentin to chemical stimuli. *Endod. Dent. Traumatol.* **1986**, *2*, 130–137. [CrossRef] [PubMed]
217. Roudaut, Y.; Lonigro, A.; Coste, B.; Hao, J.; Delmas, P.; Crest, M. Touch sense: Functional organization and molecular determinants of mechanosensitive receptors. *Channels* **2012**, *6*, 234–245. [CrossRef]
218. Christensen, A.P.; Corey, D.P. TRP channels in mechanosensation: Direct or indirect activation? *Nat. Rev.* **2007**, *8*, 510–521. [CrossRef]
219. Narhi, M.; Jyvasjarvi, E.; Virtanen, A.; Huopaniemi, T.; Ngassapa, D.; Hirvonen, T. Role of intradental A- and C-type nerve fibres in dental pain mechanisms. *Proc. Finn. Dent. Soc.* **1992**, *88*, 507–516.
220. Narhi, M.; Jyvasjarvi, E.; Hirvonen, T.; Huopaniemi, T. Activation of heat-sensitive nerve fibres in the dental pulp of the cat. *Pain* **1982**, *14*, 317–326. [CrossRef]
221. Hamilton, S.G. ATP and pain. *Pain Pract.* **2002**, *2*, 289–294. [CrossRef]
222. Wirkner, K.; Sperlagh, B.; Illes, P. P2X3 receptor involvement in pain states. *Mol. Neurobiol.* **2007**, *36*, 165–183. [CrossRef]
223. Sato, M.; Ogura, K.; Kimura, M.; Nishi, K.; Ando, M.; Tazaki, M.; Shibukawa, Y. Activation of Mechanosensitive Transient Receptor Potential/Piezo Channels in Odontoblasts Generates Action Potentials in Cocultured Isolectin B4-negative Medium-sized Trigeminal Ganglion Neurons. *J. Endod.* **2018**, *44*, 984–991. [CrossRef] [PubMed]

224. Liu, X.; Wang, C.; Fujita, T.; Malmstrom, H.S.; Nedergaard, M.; Ren, Y.F.; Dirksen, R.T. External Dentin Stimulation Induces ATP Release in Human Teeth. *J. Dent. Res.* **2015**, *94*, 1259–1266. [CrossRef]
225. Zimmermann, H.; Braun, N. Extracellular metabolism of nucleotides in the nervous system. *J. Auton. Pharmacol.* **1996**, *16*, 397–400. [CrossRef] [PubMed]
226. Liu, X.; Yu, L.; Wang, Q.; Pelletier, J.; Fausther, M.; Sevigny, J.; Malmstrom, H.S.; Dirksen, R.T.; Ren, Y.F. Expression of ecto-ATPase NTPDase2 in human dental pulp. *J. Dent. Res.* **2012**, *91*, 261–267. [CrossRef]
227. Burnstock, G. Physiology and pathophysiology of purinergic neurotransmission. *Physiol. Rev.* **2007**, *87*, 659–797. [CrossRef]
228. Kuroda, H.; Shibukawa, Y.; Soya, M.; Masamura, A.; Kasahara, M.; Tazaki, M.; Ichinohe, T. Expression of P2X(1) and P2X(4) receptors in rat trigeminal ganglion neurons. *Neuroreport* **2012**, *23*, 752–756. [CrossRef] [PubMed]
229. Xiang, Z.; Bo, X.; Burnstock, G. Localization of ATP-gated P2X receptor immunoreactivity in rat sensory and sympathetic ganglia. *Neurosci. Lett.* **1998**, *256*, 105–108. [CrossRef]
230. Jiang, J.; Gu, J. Expression of adenosine triphosphate P2X3 receptors in rat molar pulp and trigeminal ganglia. *Oral Surg. Oral Med. Oral Pathol. Oral Radiol.* **2002**, *94*, 622–626. [CrossRef]
231. Alavi, A.M.; Dubyak, G.R.; Burnstock, G. Immunohistochemical evidence for ATP receptors in human dental pulp. *J. Dent. Res.* **2001**, *80*, 476–483. [CrossRef]
232. Cook, S.P.; Vulchanova, L.; Hargreaves, K.M.; Elde, R.; McCleskey, E.W. Distinct ATP receptors on pain-sensing and stretch-sensing neurons. *Nature* **1997**, *387*, 505–508. [CrossRef]
233. Kawaguchi, A.; Sato, M.; Kimura, M.; Ichinohe, T.; Tazaki, M.; Shibukawa, Y. Expression and function of purinergic P2Y12 receptors in rat trigeminal ganglion neurons. *Neurosci. Res.* **2015**, *98*, 17–27. [CrossRef] [PubMed]
234. Nishiyama, A.; Sato, M.; Kimura, M.; Katakura, A.; Tazaki, M.; Shibukawa, Y. Intercellular signal communication among odontoblasts and trigeminal ganglion neurons via glutamate. *Cell Calcium* **2016**, *60*, 341–355. [CrossRef] [PubMed]
235. Cho, Y.S.; Ryu, C.H.; Won, J.H.; Vang, H.; Oh, S.B.; Ro, J.Y.; Bae, Y.C. Rat odontoblasts may use glutamate to signal dentin injury. *Neuroscience* **2016**, *335*, 54–63. [CrossRef] [PubMed]
236. Horst, O.V.; Tompkins, K.A.; Coats, S.R.; Braham, P.H.; Darveau, R.P.; Dale, B.A. TGF-β1 Inhibits TLR-mediated odontoblast responses to oral bacteria. *J. Dent. Res.* **2009**, *88*, 333–338. [CrossRef] [PubMed]
237. Jiang, H.W.; Zhang, W.; Ren, B.P.; Zeng, J.F.; Ling, J.Q. Expression of toll like receptor 4 in normal human odontoblasts and dental pulp tissue. *J. Endod.* **2006**, *32*, 747–751. [CrossRef] [PubMed]
238. Mutoh, N.; Tani-Ishii, N.; Tsukinoki, K.; Chieda, K.; Watanabe, K. Expression of toll-like receptor 2 and 4 in dental pulp. *J. Endod.* **2007**, *33*, 1183–1186. [CrossRef] [PubMed]
239. Staquet, M.J.; Carrouel, F.; Keller, J.F.; Baudouin, C.; Msika, P.; Bleicher, F.; Kufer, T.A.; Farges, J.C. Pattern-recognition receptors in pulp defense. *Adv. Dent. Res.* **2011**, *23*, 296–301. [CrossRef]
240. Rechenberg, D.K.; Galicia, J.C.; Peters, O.A. Biological Markers for Pulpal Inflammation: A Systematic Review. *PLoS ONE* **2016**, *11*, e0167289. [CrossRef]
241. Hahn, C.L.; Liewehr, F.R. Innate immune responses of the dental pulp to caries. *J. Endod.* **2007**, *33*, 643–651. [CrossRef]
242. Lundy, F.T.; Linden, G.J. Neuropeptides and Neurogenic Mechanisms in Oral and Periodontal Inflammation. *Crit. Rev. Oral Biol. Med.* **2004**, *15*, 82–98. [CrossRef]
243. Caviedes-Bucheli, J.; Munoz, H.R.; Azuero-Holguin, M.M.; Ulate, E. Neuropeptides in dental pulp: The silent protagonists. *J. Endod.* **2008**, *34*, 773–788. [CrossRef] [PubMed]
244. Mickle, A.D.; Shepherd, A.J.; Mohapatra, D.P. Sensory TRP channels: The key transducers of nociception and pain. *Prog. Mol. Biol. Transl. Sci.* **2015**, *131*, 73–118. [PubMed]
245. Hossain, M.Z.; Unno, S.; Ando, H.; Masuda, Y.; Kitagawa, J. Neuron-Glia Crosstalk and Neuropathic Pain: Involvement in the Modulation of Motor Activity in the Orofacial Region. *Int. J. Mol. Sci.* **2017**, *18*, 2051. [CrossRef] [PubMed]
246. Patapoutian, A.; Tate, S.; Woolf, C.J. Transient receptor potential channels: Targeting pain at the source. *Nat. Rev. Drug Discov.* **2009**, *8*, 55–68. [CrossRef] [PubMed]
247. Rodd, H.D.; Boissonade, F.M. Substance P expression in human tooth pulp in relation to caries and pain experience. *Eur. J. Oral Sci.* **2000**, *108*, 467–474. [CrossRef] [PubMed]

248. Killough, S.A.; Lundy, F.T.; Irwin, C.R. Substance P expression by human dental pulp fibroblasts: A potential role in neurogenic inflammation. *J. Endod.* **2009**, *35*, 73–77. [CrossRef]
249. Sacerdote, P.; Levrini, L. Peripheral mechanisms of dental pain: The role of substance P. *Mediat. Inflamm.* **2012**, *2012*, 951920. [CrossRef]
250. Nakanishi, T.; Takegawa, D.; Hirao, K.; Takahashi, K.; Yumoto, H.; Matsuo, T. Roles of dental pulp fibroblasts in the recognition of bacterium-related factors and subsequent development of pulpitis. *Jpn. Dent. Sci. Rev.* **2011**, *47*, 161–166. [CrossRef]
251. Suvas, S. Role of Substance P Neuropeptide in Inflammation, Wound Healing, and Tissue Homeostasis. *J. Immunol.* **2017**, *199*, 1543–1552. [CrossRef]
252. Goehler, L.E.; Gaykema, R.P.; Opitz, N.; Reddaway, R.; Badr, N.; Lyte, M. Activation in vagal afferents and central autonomic pathways: Early responses to intestinal infection with Campylobacter jejuni. *Brain Behav. Immun.* **2005**, *19*, 334–344. [CrossRef]
253. Chiu, I.M.; Heesters, B.A.; Ghasemlou, N.; Von Hehn, C.A.; Zhao, F.; Tran, J.; Wainger, B.; Strominger, A.; Muralidharan, S.; Horswill, A.R.; et al. Bacteria activate sensory neurons that modulate pain and inflammation. *Nature* **2013**, *501*, 52–57. [CrossRef] [PubMed]
254. Boonen, B.; Alpizar, Y.A.; Meseguer, V.M.; Talavera, K. TRP Channels as Sensors of Bacterial Endotoxins. *Toxins* **2018**, *10*, 326. [CrossRef] [PubMed]
255. Diogenes, A.; Ferraz, C.C.; Akopian, A.N.; Henry, M.A.; Hargreaves, K.M. LPS sensitizes TRPV1 via activation of TLR4 in trigeminal sensory neurons. *J. Dent. Res.* **2011**, *90*, 759–764. [CrossRef] [PubMed]
256. Ferraz, C.C.; Henry, M.A.; Hargreaves, K.M.; Diogenes, A. Lipopolysaccharide from Porphyromonas gingivalis sensitizes capsaicin-sensitive nociceptors. *J. Endod.* **2011**, *37*, 45–48. [CrossRef] [PubMed]
257. Lin, J.J.; Du, Y.; Cai, W.K.; Kuang, R.; Chang, T.; Zhang, Z.; Yang, Y.X.; Sun, C.; Li, Z.Y.; Kuang, F. Toll-like receptor 4 signaling in neurons of trigeminal ganglion contributes to nociception induced by acute pulpitis in rats. *Sci. Rep.* **2015**, *5*, 12549. [CrossRef] [PubMed]
258. Meseguer, V.; Alpizar, Y.A.; Luis, E.; Tajada, S.; Denlinger, B.; Fajardo, O.; Manenschijn, J.A.; Fernandez-Pena, C.; Talavera, A.; Kichko, T.; et al. TRPA1 channels mediate acute neurogenic inflammation and pain produced by bacterial endotoxins. *Nat. Commun.* **2014**, *5*, 3125. [CrossRef] [PubMed]
259. Boonen, B.; Alpizar, Y.A.; Sanchez, A.; Lopez-Requena, A.; Voets, T.; Talavera, K. Differential effects of lipopolysaccharide on mouse sensory TRP channels. *Cell Calcium* **2018**, *73*, 72–81. [CrossRef] [PubMed]
260. Alpizar, Y.A.; Boonen, B.; Sanchez, A.; Jung, C.; Lopez-Requena, A.; Naert, R.; Steelant, B.; Luyts, K.; Plata, C.; De Vooght, V.; et al. TRPV4 activation triggers protective responses to bacterial lipopolysaccharides in airway epithelial cells. *Nat. Commun.* **2017**, *8*, 1059. [CrossRef]
261. Pashley, D.H. How can sensitive dentine become hypersensitive and can it be reversed? *J. Den.* **2013**, *41*, S49–S55. [CrossRef]
262. Stenvik, A.; Iversen, J.; Mjor, I.A. Tissue pressure and histology of normal and inflamed tooth pulps in macaque monkeys. *Arch. Oral Biol.* **1972**, *17*, 1501–1511. [CrossRef]
263. Hossain, M.Z.; Daud, S.; Nambiar, P.; Razak, F.A.; Ab-Murat, N.; Saub, R.; Bakri, M.M. Correlation between numbers of cells in human dental pulp and age: Implications for age estimation. *Arch. Oral Biol.* **2017**, *80*, 51–55. [CrossRef] [PubMed]
264. Daud, S.; Nambiar, P.; Hossain, M.Z.; Rahman, M.R.; Bakri, M.M. Changes in cell density and morphology of selected cells of the ageing human dental pulp. *Gerodontology* **2016**, *33*, 315–321. [CrossRef] [PubMed]
265. Yang, L.C.; Huang, F.M.; Lin, C.S.; Liu, C.M.; Lai, C.C.; Chang, Y.C. Induction of interleukin-8 gene expression by black-pigmented Bacteroides in human pulp fibroblasts and osteoblasts. *Int. Endod. J.* **2003**, *36*, 774–779. [CrossRef] [PubMed]
266. Coil, J.; Tam, E.; Waterfield, J.D. Proinflammatory cytokine profiles in pulp fibroblasts stimulated with lipopolysaccharide and methyl mercaptan. *J. Endod.* **2004**, *30*, 88–91. [CrossRef]
267. Patel, T.; Park, S.H.; Lin, L.M.; Chiappelli, F.; Huang, G.T. Substance P induces interleukin-8 secretion from human dental pulp cells. *Oral Surg. Oral Med. Oral Pathol. Oral Radiol.* **2003**, *96*, 478–485. [CrossRef]
268. Park, S.H.; Hsiao, G.Y.; Huang, G.T. Role of substance P and calcitonin gene-related peptide in the regulation of interleukin-8 and monocyte chemotactic protein-1 expression in human dental pulp. *Int. Endod. J.* **2004**, *37*, 185–192. [CrossRef] [PubMed]

269. Yamaguchi, M.; Kojima, T.; Kanekawa, M.; Aihara, N.; Nogimura, A.; Kasai, K. Neuropeptides stimulate production of interleukin-1β, interleukin-6, and tumor necrosis factor-α in human dental pulp cells. *Inflamm. Res.* **2004**, *53*, 199–204. [CrossRef]
270. Binshtok, A.M.; Bean, B.P.; Woolf, C.J. Inhibition of nociceptors by TRPV1-mediated entry of impermeant sodium channel blockers. *Nature* **2007**, *449*, 607–610. [CrossRef]
271. Zakir, H.M.; Mostafeezur, R.M.; Suzuki, A.; Hitomi, S.; Suzuki, I.; Maeda, T.; Seo, K.; Yamada, Y.; Yamamura, K.; Lev, S.; et al. Expression of TRPV1 channels after nerve injury provides an essential delivery tool for neuropathic pain attenuation. *PLoS ONE* **2012**, *7*, e44023. [CrossRef]
272. Kim, H.Y.; Kim, K.; Li, H.Y.; Chung, G.; Park, C.K.; Kim, J.S.; Jung, S.J.; Lee, M.K.; Ahn, D.K.; Hwang, S.J.; et al. Selectively targeting pain in the trigeminal system. *Pain* **2010**, *150*, 29–40. [CrossRef]
273. Badral, B.; Davies, A.J.; Kim, Y.H.; Ahn, J.S.; Hong, S.D.; Chung, G.; Kim, J.S.; Oh, S.B. Pain fiber anesthetic reduces brainstem Fos after tooth extraction. *J. Dent. Res.* **2013**, *92*, 1005–1010. [CrossRef] [PubMed]
274. Yang, B.H.; Piao, Z.G.; Kim, Y.B.; Lee, C.H.; Lee, J.K.; Park, K.; Kim, J.S.; Oh, S.B. Activation of vanilloid receptor 1 (VR1) by eugenol. *J. Dent. Res.* **2003**, *82*, 781–785. [CrossRef] [PubMed]
275. Mickle, A.D.; Shepherd, A.J.; Mohapatra, D.P. Nociceptive TRP Channels: Sensory Detectors and Transducers in Multiple Pain Pathologies. *Pharmaceuticals* **2016**, *9*, 72. [CrossRef] [PubMed]
276. Gavva, N.R.; Treanor, J.J.; Garami, A.; Fang, L.; Surapaneni, S.; Akrami, A.; Alvarez, F.; Bak, A.; Darling, M.; Gore, A.; et al. Pharmacological blockade of the vanilloid receptor TRPV1 elicits marked hyperthermia in humans. *Pain* **2008**, *136*, 202–210. [CrossRef] [PubMed]
277. Quiding, H.; Jonzon, B.; Svensson, O.; Webster, L.; Reimfelt, A.; Karin, A.; Karlsten, R.; Segerdahl, M. TRPV1 antagonistic analgesic effect: A randomized study of AZD1386 in pain after third molar extraction. *Pain* **2013**, *154*, 808–812. [CrossRef] [PubMed]

© 2019 by the authors. Licensee MDPI, Basel, Switzerland. This article is an open access article distributed under the terms and conditions of the Creative Commons Attribution (CC BY) license (http://creativecommons.org/licenses/by/4.0/).

Review

Potential Causes of Titanium Particle and Ion Release in Implant Dentistry: A Systematic Review

Rafael Delgado-Ruiz [1],* and Georgios Romanos [2,3]

1. Department of Prosthodontics and Digital Technology, School of Dental Medicine, Stony Brook University, New York, NY 11794, USA
2. Department of Periodontics, School of Dental Medicine, Stony Brook University, New York, NY 11794, USA; georgios.romanos@stonybrookmedicine.edu
3. Department of Oral Surgery and Implant Dentistry, Dental School, Johann Wolfgang Goethe University, 60323 Frankfurt, Germany
* Correspondence: rafael.delgado-ruiz@stonybrookmedicine.edu

Received: 25 October 2018; Accepted: 11 November 2018; Published: 13 November 2018

Abstract: Implant surface characteristics, as well as physical and mechanical properties, are responsible for the positive interaction between the dental implant, the bone and the surrounding soft tissues. Unfortunately, the dental implant surface does not remain unaltered and changes over time during the life of the implant. If changes occur at the implant surface, mucositis and peri-implantitis processes could be initiated; implant osseointegration might be disrupted and bone resorption phenomena (osteolysis) may lead to implant loss. This systematic review compiled the information related to the potential sources of titanium particle and ions in implant dentistry. Research questions were structured in the Population, Intervention, Comparison, Outcome (PICO) framework. PICO questionnaires were developed and an exhaustive search was performed for all the relevant studies published between 1980 and 2018 involving titanium particles and ions related to implant dentistry procedures. Preferred Reporting Items for Systematic Reviews and Meta-Analyses (PRISMA) guidelines were followed for the selection and inclusion of the manuscripts in this review. Titanium particle and ions are released during the implant bed preparation, during the implant insertion and during the implant decontamination. In addition, the implant surfaces and restorations are exposed to the saliva, bacteria and chemicals that can potentially dissolve the titanium oxide layer and, therefore, corrosion cycles can be initiated. Mechanical factors, the micro-gap and fluorides can also influence the proportion of metal particles and ions released from implants and restorations.

Keywords: dental implants; titanium particles; wear; corrosion

1. Introduction

Titanium implants have been used for dental, orthopedic and other medical applications since the early 1980s [1,2]. Implant-related factors, such as the surface characteristics, material composition and chemistry, are responsible for the osseointegration. Specifically, the presence of a titanium oxide layer on the implant surface is considered crucial for the maintenance of the osseointegration and the prevention of the corrosion of the titanium surface [3]. The dental implant surface does not remain unaltered and, if changes occur at the implant surface, mucositis and peri-implantitis processes could be initiated; osseointegration might be disrupted, and bone resorption phenomena (osteolysis) may lead to implant loss [1–4]. Particularly, at the moment of implant insertion, the implant surface can incur changes in its chemical and topographic structures, which is sometimes irreversible [1] and also titanium particles of different sizes and characteristics can be released from the dental implant surface [5–9].

Subsequently, if the implant surface is altered, degraded or dissolved by the effects of acidic substances or an acidic environment, the titanium oxide layer can be lost, and corrosion phenomena can begin [10–14]. Furthermore, during functional loading, the combination of mechanical and chemical corrosion (fretting) between the implant surface and the adjacent bone can facilitate the release of more metal particles and ions and, if friction occurs between the internal implant walls and the prosthetic abutment, additional particles and ions can be released into the surrounding tissues [2–4,15]. Lastly, cleaning and disinfection of the implant surface with mechanical, physical or chemical methods have been shown to induce changes of different magnitudes at the implant and abutment, with subsequent titanium oxide layer loss, titanium particle or ion loss and corrosion initiation [16–20].

However, the sources of titanium particle or ions in implant dentistry procedures have not been studied in depth and methods for its reduction are currently unknown. Therefore, all the information related to the potential sources of titanium ions and particles and suggestions of methods for their control in implant dentistry is fundamental for the long-term survival of dental implants to improve clinical practice.

The purpose of this systematic review is to compile the information related to the potential sources of titanium particles and ions in dental implantology. Taken together, the findings of this review, including the percentages of particles, their characteristics, and detection methods, suggest ways for reducing titanium particle and ion release in implant dentistry.

2. Results

The initial search returned 635 articles. A total of 22 articles were removed because they were duplicates; the remaining 613 articles abstracts were read in full, and 378 articles were excluded because they did not fulfil the inclusion criteria. The full texts of the remaining 235 articles were read in full for eligibility, and 29 articles were removed because they presented reviews or expert opinions or were duplicated. Finally, 206 articles were included in this systematic review (Figure 1).

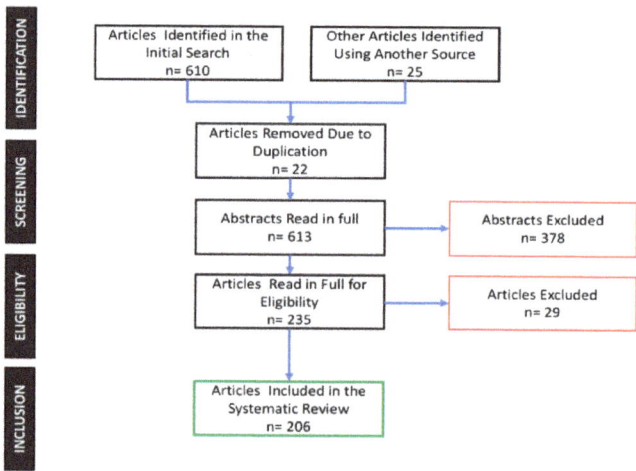

Figure 1. Preferred Reporting Items for Systematic Reviews (PRISMA) flow diagram of the screening and selection process.

The findings were grouped as follows: sources of titanium release during the surgical phase, during the prosthetic phase and during maintenance (Figure 2).

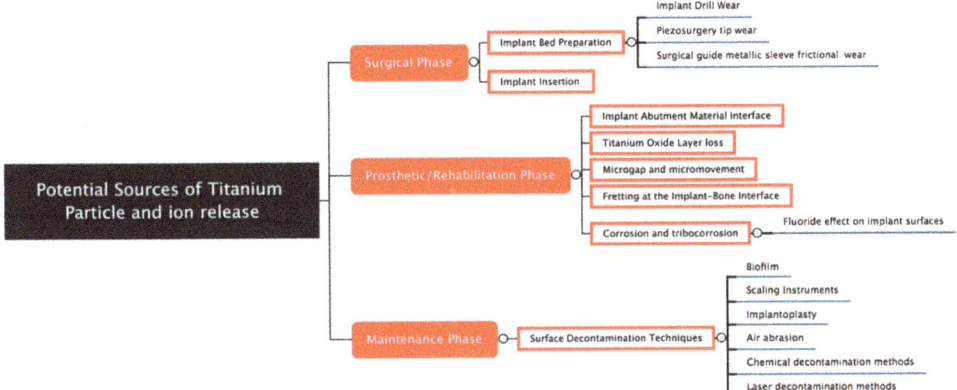

Figure 2. This systematic review found that titanium particles and ions can be released during the surgical, prosthetic and maintenance phases due to different causes during the life span of a dental implant. The rectangles filled in red are the three phases in implant dentistry procedures in which titanium particles and ions can be released. The rectangles with the red frame are procedures within the previous phases which resulted in titanium particles and ions release. The underlined sentences are the detailed sources or initiators of titanium particles and ions release.

2.1. Potential Causes of Titanium Particle Release during the Surgical Phase

2.1.1. Implant Bed Preparation

Bone-cutting instruments, when subjected to frictional forces, can suffer variable levels of metal attrition, wear and corrosion. Traces of different metallic elements have been observed after osteotomy in regional ganglions, kidneys and lungs [21,22]. In addition, it has been demonstrated that the irrigation liquid collected from the implant bed preparation contains metallic debris and ions. Rashad et al. [23] performed implant osteotomies with rotatory instruments and piezosurgery devices to detect drill deposits in bone or in the recovered irrigation liquid. After implant bed preparation, bone samples were examined by (Scanning Electron Microscope) SEM/energy dispersion X-ray spectroscopy (EDX), the irrigation liquid was collected and filtered with polycarbonate membranes with a specific pore size. The membranes were further analysed by SEM/EDX for the detection of metal content. The results showed different residual metals from the drills in the irrigation liquid (Ag, Si, Fe, Ti, V, Cu, Mn, Zr, Cr, Bi, Mg) for both osteotomy methods. The authors recommended the use of copious irrigation to reduce the metallic particle content in the implant bed area [23].

Implant drill bits and piezosurgery tips suffer characteristic patterns of wear after osteotomy procedures. Drills showed abrasive wear, plastic deformation, blunting, coating damage and material loss, mainly at the tip and cutting edges of the drill flank [24]. Meanwhile, piezosurgery showed abrasive wear of the tip and flanks and plastic deformation of the cutting point. The characteristics of the wear suffered by osteotomy instruments have been related to the drill material, drill and drill mechanical properties [24].

Carvalho et al. [25,26] demonstrated that substance loss, steel melting and condensation of particles detached from the active point increased proportionally to the number of uses; indeed, drills used 10 times showed 17.86% deformation and, drills used 50 times showed 33.97% deformation. These findings were also confirmed in drills used 50 or more times, on which increased areas of metal subtraction, the addition of loosened metal, and metal surface abrasion at the lateral surface of the cutting tip and drill edges were detected [27–29].

Another factor that seems to increase the wear of implant drills is the use of guided surgery techniques. Bone heating, drill deformation and roughness were evaluated after osteotomy with guided surgery and conventional osteotomy; greater drill deformation and wear were recorded after the 40th osteotomy in the guided surgery group produced by the drill friction against the metal sleeve and the bone [30].

Sterilization and irrigation also increase the release of metal particles and ions as well as drill wear. Allsobrook et al. [31] evaluated Straumann, Nobel Biocare and Neoss drills. The authors performed 50 osteotomies per drill. After 20 osteotomies, the drills suffered 30 µm to 100 µm of wear at the tip and edges and loss of the surface coating. Furthermore, analysis of corrosion after sterilization processes and irrigation with saline showed increased pitting corrosion of the drills after 20 cycles of sterilization [24,31,32]. Other authors also found that the autoclave sterilization of implant drills increased the surface corrosion of the drills and therefore the ion and particle release [25,33,34]. Similar results were obtained when implant drills with three different characteristics (smooth stainless steel drills, coated drills and smooth zirconia drills) were compared. After repeated drilling and sterilization, it was found that drills in all groups suffered wear, deformation, coating delamination and surface roughness changes [35].

The material of the drill seems to influence the amount of particle loosening of the drill surface. Zirconia drills showed less qualitative substrate loss than titanium drills when drilling 20 implant beds at a standard drilling speed of 800 rpm [36].

Remarks

During osteotomy procedures, surgical drills, implant drills and piezosurgery tips suffer deformation, surface wear, microfractures, delamination, and metal particle release (Figures 3–5). Additionally, sterilization of the cutting tools can initiate corrosion and can increase the particle and ion release. Sufficient irrigation, adequate suction of metallic debris, the use of single-use drills, adequate control of the number of sterilization procedures, inspection of cutting tool integrity, timely replacement of worn drills and the use of harder and more resistant cutting tools might reduce the deposition of metallic particles and ions released during implant bed preparation.

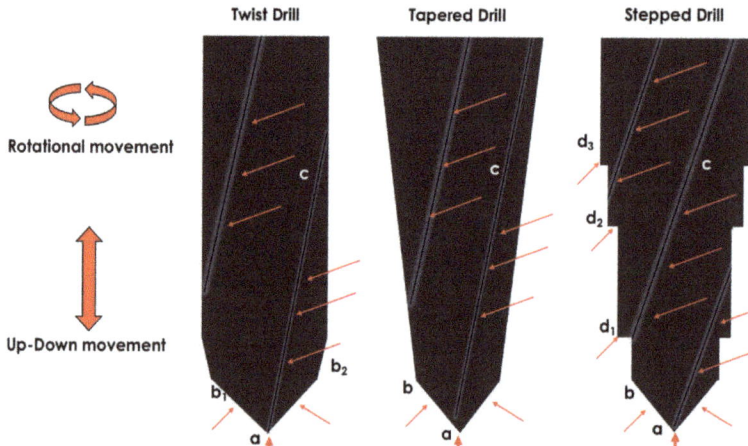

Figure 3. Areas of wear of implant drills. Different drill designs present wear and particle loosening at different levels under the effects of axial and rotational forces. Twist drills suffer deformation, blunting and delamination of the drill tip (a), tip angles (b1 and b2), and cutting blades (c). Tapered drills suffer deformation, blunting and delamination of the drill tip (a), tip angle (b), and cutting blades (c). Stepped drills suffer deformation, blunting and delamination of the drill tip (a), tip angles (b), cutting blades (c) and step angles (d1, d2, d3). Thin red arrows are showing the blade areas suffering wear as well as the tip angles. Thick arrows are showing the tip of the drills suffering wear and deformation.

Int. J. Mol. Sci. **2018**, *19*, 3585

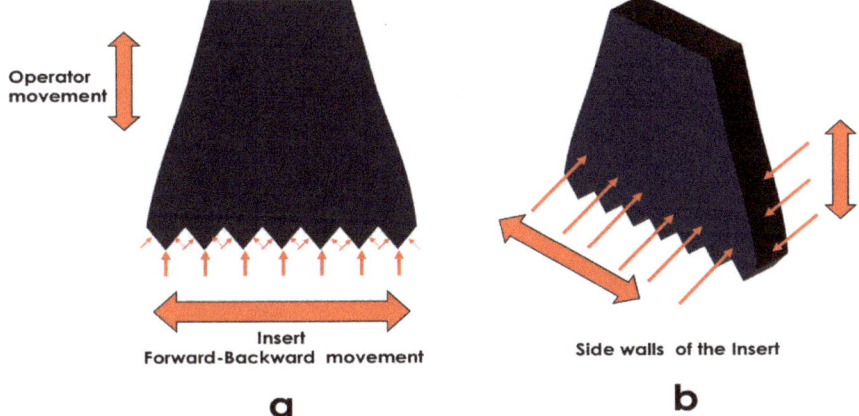

Figure 4. Areas of wear of piezosurgery inserts. The piezosurgery insert directions of movement: a vertical movement induced by the operator, and a minimal vertical displacement during the ultrasonic movement in conjunction with a horizontal component produced by oscillation of the tip. (**a**) The piezosurgery insert oscillates in a forward-backward movement. The insert tip and the sides of the tip suffer deformation, wear and particle detachment. (**b**) Additionally, the sidewalls and borders of the insert suffer deformation, wear and particle detachment. Short and long thing arrows are showing areas of angle and lateral wear and deformation. Thick arrows are showing the vertex of the tips suffering wear and deformation.

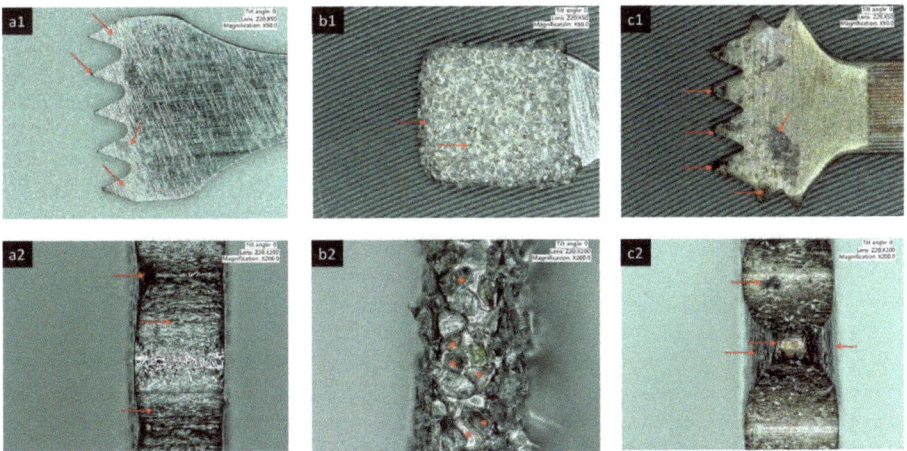

Figure 5. Piezosurgery tips used 1 time. The tips are made from different materials. (**a1,a2**) Stainless steel insert showing lateral wear of the active tip, abrasion and particle loosening at the flanks, deformation of the tip edges and material delamination. (**b1,b2**). Diamond-coated insert showing wear at the sides and empty spaces representing particle loosening. (**c1,c2**) Gold nitride-coated insert with several areas at the flanks, tips and sides of the tips showing excessive wear, particle delamination and deformation. The red arrows are showing the areas of wear, delamination and deformation suffered by piezosurgery inserts. The red asterisk is showing particle detachment from diamond coated piezosurgery inserts.

2.1.2. Implant Insertion

When the dental implant is inserted, frictional moments occur at the bone-implant interface and the initial mechanical interlocking results from interactions between the bone and the implant threads and walls [37]. As a consequence of implant insertion, microfractures and compression occur at the bone side and the implant surface is simultaneously subjected to a combination of torsional and frictional forces, which may alter the original implant surface. Wawrzinek et al. [38] described that shear forces originating from the friction of implants inserted against the surrounding bone tissue can produce a shifting of stresses at different locations along the implant surface related to the heterogeneity of the bone tissue (cortical or cancellous) and the implant geometry. Therefore, localized areas of stress concentration can appear and titanium particles can be released into the bone tissue [37]. A harder implant surface should be able to retain its characteristics more than a softer surface and should demonstrate increased abrasion resistance [39], however, the abrasive wear phenomena during implant insertion are influenced by additional factors that cannot be controlled simultaneously; these factors include the properties of the specimens, interactions with the environment and the experimental conditions [40–42].

Schliephake et al. [43] showed that the placement of self-tapping titanium implants abraded particles from the implant surface, as evidenced by deposits of these particles at the implant-bone interface. Seki et al. [44] and Kim et al. [45] revealed titanium particles in the tissue interposed between the bone and the titanium plate used for oral and maxillofacial fracture fixation. The authors speculated that the particles originated from two causes: first, the physical-mechanical removal of the oxidation layer during insertion; and second, the dissolution of the subjacent titanium into the surrounding tissues (corrosion), resulting in detection of the titanium particles and ions [44,45].

Different observations also showed that particles abraded from the implant surface ranged between 1.8–3.2 µm in diameter and were located at the bone surface and up to 100 µm inside the surrounding bone [5]. Other authors found titanium particles ranging widely in size from 20 nm to 20 µm at the implantation site concentrated at the cortical layer. Apparently, the process of implant insertion alone can release up to 0.5 mg of metallic debris at the implant-bone interface. Previous studies in orthopaedics have shown that aseptic osteolysis can be induced by 0.220 mg to 3.0 mg of titanium particles in the implant area as well as inside the medullary spaces [6]. In the case of titanium plasma-sprayed (TPS) implants, the particle dimensions decreased with increasing distance from the implant surface, probably due to gradual and passive dissolution, fretting and wear [7–9].

Martini et al. [46] found that titanium particles released from TPS implants can be present within 200–250 µm of the implant surface and some debris could be observed at 500 µm. It was speculated that detachment of titanium particles from the TPS implants resulted from the implant surface morphology, frictional forces during implant insertion and frictional forces between the titanium (Ti) coating and the pre-existing bone. These released metallic particles impeded bone formation and created gaps of 180–260 µm between the pre-existing bone and implant surface compared to non-TPS surfaces, which showed bone matrix on the implant surface after 14 days [47–49].

Wennerberg et al. [50] performed in vitro and in vivo studies to investigate the titanium concentration around titanium implants with different surface roughness inserted in rabbit tibias. They found that moderately rough surfaces (Sa 2.21 µm) presented more titanium release up to 400 µm from the implant surface than smooth surfaces (Sa < 1.43 µm). The authors recommended further chemical analysis of the implant surfaces to evaluate potential chemical changes at the implant surfaces and their biological consequences [51].

Mints et al. [52] compared acid-etched, anodized and machined implant surfaces after their insertion into bone blocks. They found that all the implants suffered surface damage, material removal from the implant surface, oxide layer breakdown, metal debris transportation (due to the insertion and friction) from the apical and middle third to the coronal area and cracks at the implant surface as result of the insertion process. The titanium debris ranged from nano- to microparticles; the nanoparticles were located in the coronal area and the microparticles were located in the mid and apical regions [52].

It seems that although almost all of an implant surface is exposed to wear and particle detachment during insertion, the tip of the threads and the lower flank of the threads are more exposed, while the apex microstructure is least exposed in conventional implant bed preparation [53]. It was also concluded that surfaces with subtractive modifications appeared to suffer less wear and particle loosening than surfaces with additive modifications and that re-establishment of the damaged TiO_2 layer was superior for surfaces with subtractive modifications [54].

Particles and ions released during titanium implant insertion can also be retained at the soft tissue level. Concentrations of specific metallic elements retained in the gingival cuff, surrounding the implant neck were detected in an analysis of histological sections processed with the laser ablation detection technique. The elemental mapping showed titanium from 0.4 mm up to 4 mm from the implant, with additional contents of aluminium and vanadium [55].

One in vitro study did not find metal particles released after the insertion and removal of sandblasted and acid-etched implants. The study was performed in polyurethane blocks with different densities and the methods of analysis were SEM observation of the implant surface and X-ray diffraction analysis. Albeit authors observed surface deformation, changes and wear, no traces of metals were observed in the samples [56]. Titanium particles and ions are not always detected; apparently, the irrigation of the surgical site could dilute and remove metallic particles and ions, thus reducing the overall metal content. However, the presence of measurable metals and ions in intraoperative fluid samples indicates that metal particles and ions are certainly released at the time of implantation [57].

The Food and Drug Administration (FDA), explained that variability in metals detection is possible given test interpretation differences, accuracy and precision of the detection method, variability in the test specimen, contamination by metal ions, variance between laboratories, interfering substances and lack of proficiency testing [58,59]. These factors should be considered in future comparisons of experiments.

Remarks

Implant insertion produces changes at the implant surface, as one or more of the following factors were observed: wear, deformation, particle delamination, scratches and cracks at the lower flank of the threads, thread tips and implant apex in different proportions. Released metallic particles with variable sizes and metallic ions can be located adjacent to the implant surface or can be displaced to other distant locations (Table 1). Factors, such as higher bone density, additive implant surfaces and lack of irrigation might lead to the detection of higher percentages of metallic particles and ions, while lower bone density, subtractive implant surfaces, and abundant irrigation might reduce the number of detectable metallic particles and ions produced during implant insertion. It is recommended to standardize the methods of metals particle and ion detection, as suggested by the FDA [59].

Table 1. Titanium/metal particles released during implant insertion. The table summarizes the particle size, locations and detection methods. The particles ranged from a few nanometers to micrometers in size.

Author and Year of Publication	Original Implant Surface	Animal Model and Area of Implant Insertion	Localization of the Metal Particles	Method of Detection	Metal Detected	Particle Size/Recovered Particle Weight	Particle Geometry
Schliephake et al. [43]	Titanium, machined	Minipig mandible	- Peri-implant bone and implant surface - Lungs, liver and kidneys	- Scanning Electron Microscopy (SEM) - Energy-Dispersive X-ray Spectroscopy (EDX) - Flameless atomic absorption spectroscopy (FAAS)	Titanium particles Titanium concentration as ng/mg dry weight of the organ	5-30 μm Kidneys: 2.92 ± 0.69 ng/mg Liver: 11.5 ± 1.35 ng/mg Lungs: 135.7 ± 12.42 ng/mg	Solid and leaf-like particles -
Tanaka et al. [5]	TPS	Dog mandible	Implant-bone interface and surrounding bone tissue	- SEM - Transmission Electron Microscopy (TEM) - X-ray microanalyzer - Electron diffraction	Titanium particles	1.8-3.2 μm	-
Martini et al. [46]	- TPS - TPS + coating of fluorohydroxyapatite	Mongrel sheep femoral and tibial diaphysis	- Surface of TPS implants - Inside the new bone - Inside the medullary spaces near the TPS surface	(EDS)	Titanium particles	-	-
Franchi et al. [47]	- Titanium, machined - TPS - Alumina oxide, sandblasted and acid-etched - Zirconium oxide, sandblasted + acid-etched	Sheep femur and tibia	- Peri-implant tissue - Inside the new bone - Near blood vessels of peri-implant connective tissue around TPS implants	SEM	Titanium granules	3-60 μm	-
Wennerberg et al. [50]	- Titanium, turned - Titanium, sandblasted	New Zealand rabbit tibia	Titanium detection was related to the distance of evaluation	- X-ray fluorescence spectroscopy (SRXRF) - Secondary ion mass spectrometry (SIMS)	Titanium concentration as ng/mg dry weight of implant	Turned: 206.7 ± 25.2 ng/mg Sandblasted: 210 ± 35.2 ng/mg	-

Table 1. Cont.

Author and Year of Publication	Original Implant Surface	Animal Model and Area of Implant Insertion	Localization of the Metal Particles	Method of Detection	Metal Detected	Particle Size/Recovered Particle Weight	Particle Geometry
Meyer et al. [60]	- Titanium, sandblasted + acid-etched - TPS - Machined	- Minipig mandible	- Titanium particles detected at the crestal bone	- SEM - EDS	- Titanium particles and nanoparticles	20 nm to a few microns	- Angular or round elongated particles - Large and oval-shaped particles
Flatebø et al. [55]	- Titanium, anodized	- Humans	- Titanium particles detected in the gingival mucosa around cover screws	- Laser ablation inductively coupled plasma mass spectrometry (LA-ICP-MS) - High-resolution optical darkfield microscopy (HR-ODM) - SEM	- Titanium particles - Titanium isotopes	140–2300 nm	-
Senna et al. [6]	- Titanium, anodized - Titanium, sandblasted + acid-etched	- In vitro bovine ribs	- Titanium particles detected along the implantation site bone walls and cortical layer - Implant surface damage	- SEM with backscattered electron detection (BSD)	- Titanium particles	10 nm to 20 μm	-
Deppe et al. [54]	- Titanium, sandblasted + acid-etched	- Human cadaver edentulous jaws	- Implant surface damage at the apical thread flanks	- SEM	- Areas of the implant surface with loose material, lack of surface characteristics (delamination)	-	-
Sridhar et al. [56]	- Titanium, sandblasted + acid-etched	- Polyurethane foam blocks with different densities	- No particles were detected	- Digital optical microscopy - SEM - X-ray diffraction (XRD)	-	-	-

223

Table 1. *Cont.*

Author and Year of Publication	Original Implant Surface	Animal Model and Area of Implant Insertion	Localization of the Metal Particles	Method of Detection	Metal Detected	Particle Size/Recovered Particle Weight	Particle Geometry
Deppe et al. [51]	- Four different implants with different surface treatments were compared - Titanium, sandblasted + acid-etched (Ankylos and Straumann) - Acid-etched (Frialit) - Titanium, anodized (Nobel)	- Porcine mandible	- Evaluation of the implant surface damage/changes	- 3D confocal microscopy	- Changes in the surface topography were detected along all the implant surfaces - Major changes were observed at the apical and cervical areas - Significant destruction of the surface of anodized implants was recorded	-	-
Pettersson et al. [49]	- Titanium, machined implants - Titanium, anodized implants	- Pig jaw bone	- Peri-implant bone	- SEM for the evaluation of implant surface changes - Coupled plasma atomic emission spectroscopy (ICP-AES) for analysis of the released titanium particles	Titanium particles were detected	- Anodized titanium implant with parallel walls 2.80 ± 0.85 µg - Anodized titanium implant, slightly tapered 2.00 ± 0.56 µg - Machined titanium implant, slightly tapered 0.91 ± 0.36 µg	-

2.2. Potential Causes of Titanium Particle and Ion Release in the Prosthetic Phase

The implant-abutment interface involves the interaction of the internal walls of the implant connection and the surface of the abutment connection. The material properties, the magnitude, direction and duration of the forces, the composition and saliva pH and the microflora can all influence the amount of titanium particles and ions released from the implant-abutment interface into the surrounding environment [61–70].

2.2.1. Implant-Abutment Material Interface

Titanium and zirconia abutments induce different levels of wear to the implant connection. Klotz et al. [61] applied cyclic load of up to 1,000,000 cycles with forces from 20 N to 200 N to titanium and zirconia abutments connected to titanium implants. They found that implants with zirconia abutments suffered greater wear and more titanium particle release than implants with titanium abutments. The authors explained that these differences were produced because the hardness of the zirconia is approximately 10 times higher than the hardness of grade 4, commercially pure titanium (1600–2000 Vickers hardness (HV) for zirconia vs. 258 HV for titanium) [61,66,67].

The amount of wear at the implant-abutment connection was quantified by Stimmelmayr et al. [62], who demonstrated significantly more wear at the shoulder of implants connected to zirconia abutments (10.2 ± 1.5 μm) than that of implants connected to titanium abutments (0.7 ± 0.3 μm) [62]. When materials with different mechanical properties interact, more wear and deformation are expected in the weakest material. Indeed, the zirconia flexural strength is greater than 1000 MPa, and its elastic modulus is greater than 200 GPa, making it a more rigid material than titanium [63,64].

In similar studies comparing the zirconia abutment-titanium implant interface with the titanium-titanium interface, small regions of scratching and crushing after dynamic loading were observed in the titanium-titanium interface. In contrast, the zirconia-titanium implant interface was dramatically affected after dynamic loading. The interface between materials with very different Young's moduli could suffer wear, micro separations, and consequently, mechanical failure of the connection [65].

Furthermore, interactions between pure titanium and titanium alloys with greater hardness can result in more deformation and wear at the implant connection. For instance, an abutment made in titanium alloy (Ti6Al4V) is characterized by a mean hardness value of approximately 350–370 HV, while an implant synthesized from commercially pure titanium has a hardness of approximately 200–280 HV. Consequently, plastic deformation and wear often occur within implant connection surfaces, resulting in loosening of the joint and a reduction in its mechanical integrity [68–70].

Remarks

Implant-abutment connection wear is influenced by the characteristics of the coupled materials. The combination of an implant with an abutment (harder or softer) will result in wear of the weakest involved material and potentially in metallic particle release. The released particles can remain inside the connection area (thus increasing the frictional wear) or can be displaced to adjacent tissues, potentially favoring a foreign body reaction.

It is recommended to utilize materials with similar hardness and mechanical properties to reduce the wear of the abutment and the inner walls of the implant at the implant-abutment connection.

2.2.2. Microgap and Micromovement

The mismatch between implant and abutment components (micro-gap) can be increased by micromotion phenomena under functional loading, which could result in increased friction, microleakage, material wear, titanium particle release and screw loosening [68,71–80]. Braian et al. [71] evaluated the horizontal micro-gap between abutments and implants with external and internal hexagonal connections. They found the smallest micro-gap for prefabricated gold abutments (less than

50 µm) placed on implants with external hexagon compared to prefabricated plastic cylinders (less than 130 µm) placed on implants with internal hexagonal connections.

Meanwhile, Morse taper connections with no or minimal micro-gap and titanium abutments instead of zirconium abutments can reduce the wear and micromovement at the implant-abutment interface [72–74]. Conical connections also showed micro-gaps. Fatigue changes of conical connections were evaluated after functional loading by SEM and EDX. The results showed that the micro-gap between dental implant-abutment assemblies with conical connections exists prior to cyclic loading, that titanium and metallic particles (ranging from 2 µm to 80 µm in diameter) were released within the interface and outside the interface and that the effects of the wear and the micro-gap increased with the number of cycles [75].

The physicochemical and microscopic characteristics of different implant-abutment configurations were evaluated, and without exception, defects from 0.5 to 5.6 µm were present in all the samples. Therefore, the original micro-gap between the parts can lead to micromotion and wear, the release of more particles, the penetration of oral fluids and bacteria into the connection, the initiation of corrosion, as well as potential late failures [68,76–80].

Remarks

The micro-gap potentially exists in all implant-abutment connections, given unavoidable discrepancies in the fabrication process. Its dimensions are variable and it has been demonstrated that smaller micro-gaps are present in Morse and conical connections and that larger micro-gaps exist in external connections than internal connections. Larger micro-gaps result in increased micromovements, which under functional loading can increase the friction between the parts, the wear and the particle release. In addition, under functional loading, saliva and bacteria will flow, producing an additional effect, along with the displacement of titanium particles to peri-implant tissues and peri-implant bacteria to the gap of the implant-abutment connection. The presence of saliva, bacteria and their sub-products and the wear and the corrosion initiation of the metallic parts inside the implant-abutment connection might induce mechanical (screw loosening, corrosion, fatigue, fracture) and biological (mucositis, peri-implantitis) implant failures.

Therefore, it is recommended to use internal connections and Morse or conical connections that possess smaller micro-gaps. Additionally, implant companies should provide information about the micro-gap that exists in their implant-abutment connections to clinicians.

2.2.3. Titanium Oxide Layer Loss

When the titanium implant surface is exposed to air, a stable titanium oxide film is spontaneously formed at the implant surface. This thin layer (1.5–10 nm thickness) is formed due to the high affinity of Ti for oxygen [81,82]. The oxide layer protects the bulk material from reactive species and consists of TiO_2 coexisting with other titanium oxides, such as TiO and Ti_2O_3 [83,84]. The resistance to corrosion of titanium implants originates in this titanium oxide layer [85–87].

Once the oxide layer is formed (passivation), it can be altered and damaged by various environmental and functional factors and in the event of damage, the oxide layer can spontaneously reform under normal physiological conditions (re-passivation) [86]. Examples of environmental and functional factors that can alter the oxide layer are abnormal cyclic loads (overloading or continued loading), micromotion of the implant, micromovement of the implant-abutment interface, acidic environments and the combined effects of these factors [88]. Continued attack of the implant surface by these factors can result in permanent breakdown of the oxide film, leaving exposed the bulk metal to electrolytes. Varying pH conditions can turn the implant environment into a more acidic environment and active dissolution of metallic ions can occur (corrosion) [89–91].

Regarding the effects of an acidic environment on the integrity of the titanium oxide layer and bulk titanium, metabolites, such as lactic acid can induce the depletion of oxygen sources required for re-passivation, thus hindering the capability of the titanium surface to reform the oxide layer [11,89,91–97]. Therefore, metallic Ti ions are released into the surrounding tissues. Simultaneously, metallic debris is also loosened from the weakened implant surfaces exacerbating inflammatory conditions and facilitating further surface corrosion [11,89,91–97].

Saliva also plays a role in the corrosion of dental implants [98]. The dental implant interface is exposed continuously to saliva via the gingival sulcus. Saliva can act as a weak electrolyte, and the oral cavity can simulate an electrochemical cell facilitating dissolution of the oxide layer. Further electrochemical corrosion of titanium and its alloys may lead to crevice corrosion and ultimately the release of corrosion products [90,99,100].

Remarks

The titanium oxide layer appears on the implant surface as soon as the implant comes into contact with air, and although the titanium oxide layer has the capability to regrow (re-passivation), the action of continued wear, exposure to chemicals, bacteria and their sub-products, and the presence of an acidic environment can deteriorate and degrade the titanium oxide layer. To preserve the oxide layer integrity, the use of non-aggressive chemicals to disinfect the titanium surface and reduce the bacterial content and environmental acidity is recommended.

2.2.4. Corrosion

Degradation of the implant surface in the human body can be induced by two main events, i.e., wear (a mechanical degradation producing particles) and corrosion (a chemical degradation that mainly produces soluble metal ions) [101]. The term "corrosion" is generally used for metals and consists of material degradation induced by actions of the environment [102–104].

Over time and by the effects of implant function, the implant surface can experience corrosion and can release ions and particulate debris [4,105]. The metallic debris released from titanium implants can be present in various forms, including nanometric and micrometric particles, colloidal and ionic forms (bonded to proteins) [106,107], organic forms (iron-storage complexes), inorganic metal oxides and salts [106].

In the oral cavity, fluctuations in temperature, pH, oxygen, bacteria and food decomposition are attacking continuously the implant surface [91]. The TiO_2 layer is disrupted, and ions and debris generated by the physicochemical degradation of the surface are removed [108]. Through the abrasion produced by food, liquids and toothbrushes, the cycle continues [109].

Factors that can alter the corrosion resistance of the titanium surface are inflammation of the surrounding tissues (which can produce local acidification) and the acidic environment created by lactic acid released by bacteria [94,110]. The lipopolysaccharide (LPS) of gram-negative bacteria can increase the inflammation of peri-implant tissues by its marked effects on macrophages, lymphocytes, fibroblasts, and osteoblasts [111,112]. LPS chains attack the oxide film of the titanium surface (by adsorption phenomena), inducing voids in the oxide film. The Ti surface is exposed and ion exchange between the exposed surface (metal ions, M^+) and the saliva (electrons, e^-) initiates the corrosion process [113].

Furthermore, the interactions between the current flows of the dental implant and the prosthetic superstructures (produced by the differences in the electric potential of the materials) can create crevice, pitting and galvanic corrosion and the subsequent dissolution of the pure metal and alloy components [100]. Under mechanical loading, the corrosion resistance of the metal alloy decreases in different proportions, with cast and machined titanium having the most passive current density at a given potential and chromium-nickel alloys having the most active critical current density values. High-gold-content alloys have excellent corrosion resistance, and palladium alloys have a low critical current density due to the presence of gallium [114].

Chromium-cobalt framework and implant interactions were analysed in vitro, and the results showed that both the implants and the frameworks suffered active degradation processes, ions of all the materials were released and leakage of cobalt ions was greater than the leakage of titanium and chromium ions. In addition, the surfaces of the implants and frameworks became rougher after exposure to saliva [115]. These differences in percentages could be explained by the nature of metallic elements (crystalline structure, surface energy, solubility and exposed area) [116,117] and the quantity and duration of exposure [118]. Ion leakage will occur for all solid surfaces in contact with liquids until the solubility constant is reached [93].

Corrosion could occur also at the intraosseous portion of the implant surrounded by bone. This occurs by a phenomenon in which two surfaces (the implant surface and bone) under mechanical loading can have an oscillatory relative motion of a small, amplitude (fretting) [93], in which chemical reactions are prevalent. This type of corrosion is characterized by particle removal, oxide formation and increased abrasion, which increase the wear of the implant surface [119].

2.2.5. Tribocorrosion

This mechanism involves a combination of tribological (wear and fretting) and corrosive (chemical or electrochemical) events and is influenced by variations in mechanical contact conditions (loading and relative velocity) and in the nature of the environment (pH, humidity and biochemistry) [120–122].

Revathi et al. [120] described the sequence of tribocorrosion as follows: a given load is applied between two surfaces; in the presence of lubricant particles, the load will allow a sliding movement that will produce oxide layer fractures, microcracks, diffusion and re-passivation, wear debris release and finally material dissolution through five types of corrosion, i.e., microbial, galvanic, uniform, crevice and fretting corrosion [121].

Specifically, for titanium implants and the salivary pH, the titanium showed inferior performance in tribocorrosion at pH 6.0, which manifested as greater weight loss and increased cracking [113]. The normal pH of saliva ranges between 6.3 and 7.0 [11,123], various conditions can lower the pH of saliva to below 6.0 (infection, food, oral hygiene products, age, periodontitis, smoking, systemic disease and salivary gland radiation) [123], favoring the corrosion process [124–126]; under these simultaneous actions, the total material loss may be significantly greater than that under mechanical wear or corrosion individually [127].

2.2.6. Fluoride and Titanium Corrosion

Fluoride is one of the main methods for dental caries prevention and is present in many toothpastes and gels. Its percentages range from 0.1 to 2.0 wt%. At these concentrations, fluoride can reduce the corrosion resistance of metallic implants [128]. Apparently, the presence of fluoride ions in the electrolytic environment of the oral cavity attacks the titanium oxide layer and facilitates its dissolution [129].

The interaction between fluoride and titanium surface was described by Kaneko et al. [130]; the oxide film reacts in the presence of fluoride solutions by forming complexes of molecules on the implant surface (titanium fluoride, titanium oxide fluoride and sodium titanium fluoride). These soluble molecules replace the titanium oxide film, allowing corrosion initiation [130]. Once the titanium oxide film is lost, the compound films formed on the surface will undergo rapid dissolution. The rate of dissolution depends on the formation of a new oxide compound at the metal oxide interface, the flow of electrons to fill the metal or oxygen vacancies in the film, the generation or consumption of vacancies at the oxide/electrolyte interface and the chemical or electrochemical dissolution process itself [10–13].

The effect of the time of immersion of titanium implants with different surface treatments (sandblasted and acid-etched, micro-sanded with calcium phosphate and acid-etching, saline solution and saliva) on the percentage of metallic particle and ion release were evaluated by Barbieri et al. [14]. Different time points were selected, and mass spectrometry was used for the detection of particles and ions in the solution. The authors found that all implant surfaces released titanium, nickel and vanadium after seven days and that these percentages increased in proportional to the elapsed time until the sixth month [14].

Remarks

Fluoride ions have the capability to bond to the titanium oxide layer and create compounds that dissolve easily in acidic media, thereby facilitating titanium or metal dissolution and particle and ion release. Although the behavior of fluoride in vitro may vary slightly compared to that in clinical settings (based on the buffer capability of food and saliva), the noxious effect of fluoride on titanium corrosion cannot be denied. It is recommended to utilize non-fluoride rinses or gels for daily care and use alternative (non-acidic) substances in patients with titanium dental implants.

2.3. Potential Causes of Titanium Particle and Ion Release during the Maintenance Phase

The maintenance phase of dental implants and restorations involves the control of biological (i.e., biofilm and plaque) and mechanical risk factors. There are no defined protocols for dental implant maintenance, cleaning or disinfection, and apparently, all existing methods have negative effects of different magnitudes on the implant surface.

2.3.1. Biofilms

The extraosseous surface of dental implants (i.e., polished neck) and restorations (i.e., abutments, metal frameworks) will develop a biofilm once exposed to the oral cavity environment [131,132]. Different surface characteristics, such as chemistry, energy and topography, can influence biofilm formation [133,134].

The first step in biofilm formation is the adsorption of a layer of organic molecules (i.e., salivary proteins) to the material surface [133]. The subsequent step is the adhesion of cells and bacteria mediated by membrane binding sites, such as glucan-binding sites or specific protein-binding sites, such as those for proline-rich proteins [134]. Then, different bacterial populations attach to the pellicle, allowing accumulation of the biofilm, which depends on additional surface characteristics, such as the surface tension. These steps were demonstrated in experimental studies showing that the initial retention of microorganisms to surfaces was strongly related to the forces required for mechanical removal and to the energy of the exposed surface [135].

Within the oral cavity, there is a continuous introduction and removal of microorganisms and nutrients; for these microorganisms to survive they must be adhered to soft or hard tissues to resist, shear forces [135,136]. Bacterial adhesion to soft tissues is reduced by the turnover of the oral epithelia, however, hard, solid structures (teeth, dentures, implants) provide non-shedding surfaces, which allow the formation of thicker and more stable biofilms [137]. In general, established biofilms maintain equilibrium with the host, but uncontrolled accumulation or metabolism of bacteria on hard surfaces can cause dental caries, gingivitis, periodontitis, peri-implantitis and stomatitis [137].

Oral bacteria will adsorb, organize and group on exposed surfaces, forming plaques. These microenvironments of bacteria and their sub-products require efficient removal from the contaminated surface. The main problem associated with plaque removal from an implant surface is potential damage to the surface, which can be permanent [138,139].

Therefore, methods of implant surface detoxification and plaque and calculus removal causing little or no damage to the surface should be preferred. Current methods used for the decontamination of implant surfaces include mechanical instruments, chemical agents and lasers [140]. Different advantages, disadvantages and limitations have been correlated with these methods, thus, there is no defined gold standard for implant surface decontamination [141].

2.3.2. Scaling Instruments

Non-metal instruments (plastic and Teflon tips) were found to cause minimal damage to both smooth and rough titanium surfaces. Meanwhile, hard instruments (metallic) cause major damage to smooth and rough surfaces [16–20,142,143]. Burs seemed to be the instruments of choice if the smoothening of a rough surface was required, but they led to increased metal particle release [16].

Non-metal instruments and rubber cups were shown to be adequate for smooth and rough implant surfaces, air abrasives should be preferred if the surface integrity must be maintained. However, these approaches involve two events that alter the original surface: first, the release of titanium particles produced by the cleaning method; and second, the deposition of instrumentation materials, and residues of the air-abrasive (cleaning powders) [17].

Instruments used for the mechanical cleaning of the implant surface, such as metal curettes and conventional sonic and ultrasonic scalers can damage the implant surface (particle release, modification of the original surface topography and chemical changes). Meanwhile, non-metal instruments and air abrasives produce less damage and fewer alterations to the surface but have been associated with incomplete plaque removal and the generation of sub-products [16]. If the surface roughness is modified, biofilm formation or cell re-attachment can also be influenced, thus altering the healing process [17].

Hallmon et al. [18] and Homiak et al. [19] also found that after multiple uses, plastic curettes changed the structure of the titanium surface. Similarly, Cross-Poline et al. [20] studied the effects of instrumentation on titanium surfaces. They found that the surfaces changed compared to the original control surfaces and observed both particle detachment from the surface and traces of instrumentation materials on the treated surface [20].

The most external areas of the implant are more exposed to damage and deformation than internal areas during mechanical instrumentation. Augthun et al. [142] found roughening of the original implant surface at the implant thread edges after the use of a steel curette for 60 s. Differences can be expected given the multiple factors that influence the extent of surface damage, i.e., the number of strokes, the pressure, the number of treatments and the cleaning instrument [143].

2.3.3. Implantoplasty

This procedure aims to reduce the adherence of plaque by eliminating the contaminated titanium surface, removing inaccessible areas below the contaminated threads, and smoothing the surface topography, thus facilitating implant cleaning [144]. The most efficient drills for implantoplasty regarding the smoothness of the surface achieved were conical carbide drills (Ra < 1 µm) compared with round carbide drills (Ra > 1 µm) [18,145].

Intentionally changing the implant surface by implantoplasty with diamond burs and polishing devices was evaluated; originally, smooth titanium surfaces suffered severe damage and increased surface roughness, together with metallic traces and deposits on the titanium surfaces [146]. When performing implantoplasty, titanium or metallic particles are released, and extraordinary care must be taken to completely remove all titanium particle deposits from the surrounding tissue [147]. Carbide and diamond burs for implantoplasty were compared in vitro; the burs were used alone and in sequence. The original roughness of titanium implants with TPS and SLA surfaces were compared with that of the surfaces after implantoplasty, and both drills changed the roughness of both surfaces [148]. Implantoplasty can reduce bacterial adhesion (through reducing the surface roughness and eliminating non-cleansable areas), but some risks exist with this treatment, including high temperatures, released particles, implant surface damage, implant diameter reduction and implant fracture [149].

Implants immersed in acrylic blocks, were treated by implantoplasty with one of the following methods: diamond burs and silicone polishers; diamond burs and Arkansas stones; diamond drill short sequence; diamond drill short sequence and silicone polishers; diamond drill complete sequence; and diamond drill complete sequence and silicone polishers [150]. The authors found that all methods reduced the surface roughness, but pollution of the operative field was observed in the groups using silicone polishers. The authors concluded that the use of diamond burs and Arkansas stones resulted in a smoother surface with less debris and recommended studies investigating the bio-toxicity of the different types of debris that can be generated during implantoplasty procedures [150].

One potential drawback of implantoplasty procedures is that titanium particles removed from a contaminated implant surface are also contaminated and may not be fully removed from the surrounding environment, which could disseminate bacteria into the surrounding tissues [151].

Remarks

Implantoplasty is the implant surface decontamination procedure releasing relatively more titanium and metal debris. The surface topography is completely modified after implantoplasty, and the external geometry of the implant is removed (using different drills) to facilitate decontamination. A second procedure (polishing) is performed to reduce the surface roughness. As a consequence, particles of different sizes (nanometric and micrometric) and weights are released. It is recommended to use sufficient irrigation during and after the implantoplasty procedure, as well as powerful surgical suction that will be useful for removing released particles from the peri-implant tissues.

2.3.4. Air-Abrasion

Air-abrasion or polishing is a mechanical method for cleaning teeth and implants surfaces. The method uses powders with different particle sizes contained in a waterjet [152,153] that can remove contaminants and clean and polish the surface [154]. Low-abrasive powders (glycine and erythritol) are used in the waterjet, and a specially designed subgingival nozzle allows application to contaminated implant areas [155–157].

The cleaning efficiency of the air-abrasion method was investigated by Tastepe et al. [158]. The authors applied air-abrasion to the simulated subgingival area of titanium implant surfaces in 48 titanium discs. The most efficient decontamination was obtained when the parameters of cleaning were adjusted as follows: application of higher air pressure (while avoiding air emphysema), better insertion of the nozzle tip into the subgingival area, increased movement of the nozzle tip in the subgingival area (up-down, rotation, and slow upward movements) and sufficient water flow [158].

Ronay et al. [159] compared three implant surface debridement methods: curettes, ultrasonic scaling and air-powder abrasion. The authors found that air-abrasion cleaned more surface area than the other methods and did not alter the titanium surface [159]. Duarte et al. [160] evaluated bacterial adhesion on smooth and rough surfaces after decontamination with one of the following procedures: erbium-doped:yttrium, aluminium, and garnet (Er:YAG) laser, metal and plastic curettes and air-powder abrasion. They found that smooth implant surface roughness was increased when metal curettes were used, and rough surfaces showed reduced roughness and bacterial adhesion when metal curettes followed by air-abrasion were used; the authors did not evaluate the presence of metallic debris [160].

Kreisler et al. [161] compared Er:YAG laser and an air-powder system to remove *Porphyromonas gingivalis* from titanium plates. After the surface treatment, fibroblasts were incubated on the specimens, and the proliferation rate was evaluated. They found that both treatments (laser and air powder) supported comparable cell growth, but the air-powder treatment produced slight changes on the implant surface, whereas the laser-treated surfaces remained unchanged [162]. Apparently, in air-abrasion, the optimal air pressure for decontaminating a surface without causing significant alterations is 60–90 psi for 60 s [163].

Remarks

Air-abrasion methods with soft powders seem to provide adequate decontamination of titanium discs and titanium implant surfaces. The surfaces treated with this method show minimal or no titanium particle release when the air pressure is adequate (60–90 psi) and sufficient water flow is provided. To reach subgingival areas, specially designed tips are required. This method is apparently safe for dental implant surfaces.

2.3.5. Chemical Decontamination Methods

Chemical methods can be used alone or in combination with mechanical methods for more efficient implant surface decontamination. Chemical methods reduce bacterial adhesion and eliminate bacterial toxins or sub-products present at the implant surface. Among these chemicals are citric acid, tetracycline, saline, chlorhexidine, hydrogen peroxide, tetracycline and doxycycline [164–167]. Usually, a carrier (cotton swab) is immersed or soaked in the chemical solution and applied to the implant surface with a rubbing movement to clean all the contaminated implant surfaces [168].

Wheelis et al. [168] observed that the combination of rubbing, treatment with a carbon dioxide laser, and any of the following chemicals produced different levels of surface corrosion: citric acid, 15% hydrogen peroxide, chlorhexidine gluconate, tetracycline, doxycycline, sodium fluoride and peroxyacetic acid. Surface corrosion and pitting were presented when more acidic solutions were used (pH < 3); mildly acidic solutions caused surface discoloration, and neutral solutions did not cause signs of corrosion. However, EDS-analysis of all cotton swabs showed the presence of titanium particles [168].

This findings can be explained as follows: acidic solutions (pH < 3) and/or solution with high fluoride concentrations (greater than 0.2%) disrupt the oxide layer on the titanium surface and may inhibit re-passivation (resulting in corrosion), causing localized dissolution of the bulk titanium (pitting), discoloration and the release of ions and metallic debris into the surrounding medium [92,96,129,169–173].

Chemicals for decontamination have also been used in combinations; Wiedmer et al. [174] used combinations of hydrogen peroxide and titanium oxide (H_2O_2 + TiO_2) compared to H_2O_2 alone and chlorhexidine (CHX) for the decontamination of titanium surfaces contaminated with *Staphylococcus epidermidis* biofilms. The authors found that surface treatment with a H_2O_2 + TiO_2 suspension was superior to that with H_2O_2 and CHX for the decontamination of dental implants. This antimicrobial effect is produced by the chemical interaction of TiO_2 particles with H_2O_2, producing ROS (hydroperoxyl and hydroxyl radicals) and rupturing the bacterial membrane. Unfortunately, its effects on the titanium oxide layer have not been confirmed [174].

Among chemical methods, the application of citric acid is considered slightly superior to that of saline for biofilm removal from titanium surfaces [175,176]. However, studies showed that citric acid at a 40% concentration applied by rubbing to the titanium surface induced changes in the topography and potentially increased the surface roughness; because of its acidic nature, it also potentially dissolves the titanium oxide layer [96,177–179].

Ungvari et al. [180], compared three chemical cleaning methods (3% H_2O_2 for 5 min; saturated citric acid at pH 1 for 1 min, and chlorhexidine gel at 0.12% for 5 min). After treatment, the samples were evaluated by atomic force microscopy (AFM) and X-ray photoelectron spectroscopy (XPS). AFM showed a slight increase in the surface roughness but did not reveal differences in roughness among the groups, and XPS confirmed the presence of an intact TiO_2 layer on all the surfaces, thus demonstrating that these chemicals applied under these conditions will not damage the titanium oxide layer of the titanium surface [180].

Remarks

Chemical methods facilitate the removal of plaque, elimination of bacteria and reduction of toxins deposited on the implant surface. As a consequence of their pH, some chemical methods can damage the titanium oxide layer and produce corrosion of the titanium surface. When applying chemical substances to an implant surface, the friction removes existing corrosion products, releasing titanium particles from the implant surface and leaving the bulk titanium exposed.

Among the chemical decontamination methods, it seems that saline, chlorhexidine, hydrogen peroxide and saturated citric acid (less than 1 min) produce minimal alterations on the implant surface. There is a lack of information about the effects of tetracycline and doxycycline on titanium particle release from implant surfaces.

2.3.6. Laser Decontamination Methods

The antimicrobial activity of laser light is based on its photothermal effects [181] and its capability to denature proteins [182], and these effects on titanium surfaces are related to the wavelength and operation settings.

When the CO_2-laser beam is directed onto a titanium implant surface, the laser light is reflected by rougher surfaces [183].

Different power settings for CO_2 lasers might have different impacts on the titanium surface. The effect of laser irradiation on four titanium surfaces at a power of 1.0 W and 4.0 W with 50 pulses per second (pps) and energies from 15.2 to 60.8 Joules per pulse with a laser beam diameter of 200 µm was compared and surface alterations were found both macroscopically (dark spots) and microscopically (melted and glazed Ti) at power outputs above 2 W [183–186]. Shibli et al. [187], who evaluated the effects of laser decontamination of the surface of failed titanium implants. A CO_2-laser at a power of 1.2 W and energy of 40 J for 40 s at a distance of 30 mm from the implant surface was used. No alterations in the implant surface were detected by SEM or EDS analysis [187].

Also Park et al. [188,189] showed that the CO_2-laser at a low power (1.0 or 2.0 W) did not alter the implant surface, regardless of implant type, while at of 3.5 and 5 W, the laser produced surface alterations and gas. Romanos et al. [190] evaluated the effects of CO_2 lasers for the treatment of peri-implantitis and concluded that the application of a CO_2 laser does not alter the implant surface, decreases bacterial contamination, and enhances osseointegration. Finally, Stuebinger et al. [191] compared different lasers for titanium surface decontamination. The CO_2 laser at a power of 2, 4 and 6 W for 10 s in continuous-wave mode did not modify surfaces [191–193].

Schmage et al. [194] compared 10 methods for the surface decontamination of titanium discs. An Er:YAG laser was used in pulsed mode at a distance of 2 mm from the implant surface and the authors found that it produced slight surface damage and only partial bacteria removal. Unfortunately, laser energy, power pulse rate were not reported [194].

The Er:YAG-laser can induce different effects on different implant surfaces. Shin et al. [195] evaluated the effect of Er:YAG-laser irradiation on the microscopic structure and surface roughness of different implant surfaces. Titanium implants with anodized and SLA surfaces were irradiated with an Er:YAG-laser with a 60, 100, 120 and 140 mJ/pulse at 10 Hz. No significant surface changes were observed in SLA surface implants, but severe changes were observed in anodic-oxidized implants with only 100 mJ/pulse irradiation [195].

Er:YAG-lasers seem to be safe for titanium surface at a power of 1 W. Park et al. [189] used higher powers of 1, 2, 3, 4 and 5 W at 20 pulses per second on pure titanium discs (machined and anodized). They observed that 2, 3, 4 and 5 W of power generated melting, coagulation and microfractures of the titanium surface in proportion to the used power.

The time of laser exposure and the energy applied are also responsible for the alterations observed in different titanium implant surfaces. Settings of 100 mJ/pulse, 10 Hz and 1 min preserve the titanium surface structure [196,197]. Meanwhile, higher energies and longer times produce surface melting, particle loosening, and surface fractures on hydroxyapatite (HA)-coated surfaces and TiO_2 surfaces [198,199]. Finally, Takagi et al. [200] analysed the efficacy of implant decontamination (calcified deposit removal) and the surface alterations of treatment with erbium lasers (Er:YAG and Er,Cr:YSGG), cotton pellets and titanium curettes. The authors found that Er:YAG and Er,Cr:YSGG lasers at 40 mJ/pulse (ED 14.2 J/cm^2/pulse) and 20 Hz with water spray in non-contact mode were superior to cotton pellets or titanium curettes and the surfaces suffered minimal damage [200].

Romanos et al. [201] analyzed the effect of a diode laser (980 nm) with different power settings applied to titanium discs in continuous-wave mode in comparison with an Nd:YAG laser. Castro et al [202] also showed intact titanium surfaces after treatment with the diode laser but extensive melting and damage after irradiation with the Nd:YAG laser [201,202].

Gonçalves et al. [203] used a 980-nm diode laser with continuous emission for 5 min at 2.5 and at 3 W to irradiate implants contaminated with *P. gingivalis* and *E. faecalis* and evaluated the bacteria reduction and the changes to the implant surfaces. The authors demonstrated that while the used parameters did not change the treated surfaces, 3 W was 100% effective for surface decontamination [203]. The thermal effects of diode lasers (810 nm and 980 nm) when used at low energy (1 W) in pulsed mode with air/water cooling maintained the temperature below the critical threshold of 47 °C [204–206].

Remarks on Laser-Assisted Decontamination Methods

Lasers used for titanium surface decontamination produce titanium surface alterations depending on the laser energy, radiation time and titanium surface characteristics. There are no studies measuring titanium particles released during or after laser decontamination. The changes in the titanium surface induced by lasers are mainly produced by the temperature increase at the irradiated spot. Each laser behaves differently on rough or machined surfaces. In general, surfaces that reflect the laser (machined surfaces) produce smaller local temperature increases but can affect the surrounding areas. On the other hand, surfaces that absorb the laser (rough surfaces) show significant implant temperature increments. Safe operation settings for each laser and surface should be carefully evaluated before the laser decontamination of implant surfaces to avoid thermal damage and surface changes. To reduce the surface damage during laser decontamination, it is recommended to use pulsed mode, short periods of irradiation, cooling with proper air-water ratios and low energies. This section may be divided by subheadings. It should provide a concise and precise description of the experimental results, their interpretation as well as the experimental conclusions that can be drawn.

3. Materials and Methods

PICO questionnaires were developed and problems, interventions, comparisons and outcomes were organized for the surgical, prosthetic and maintenance phases in implant dentistry procedures (Table 2). An exhaustive search of the literature was performed in PubMed, Medline and Google Scholar for all the relevant studies in the literature published between 1980 and 2018 involving titanium particles and ions related to dental implants and implant dentistry procedures. The following search keywords were used in this systematic review.

Table 2. PICO questions used for preparation of the systematic review to identify sources and aetiological factors for titanium particle and ion release.

P (Probelem)	Subgroup		P (Population)	I (Interventions)	C (Comparisons)	O (Outcomes)
The potential sources of titanium particle and ion release are not known or compiled in the literature	Surgical phase	Implant bed preparation	Experimental, animal and human studies	Bone drilling	Implant drills and other implant bed preparation methods	- Metal content in the adjacent bone or soft tissues after implant bed preparation - Tool wear, damage and corrosion
		Implant placement	Experimental, animal and human studies	Implant Insertion	Implant after Insertion	- Implant surface, alterations after insertion - Metal content in the adjacent bone or soft tissues after implant insertion
		Implant removal	Experimental, animal and human studies	Bone drilling	Other methods for implant removal	- Metal content in the adjacent bone or soft tissues after implant removal - Implant surface alterations and corrosion evaluated after implant removal
	Prosthetic phase	Implant abutment connection	Experimental, animal and human studies	Functional load at the implant abutment connection	Type of connection, misfit gap material	- Implant connection frictional damage - Corrosion and particle/ion release at the implant-abutment connection - Metal content in the adjacent bone or soft tissues
	Maintenance phase	Implant cleaning and decontamination techniques	Experimental, animal and human studies	Implant cleaning, disinfection and polishing	Scaling Ultrasonication Rubber cups and brushes Air-polishing Lasers Cleaning and antibacterial substances	- Metal content in the adjacent bone or soft tissues after implant cleaning, decontamination or polishing - Implant surface alterations and corrosion evaluated after implant cleaning disinfection and polishing

3.1. For the Surgical Phase

Implant bed preparation AND titanium particles; OR implant bed preparation AND metal release; OR bone drilling AND titanium particles; OR bone drilling and metal release; metal debris AND drills; OR implant drill AND wear; OR implant drill and corrosion; OR implant drill and damage; OR bone piezosurgery AND titanium particles/ions; OR osteotomy and metal particles/ions; OR osteotomy and metal debris.

Implant insertion AND metal ions; implant insertion AND titanium ions; Implant insertion AND titanium particles; OR implant insertion and implant surface alterations; OR implant insertion AND titanium release; implant insertion and metal release; OR implant insertion AND titanium particles detachment; OR implant insertion and metal debris; OR implant insertion AND bone contamination; OR implant insertion and bone metal content; implant insertion and tissue metal content.

Implant removal AND metal ions release; implant removal and titanium release; Implant removal AND titanium particles; implant removal and surface alterations; OR implant removal AND metal release; OR Implant removal and particles dislodgement; OR implant removal and metal debris.

3.2. For the Prosthetic Phase

Implant abutment connection AND wear; OR implant abutment connection and deformation; OR implant abutment connection AND titanium particles; implant abutment connection AND ions release; OR implant abutment connection AND corrosion; OR implant abutment connection material AND wear; OR implant abutment connection misfit; implant abutment connection misfit AND wear; dental Implants AND fretting corrosion.

3.3. For the Maintenance Phase

Implant decontamination AND wear; OR implant decontamination and corrosion; OR fluoride AND titanium corrosion; OR chlorhexidine AND titanium corrosion; OR implant scaling AND titanium wear; OR implant surface polishing AND particles release; implant surface polishing AND titanium findings in soft tissues; implant surface polishing AND titanium particles in bone; OR chemical decontamination AND implant surface OR laser decontamination AND implant surface OR laser decontamination AND titanium particles OR laser decontamination AND titanium ions.

Two investigators (RD and GR) performed the initial searches using the keyword combinations; the titles that appeared in the search containing the keywords were reviewed and these fulfilling the inclusion criteria were included for abstract review. The abstracts of the articles were read in full, and those fulfilling the inclusion criteria were included for full-text review and data extraction.

In the case of a disagreement between reviewers, a third investigator (JLC) was included for a final decision regarding the inclusion or exclusion of the articles.

The inclusion criteria were determined as follows: experimental, animal and human studies published in the English language that analyzed titanium particle or ion release in the surgical, prosthetic/restorative or maintenance phase in implant dentistry. The titanium/metal particles released during implant insertion, particle size, location and detection methods were compiled (Table 3).

The exclusion criteria were as follows: articles in languages other than English, reviews, other systematic reviews and expert opinions as well as duplicated articles were excluded.

Table 3. Titanium/metal particles released during implant insertion. The table summarizes the particle size, locations and detection methods. The particles ranged from a few nanometers to micrometers in size.

Author and Year of Publication	Original Implant Surface	Animal Model and Area of Implant Insertion	Localization of the Metal Particles	Method of Detection	Metal Detected	Particle Size/Recovered Particle Weight	Particle Geometry
Schliephake et al. [43]	Titanium, machined	Minipig mandible	- Peri-implant bone and implant surface	- Scanning Electron Microscopy (SEM) - Energy-Dispersive X-ray Spectroscopy (EDX)	Titanium particles	5–30 μm	Solid and leaf-like particles
			- Lungs, liver and kidneys	- Flameless atomic absorption spectroscopy (FAAS)	Titanium concentration as ng/mg dry weight of the organ	Kidneys: 2.92 ± 0.69 ng/mg Liver: 11.5 ± 1.35 ng/mg Lungs: 135.7 ± 12.42 ng/mg	-
Tanaka et al. [5]	TPS	Dog mandible	- Implant-bone interface and surrounding bone tissue	- SEM - Transmission Electron Microscopy (TEM) - X-ray microanalyzer - Electron diffraction	Titanium particles	1.8–3.2 μm	-
Martini et al. [46]	- TPS - TPS + coating of fluorohydroxyapatite	Mongrel sheep femoral and tibial diaphysis	- Surface of TPS implants - Inside the new bone - Inside the medullary spaces near the TPS surface	- (EDS)	Titanium particles	-	-
Franchi et al. [47]	- Titanium, machined - TPS - Alumina oxide, sandblasted and acid-etched - Zirconium oxide, sandblasted + acid-etched	Sheep femur and tibia	- Peri-implant tissue - Inside the new bone - Near blood vessels of peri-implant connective tissue around TPS implants	- SEM	Titanium granules	3–60 μm	-
Wennerberg et al. [50]	- Titanium, turned - Titanium, sandblasted	New Zealand rabbit tibia	- Titanium detection was related to the distance of evaluation	- X-ray fluorescence spectroscopy (SRXRF) - Secondary ion mass spectrometry (SIMS)	- Titanium concentration as ng/mg dry weight of implant	Turned: 206.7 ± 25.2 ng/mg Sandblasted: 210 ± 35.2 ng/mg	-
Meyer et al. [60]	- Titanium, sandblasted + acid-etched - TPS - Machined	Minipig mandible	- Titanium particles detected at the crestal bone	- SEM - EDS	- Titanium particles and nanoparticles	20 nm to a few microns	- Angular or round elongated particles - Large and oval-shaped particles

Table 3. *Cont.*

Author and Year of Publication	Original Implant Surface	Animal Model and Area of Implant Insertion	Localization of the Metal Particles	Method of Detection	Metal Detected	Particle Size/Recovered Particle Weight	Particle Geometry
Flatebø et al. [55]	- Titanium, anodized	- Humans	- Titanium particles detected in the gingival mucosa around cover screws	- Laser ablation inductively coupled plasma mass spectrometry (LA-ICP-MS) - High-resolution optical darkfield microscopy (HR-ODM) - SEM	- Titanium particles - Titanium isotopes	140–2300 nm	-
Senna et al. [6]	- Titanium, anodized - Titanium, sandblasted + acid-etched	- In vitro bovine ribs	- Titanium particles detected along the implantation site bone walls and cortical layer - Implant surface damage	- SEM with backscattered electron detection (BSD)	- Titanium particles	10 nm to 20 µm	-
Deppe et al. [54]	- Titanium, sandblasted + acid-etched	- Human cadaver edentulous jaws	- Implant surface damage at the apical thread flanks	- SEM	- Areas of the implant surface with loose material, lack of surface characteristics (delamination)	-	-
Sridhar et al. [56]	- Titanium, sandblasted + acid-etched	- Polyurethane foam blocks with different densities	- No particles were detected	- Digital optical microscopy - SEM - X-ray diffraction (XRD)	-	-	-
Deppe et al. [51]	- Four different implants with different surface treatments were compared - Titanium, sandblasted + acid-etched (Ankylos and Straumann) - Acid-etched (Frialit) - Titanium, anodized (Nobel)	- Porcine mandible	- Evaluation of the implant surface damage/changes	- 3D confocal microscopy	- Changes in the surface topography were detected along all the implant surfaces - Major changes were observed at the apical and cervical areas - Significant destruction of the surface of anodized implants was recorded	-	-

Table 3. *Cont.*

Author and Year of Publication	Original Implant Surface	Animal Model and Area of Implant Insertion	Localization of the Metal Particles	Method of Detection	Metal Detected	Particle Size/Recovered Particle Weight	Particle Geometry
Pettersson et al. [49]	- Titanium, machined implants - Titanium, anodized implants	- Pig jaw bone	- Peri-implant bone	- SEM for the evaluation of implant surface changes - Coupled plasma atomic emission spectroscopy (ICP-AES) for analysis of the released titanium particles	Titanium particles were detected	- Anodized titanium implant with parallel walls 2.80 ± 0.85 µg - Anodized titanium implant, slightly tapered 2.00 ± 0.56 µg - Machined titanium implant, slightly tapered 0.91 ± 0.36 µg	-

4. Conclusions

Dental implants have revolutionized the dentistry profession, and titanium dental implants have demonstrated their utility and safety for more than 40 years. However, over time, it has become evident that debris and sub-products can be generated during the life of the implant.

Titanium particles and ions can also be released from metallic instruments used for implant bed preparation, from the implant surfaces during insertion and from the implant-abutment interface during insertion and functional loading. In addition, the implant surfaces and restorations are exposed to the environment, saliva, bacteria and chemicals that can potentially dissolve the titanium oxide layer. If these agents attack continuously, the implant surface can permanently lose its titanium oxide layer.

The formation of soluble compounds on the titanium surface will alter the implant surface chemistry and facilitate the dissolution and degradation of exposed bulk titanium, resulting in the initiation of corrosion cycles. Implant maintenance procedures can potentially alter implant surfaces and produce titanium debris that will be released into the peri-implant tissues.

Multiple variables, such as the bone density, mechanical overloading and the use of fluorides, can also influence the proportion of metal particles and ions released from implants and restorations. The complex oral environment can also change with age and the use of medications, and these factors have not yet been studied.

This review provides, for the first time, a summary of the potential sources of titanium particles and ion release in implant dentistry and, based on the findings, suggests methods for reducing this release. The long-term local and systemic effects of titanium particles and ions released into the oral environment and their potential effects on cells, tissues and organs remain unknown due to the rapid evolution and variability of new implant surfaces, new implant-abutment connections and new restorative materials.

Author Contributions: Conceptualization: R.D.-R., G.R.; Data Curation and Resources: R.D.-R., G.R.; Formal Analysis: R.D.-R., G.R.; Funding Acquisition and Research: R.D.-R., G.R.: Methodology: R.D.-R., G.R.; R.D.-R., G.R.; Writing—original draft: R.D.-R., G.R.; Writing—review & editing: R.D.-R., G.R.; Visualization and Methodology: R.D.-R., G.R.; Supervision: R.D.-R., G.R.

Funding: This review was entirely funded by the Department of Prosthodontics and Digital Technology and the Department of Periodontology of the School of Dental Medicine at Stony Brook University, Stony Brook, NY.

Acknowledgments: The authors thank Jose Luis Calvo-Guirado for helping in solving disagreement regarding the inclusion or exclusion or articles for this systematic review.

Conflicts of Interest: The authors declare no conflict of interest.

References

1. Staroveski, T.; Brezak, D.; Udiljak, T. Drill wear monitoring in cortical bone drilling. *Med. Eng. Phys.* **2015**, *37*, 560–566. [CrossRef] [PubMed]
2. Ferguson, A.B.; Laing, P.G.; Hodge, E.S. The ionization of metal implants in living tissues. *J. Bone Jt. Surg.* **1960**, *42*, 77–90. [CrossRef]
3. Ferguson, A.B.; Akahoshi, Y.; Laing, P.G.; Hodge, E.S. Characteristics of trace ion release from embedded metal implants in the rabbit. *J. Bone Jt. Surg.* **1962**, *44*, 317–336. [CrossRef]
4. Meachim, G.; Williams, D.F. Changes in nonosseous tissue adjacent to titanium implants. *J. Biomed. Mater. Res.* **1973**, *7*, 555–572. [CrossRef] [PubMed]
5. Tanaka, N.; Ichinose, S.; Kimijima, Y.; Mimura, M. Investigation of titanium leak to bone tissue surrounding dental titanium implant: Electron microscopic findings and analysis by electron diffraction. *Med. Electron Microsc.* **2000**, *33*, 96–101. [CrossRef] [PubMed]
6. Senna, P.; Cury, A.A.D.; Kates, S.; Meirelles, L. Surface damage on dental implants with release of loose particles after insertion into bone. *Clin. Implant Dent. Relat. Res.* **2015**, *17*, 681–692. [CrossRef] [PubMed]
7. Bianco, P.D.; Ducheyne, P.; Cuckler, J.M. Titanium serum and urine levels in rabbits with a titanium implant in the absence of wear. *Biomaterials* **1996**, *17*, 1937–1942. [CrossRef]

8. Browne, M.; Gregson, P.J. Effect of mechanical surface pretreatment on metal ion release. *Biomaterials* **2000**, *21*, 385–392. [CrossRef]
9. Ferguson, A.B. Metals in living tissues. *Surg. Clin. N. Am.* **1960**, *40*, 521–529. [CrossRef]
10. Frateur, I.; Cattarin, S.; Musiani, M.; Tribollet, B. Electrodissolution of Ti and p-Si in acidic fluoride media: Formation ratio of oxide layers from electrochemical impedance spectroscopy. *J. Electroanal. Chem.* **2000**, *482*, 202–210. [CrossRef]
11. Mabilleau, G.; Bourdon, S.; Joly-Guillou, M.L.; Filmon, R.; Baslé, M.F.; Chappard, D. Influence of fluoride, hydrogen peroxide and lactic acid on the corrosion resistance of commercially pure titanium. *Acta Biomater.* **2006**, *2*, 121–129. [CrossRef] [PubMed]
12. Siirila, H.S.; Kononen, M. The effect of oral topical fluorides on the surface of commercially pure titanium. *Int. J. Oral Maxillofac. Implant.* **1991**, *6*, 50–54.
13. Toumelin-Chemla, F.; Rouelle, F.; Burdairon, G. Corrosive properties of fluoride-containing odontologic gels against titanium. *J. Dent.* **1996**, *24*, 109–115. [CrossRef]
14. Barbieri, M.; Mencio, F.; Papi, P.; Rosella, D.; di Carlo, S.; Valente, T.; Pompa, G. Corrosion behavior of dental implants immersed into human saliva: Preliminary results of an in vitro study. *Eur. Rev. Med. Pharmacol. Sci.* **2017**, *21*, 3543–3548. [PubMed]
15. Woodman, J.L.; Jacobs, J.J.; Galante, J.O.; Urban, R.M. Metal ion release from titanium-based prosthetic segmental replacements of long bones in baboons: A long-term study. *J. Orthop. Res.* **1984**, *1*, 421–430. [CrossRef] [PubMed]
16. Louropoulou, A.; Slot, D.E.; van der Weijden, F.A. Titanium surface alterations following the use of different mechanical instruments: A systematic review. *Clin. Oral Implants Res.* **2012**, *23*, 643–658. [CrossRef] [PubMed]
17. Ruhling, A.; Kocher, T.; Kreusch, J.; Plagmann, H.C. Treatment of subgingival implant surfaces with TeflonR-coated sonic and ultrasonic scaler tips and various implant curettes. An in vitro study. *Clin. Oral Implant. Res.* **1994**, *5*, 19–29. [CrossRef]
18. Hallmon, W.W.; Waldrop, T.C.; Meffert, R.M.; Wade, B.W. A comparative study of the effects of metallic, nonmetallic, and sonic instrumentation on titanium abutment surfaces. *Int. J. Oral Maxillofac. Implant.* **1996**, *11*, 96–100. [CrossRef]
19. Homiak, A.W.; Cook, P.A.; DeBoer, J. Effect of hygiene instrumentation on titanium abutments: A scanning electron microscopy study. *J. Prosthet. Dent.* **1992**, *67*, 364–369. [CrossRef]
20. Cross-Poline, G.N.; Shaklee, R.L.; Stach, D.J. Effect of implant curettes on titanium implant surfaces. *Am. J. Dent.* **1997**, *10*, 41–45. [PubMed]
21. Bertoldi, C.; Pradelli, J.M.; Consolo, U.; Zaffe, D. Release of elements from retrieved maxillofacial plates and screws. *J. Mater. Sci. Mater. Med.* **2005**, *16*, 857–861. [CrossRef] [PubMed]
22. Thomas, P.; Bandl, W.-D.; Maier, S.; Summer, B.; Przybilla, B. Hypersensitivity to titanium osteosynthesis with impaired fracture healing, eczema, and T-cell hyperresponsiveness in vitro: Case report and review of the literature. *Contact Dermat.* **2006**, *55*, 199–202. [CrossRef] [PubMed]
23. Rashad, A.; Sadr-Eshkevari, P.; Weuster, M.; Schmitz, I.; Prochnow, N.; Maurer, P. Material attrition and bone micromorphology after conventional and ultrasonic implant site preparation. *Clin. Oral Implant. Res.* **2013**, *24*, 110–114. [CrossRef] [PubMed]
24. Ercoli, C.; Funkenbusch, P.D.; Lee, H.J.; Moss, M.E.; Graser, G.N. The influence of drill wear on cutting efficiency and heat production during osteotomy preparation for dental implants: A study of drill durability. *Int. J. Oral Maxillofac. Implant.* **2004**, *19*, 335–349.
25. Carvalho, A.C.; Queiroz, T.P.; Okamoto, R.; Margonar, R.L.; Garcia, I.R., Jr.; Filho, O.M. Evaluation of bone heating, immediate bone cell viability, and wear of high-resistance drills after the creation of implant osteotomies in rabbit tibias. *Int. J. Oral Maxillofac. Implant.* **2011**, *26*, 1193–1201.
26. De-Melo, J.F.; Gjerdet, N.R.; Erichsen, E.S. Metal release from cobalt-chromium partial dentures in the mouth. *Acta Odontol. Scand.* **1983**, *41*, 71–74. [CrossRef] [PubMed]
27. Queiroz, T.P.; Souza, F.Á.; Okamoto, R.; Margonar, R.; Pereira-Filho, V.A.; Garcia, I.R.; Vieira, E.H. Evaluation of immediate bone-cell viability and of drill wear after implant osteotomies: Immunohistochemistry and scanning electron microscopy analysis. *J. Oral Maxillofac. Surg.* **2008**, *66*, 1233–1240. [CrossRef] [PubMed]

28. Hochscheidt, C.J.; Shimizu, R.H.; Andrighetto, A.R.; Pierezan, R.; Thomé, G.; Salatti, R. Comparative analysis of cutting efficiency and surface maintenance between different types of implant drills. *Implant Dent.* **2017**, *26*, 723–729. [CrossRef] [PubMed]
29. Gupta, V.; Pandey, P.M. In-situ tool wear monitoring and its effects on the performance of porcine cortical bone drilling: A comparative in-vitro investigation. *Mech. Adv. Mater. Mod. Process.* **2017**, *3*, 1–7. [CrossRef]
30. Dos Santos, P.L.; Queiroz, T.P.; Margonar, R.; de Souza Carvalho, A.C.; Betoni, W., Jr.; Rezende, R.R.; dos Santos, P.H.; Garcia, I.R., Jr. Evaluation of bone heating, drill deformation, and drill roughness after implant osteotomy: Guided surgery and classic drilling procedure. *Int. J. Oral Maxillofac. Implant.* **2014**, *29*, 51–58. [CrossRef] [PubMed]
31. Allsobrook, O.F.L.; Leichter, J.; Holborow, D.; Swain, M. Descriptive study of the longevity of dental implant surgery drills. *Clin. Implant Dent. Relat. Res.* **2011**, *13*, 244–254. [CrossRef] [PubMed]
32. Chacon, G.E.; Bower, D.L.; Larsen, P.E.; McGlumphy, E.A.; Beck, F.M. Heat production by 3 implant drill systems after repeated drilling and sterilization. *J. Oral Maxillofac. Surg.* **2006**, *64*, 265–269. [CrossRef] [PubMed]
33. Cooley, R.L.; Marshall, T.D.; Young, J.M.; Huddleston, A.M. Effect of sterilization on the strength and cutting efficiency of twist drills. *Quintessence Int.* **1990**, *21*, 919–923. [PubMed]
34. Harris, B.H.; Kohles, S.S. Effects of mechanical and thermal fatigue on dental drill performance. *Int. J. Oral Maxillofac. Implant.* **2001**, *16*, 819–826.
35. Mendes, G.C.B.; Padovan, L.E.M.; Ribeiro-Júnior, P.D.; Sartori, E.M.; Valgas, L.; Claudino, M. Influence of implant drill materials on wear, deformation, and roughness after repeated drilling and sterilization. *Implant Dent.* **2014**, *23*, 188–194. [CrossRef] [PubMed]
36. Sartori, E.M.; Shinohara, É.H.; Ponzoni, D.; Padovan, L.E.M.; Valgas, L.; Golin, A.L. Evaluation of deformation, mass loss, and roughness of different metal burs after osteotomy for osseointegrated implants. *J. Oral Maxillofac. Surg.* **2012**, *70*, e608–e621. [CrossRef] [PubMed]
37. Guan, H.; van Staden, R.C.; Johnson, N.W.; Loo, Y.-C. Dynamic modelling and simulation of dental implant insertion process—A finite element study. *Finite Elem. Anal. Des.* **2011**, *47*, 886–897. [CrossRef]
38. Wawrzinek, C.; Sommer, T.; Fischer-Brandies, H. Microdamage in cortical bone due to the overtightening of orthodontic microscrews. *J. Orofac. Orthop.* **2008**, *69*, 121–134. [CrossRef] [PubMed]
39. Cvijović-Alagić, I.; Cvijović, Z.; Mitrović, S.; Panić, V.; Rakin, M. Wear and corrosion behaviour of Ti–13Nb–13Zr and Ti–6Al–4V alloys in simulated physiological solution. *Corros. Sci.* **2011**, *53*, 796–808. [CrossRef]
40. Hokkirigawa, K.; Kato, K. An experimental and theoretical investigation of ploughing, cutting and wedge formation during abrasive wear. *Tribol. Int.* **1988**, *21*, 51–57. [CrossRef]
41. Ni, W.; Cheng, Y.-T.; Lukitsch, M.J.; Weiner, A.M.; Lev, L.C.; Grummon, D.S. Effects of the ratio of hardness to Young's modulus on the friction and wear behavior of bilayer coatings. *Appl. Phys. Lett.* **2004**, *85*, 4028–4030. [CrossRef]
42. Rigney, D.A. Some thoughts on sliding wear. *Wear* **1992**, *152*, 187–192. [CrossRef]
43. Schliephake, H.; Reiss, G.; Urban, R.; Neukam, F.W.; Guckel, S. Metal release from titanium fixtures during placement in the mandible: An experimental study. *Int. J. Oral Maxillofac. Implant.* **1993**, *8*, 502–511.
44. Seki, Y.; Bessho, K.; Sugatani, T.; Kageyama, T.; Inui, M.; Tagawa, T. Clinicopathological study on titanium miniplates. *J. Oral Maxillofac. Surg.* **1994**, *40*, 892–896. [CrossRef]
45. Kim, Y.-K.; Yeo, H.-H.; Lim, S.-C. Tissue response to titanium plates: A transmitted electron microscopic study. *J. Oral Maxillofac. Surg.* **1997**, *55*, 322–326. [CrossRef]
46. Martini, D.; Fini, M.; Franchi, M.; Pasquale, V.D.; Bacchelli, B.; Gamberini, M.; Tinti, A.; Taddei, P.; Giavaresi, G.; Ottani, V.; et al. Detachment of titanium and fluorohydroxyapatite particles in unloaded endosseous implants. *Biomaterials* **2003**, *24*, 1309–1316. [CrossRef]
47. Franchi, M.; Bacchelli, B.; Martini, D.; Pasquale, V.D.; Orsini, E.; Ottani, V.; Fini, M.; Giavaresi, G.; Giardino, R.; Ruggeri, A. Early detachment of titanium particles from various different surfaces of endosseous dental implants. *Biomaterials* **2004**, *25*, 2239–2246. [CrossRef] [PubMed]
48. Geis-Gerstorfer, J.; Sauer, K.H.; Passler, K. Ion release from Ni-Cr-Mo and Co-Cr-Mo casting alloys. *Int. J. Prosthodont.* **1991**, *4*, 152–158. [PubMed]

49. Pettersson, M.; Pettersson, J.; Thorén, M.M.; Johansson, A. Release of titanium after insertion of dental implants with different surface characteristics—An ex vivo animal study. *Acta Biomater. Odontol. Scand.* **2017**, *3*, 63–73. [CrossRef] [PubMed]
50. Wennerberg, A.; Ide-Ektessabi, A.; Hatkamata, S.; Sawase, T.; Johansson, C.; Albrektsson, T.; Martinelli, A.; Sodervall, U.; Odelius, H. Titanium release from implants prepared with different surface roughness. An in vitro and in vivo study. *Clin. Oral Implant. Res.* **2004**, *15*, 505–512. [CrossRef] [PubMed]
51. Deppe, H.; Wolff, C.; Bauer, F.; Ruthenberg, R.; Sculean, A.; Mücke, T. Dental implant surfaces after insertion in bone: An in vitro study in four commercial implant systems. *Clin. Oral Implant. Res.* **2018**, *22*, 1593–1600. [CrossRef] [PubMed]
52. Mints, D.; Elias, C.; Funkenbusch, P.; Meirelles, L. Integrity of implant surface modifications after insertion. *Int. J. Oral Maxillofac. Implant.* **2014**, *29*, 97–104. [CrossRef] [PubMed]
53. Salerno, M.; Itri, A.; Frezzato, M.; Rebaudi, A. Surface microstructure of dental implants before and after insertion: An in vitro study by means of scanning probe microscopy. *Implant Dent.* **2015**, *24*, 248–255. [CrossRef] [PubMed]
54. Deppe, H.; Grünberg, C.; Thomas, M.; Sculean, A.; Benner, K.-U.; Bauer, F.J.M. Surface morphology analysis of dental implants following insertion into bone using scanning electron microscopy: A pilot study. *Clin. Oral Implant. Res.* **2015**, *26*, 1261–1266. [CrossRef] [PubMed]
55. Flatebø, R.S.; Høl, P.J.; Leknes, K.N.; Kosler, J.; Lie, S.A.; Gjerdet, N.R. Mapping of titanium particles in peri-implant oral mucosa by laser ablation inductively coupled plasma mass spectrometry and high-resolution optical darkfield microscopy. *J. Oral Pathol. Med.* **2011**, *40*, 412–420. [CrossRef] [PubMed]
56. Sridhar, S.; Wilson, T.G.; Valderrama, P.; Watkins-Curry, P.; Chan, J.Y.; Rodrigues, D.C. In vitro evaluation of titanium exfoliation during simulated surgical insertion of dental implants. *J. Oral Implantol.* **2016**, *42*, 34–40. [CrossRef] [PubMed]
57. Cundy, W.J.; Mascarenhas, A.R.; Antoniou, G.; Freeman, B.J.C.; Cundy, P.J. Local and systemic metal ion release occurs intraoperatively during correction and instrumented spinal fusion for scoliosis. *J. Child. Orthop.* **2015**, *9*, 39–43. [CrossRef] [PubMed]
58. Sampson, B.; Hart, A. Clinical usefulness of blood metal measurements to assess the failure of metal-on-metal hip implants. *Ann. Clin. Biochem.* **2012**, *49*, 118–131. [CrossRef] [PubMed]
59. USFDA. *FDA Executive Summary Memorandum. Metal-On-Metal Hip Implant Systems*; USFDA: Washington, DC, USA, 2012.
60. Meyer, U.; Bühner, M.; Büchter, A.; Kruse-Lösler, B.; Stamm, T.; Wiesmann, H.P. Fast element mapping of titanium wear around implants of different surface structures. *Clin. Oral Implant. Res.* **2006**, *17*, 206–211. [CrossRef] [PubMed]
61. Klotz, M.W.; Taylor, T.D.; Goldberg, A.J. Wear at the titanium-zirconia implant-abutment interface: A pilot study. *Int. J. Oral Maxillofac. Implant.* **2011**, *26*, 970–975.
62. Stimmelmayr, M.; Edelhoff, D.; Güth, J.-F.; Erdelt, K.; Happe, A.; Beuer, F. Wear at the titanium–titanium and the titanium–zirconia implant–abutment interface: A comparative in vitro study. *Dent. Mater.* **2012**, *28*, 1215–1220. [CrossRef] [PubMed]
63. Aboushelib, M.N.; Kleverlaan, C.J.; Feilzer, A.J. Evaluation of a high fracture toughness composite ceramic for dental applications. *J. Prosthodont.* **2008**, *17*, 538–544. [CrossRef] [PubMed]
64. Elias, C.N.; Fernandes, D.J.; Resende, C.R.S.; Roestel, J. Mechanical properties, surface morphology and stability of a modified commercially pure high strength titanium alloy for dental implants. *Dent. Mater.* **2015**, *31*, e1–e13. [CrossRef] [PubMed]
65. Cavusoglu, Y.; Akça, K.; Gürbüz, R.; Cehreli, M.C. A pilot study of joint stability at the zirconium or titanium abutment/titanium implant interface. *Int. J. Oral Maxillofac. Implant.* **2014**, *29*, 338–343. [CrossRef] [PubMed]
66. Yüzügüllü, B.; Avci, M. The implant-abutment interface of alumina and zirconia abutments. *Clin. Implant Dent. Relat. Res.* **2008**, *10*, 113–121. [CrossRef] [PubMed]
67. Beuer, F.; Korczynski, N.; Rezac, A.; Naumann, M.; Gernet, W.; Sorensen, J.A. Marginal and internal fit of zirconia based fixed dental prostheses fabricated with different concepts. *Clin. Cosmet. Investig. Dent.* **2010**, *2*, 5–11. [CrossRef] [PubMed]
68. Gratton, D.G.; Aquilino, S.A.; Stanford, C.M. Micromotion and dynamic fatigue properties of the dental implant–abutment interface. *J. Prosthet. Dent.* **2001**, *85*, 47–52. [CrossRef] [PubMed]

69. Quek, H.C.; Tan, K.B.; Nicholls, J.I. Load fatigue performance of four implant-abutment interface designs: Effect of torque level and implant system. *Int. J. Oral Maxillofac. Implant.* **2008**, *23*, 253–262.
70. Seetoh, Y.L.; Tan, K.B.; Chua, E.K.; Quek, H.C.; Nicholls, J.I. Load fatigue performance of conical implant-abutment connections. *Int. J. Oral Maxillofac. Implant.* **2011**, *26*, 797–806.
71. Braian, M.; Bruyn, H.; Fransson, H.; Christersson, C.; Wennerberg, A. Tolerance measurements on internal- and external-hexagon implants. *Int. J. Oral Maxillofac. Implant.* **2014**, *29*, 846–852. [CrossRef] [PubMed]
72. Rack, T.; Zabler, S.; Rack, A.; Riesemeier, H.; Nelson, K. An in vitro pilot study of abutment stability during loading in new and fatigue-loaded conical dental implants using synchrotron-based radiography. *Int. J. Oral Maxillofac. Implant.* **2013**, *28*, 44–50. [CrossRef] [PubMed]
73. Karl, M.; Taylor, T. Parameters determining micromotion at the implant-abutment interface. *Int. J. Oral Maxillofac. Implant.* **2014**, *29*, 1338–1347. [CrossRef] [PubMed]
74. Alqutaibi, A.Y.; Aboalrejal, A.N. Microgap and micromotion at the implant abutment interface cause marginal bone loss around dental implant but more evidence is needed. *J. Evid. Based Dent. Pract.* **2018**, *18*, 171–172. [CrossRef] [PubMed]
75. Blum, K.; Wiest, W.; Fella, C.; Balles, A.; Dittmann, J.; Rack, A.; Maier, D.; Thomann, R.; Spies, B.C.; Kohal, R.J.; et al. Fatigue induced changes in conical implant–abutment connections. *Dent. Mater.* **2015**, *31*, 1415–1426. [CrossRef] [PubMed]
76. Pereira, J.; Morsch, C.; Henriques, B.; Nascimento, R.; Benfatti, C.; Silva, F.; López-López, J.; Souza, J. Removal torque and biofilm accumulation at two dental implant–abutment joints after fatigue. *Int. J. Oral Maxillofac. Implant.* **2016**, *31*, 813–819. [CrossRef] [PubMed]
77. Prado, A.; Pereira, J.; Henriques, B.; Benfatti, C.; Magini, R.; López-López, J.; Souza, J. Biofilm affecting the mechanical integrity of implant-abutment joints. *Int. J. Prosthodont.* **2016**, *29*, 381–383. [CrossRef] [PubMed]
78. Lopes, P.A.; Carreiro, A.F.P.; Nascimento, R.M.; Vahey, B.R.; Henriques, B.; Souza, J.C.M. Physicochemical and microscopic characterization of implant-abutment joints. *Eur. J. Dent.* **2018**, *12*, 100–104. [CrossRef] [PubMed]
79. Tsuge, T.; Hagiwara, Y.; Matsumura, H. Marginal fit and microgaps of implant-abutment interface with internal anti-rotation configuration. *Dent. Mater. J.* **2008**, *27*, 29–34. [CrossRef] [PubMed]
80. Binon, P.P. The effect of implant/abutment hexagonal misfit on screw joint stability. *Int. J. Prosthodont.* **1996**, *9*, 149–160. [PubMed]
81. Kuromoto, N.K.; Simão, R.A.; Soares, G.A. Titanium oxide films produced on commercially pure titanium by anodic oxidation with different voltages. *Mater. Charact.* **2007**, *58*, 114–121. [CrossRef]
82. Rodriguez, L.L.; Sundaram, P.A.; Rosim-Fachini, E.; Padovani, A.M.; Diffoot-Carlo, N. Plasma electrolytic oxidation coatings on γTiAl alloy for potential biomedical applications. *J. Biomed. Mater. Res. Part B Appl. Biomater.* **2014**, *102*, 988–1001. [CrossRef] [PubMed]
83. Gittens, R.A.; Scheideler, L.; Rupp, F.; Hyzy, S.L.; Geis-Gerstorfer, J.; Schwartz, Z.; Boyan, B.D. A review on the wettability of dental implant surfaces II: Biological and clinical aspects. *Acta Biomater.* **2014**, *10*, 2907–2918. [CrossRef] [PubMed]
84. Khanlou, H.M.; Ang, B.C.; Barzani, M.M.; Silakhori, M.; Talebian, S. Prediction and characterization of surface roughness using sandblasting and acid etching process on new non-toxic titanium biomaterial: Adaptive-network-based fuzzy inference System. *Neural Comput. Appl.* **2015**, *26*, 1751–1761. [CrossRef]
85. Long, M.; Rack, H.J. Titanium alloys in total joint replacement—A materials science perspective. *Biomaterials* **1998**, *19*, 1621–1639. [CrossRef]
86. Addison, O.; Davenport, A.J.; Newport, R.J.; Kalra, S.; Monir, M.; Mosselmans, J.F.W.; Proops, D.; Martin, R.A. Do 'passive' medical titanium surfaces deteriorate in service in the absence of wear? *J. R. Soc. Interface* **2012**, *9*, 3161–3164. [CrossRef] [PubMed]
87. Abey, S.; Mathew, M.T.; Lee, D.J.; Knoernschild, K.L.; Wimmer, M.A.; Sukotjo, C. Electrochemical behavior of titanium in artificial saliva: Influence of pH. *J. Oral Implantol.* **2014**, *40*, 3–10. [CrossRef] [PubMed]
88. Mathew, M.T.; Abbey, S.; Hallab, N.J.; Hall, D.J.; Sukotjo, C.; Wimmer, M.A. Influence of pH on the tribocorrosion behavior of CpTi in the oral environment: Synergistic interactions of wear and corrosion. *J. Biomed. Mater. Res. Part B Appl. Biomater.* **2012**, *100B*, 1662–1671. [CrossRef] [PubMed]
89. Souza, J.C.M.; Henriques, M.; Oliveira, R.; Teughels, W.; Celis, J.P.; Rocha, L.A. Do oral biofilms influence the wear and corrosion behavior of titanium? *Biofouling* **2010**, *26*, 471–478. [CrossRef] [PubMed]

90. Chaturvedi, T.P. An overview of the corrosion aspect of dental implants (titanium and its alloys). *Indian J. Dent. Res.* **2009**, *20*, 91–98. [CrossRef] [PubMed]
91. Mouhyi, J.; Ehrenfest, D.M.D.; Albrektsson, T. The peri-implantitis: Implant surfaces, microstructure, and physicochemical aspects. *Clin. Implant Dent. Relat. Res.* **2012**, *14*, 170–183. [CrossRef] [PubMed]
92. Bhola, R.; Bhola, S.M.; Mishra, B.; Olson, D.L. Corrosion in titanium dental implants/ prostheses—A review. *Trends Biomater. Artif. Organs* **2011**, *25*, 34–46.
93. Reclaru, L.; Meyer, J.M. Study of corrosion between a titanium implant and dental alloys. *J. Dent.* **1994**, *22*, 159–168. [CrossRef]
94. Souza, J.C.M.; Ponthiaux, P.; Henriques, M.; Oliveira, R.; Teughels, W.; Celis, J.-P.; Rocha, L.A. Corrosion behaviour of titanium in the presence of Streptococcus mutans. *J. Dent.* **2013**, *41*, 528–534. [CrossRef] [PubMed]
95. Souza, J.C.M.; Barbosa, S.L.; Ariza, E.A.; Henriques, M.; Teughels, W.; Ponthiaux, P.; Celis, J.-P.; Rocha, L.A. How do titanium and Ti6Al4V corrode in fluoridated medium as found in the oral cavity? An in vitro study. *Mater. Sci. Eng. C* **2015**, *47*, 384–393. [CrossRef] [PubMed]
96. Rodrigues, D.; Valderrama, P.; Wilson, T.; Palmer, K.; Thomas, A.; Sridhar, S.; Adapalli, A.; Burbano, M.; Wadhwani, C. Titanium corrosion mechanisms in the oral environment: A retrieval study. *Materials* **2013**, *6*, 5258–5274. [CrossRef] [PubMed]
97. Olmedo, D.G.; Tasat, D.R.; Duffo, G.; Guglielmotti, M.B.; Cabrini, R.L. The issue of corrosion in dental implants: A review. *Acta Odontol. Latinoam.* **2009**, *22*, 3–9. [PubMed]
98. Nikolopoulou, F. Saliva and dental implants. *Implant Dent.* **2006**, *15*, 372–376. [CrossRef] [PubMed]
99. Henry, P.J. Clinical experiences with dental implants. *Adv. Dent. Res.* **1999**, *13*, 147–152. [CrossRef] [PubMed]
100. Gittens, R.A.; Olivares-Navarrete, R.; Tannenbaum, R.; Boyan, B.D.; Schwartz, Z. Electrical implications of corrosion for osseointegration of titanium implants. *J. Dent. Res.* **2011**, *90*, 1389–1397. [CrossRef] [PubMed]
101. Jacobs, J.J.; Gilbert, J.L.; Urban, R.M. Corrosion of metal orthopaedic implants. *J. Bone Jt. Surg.* **1998**, *80*, 268–282. [CrossRef]
102. Geis-Gerstorfer, J. In vitro corrosion measurements of dental alloys. *J. Dent.* **1994**, *22*, 247–251. [CrossRef]
103. Branemark, P.I.; Hansson, B.O.; Adell, R.; Breine, U.; Lindstrom, J.; Hallen, O.; Ohman, A. Osseointegrated implants in the treatment of the edentulous jaw. Experience from a 10-year period. *Scand. J. Plast. Reconstr. Surg. Suppl.* **1997**, *16*, 1–132.
104. Adell, R.; Lekholm, U.; Rockler, B.; Brånemark, P.I. A 15-year study of osseointegrated implants in the treatment of the edentulous jaw. *Int. J. Oral Surg.* **1981**, *10*, 387–416. [CrossRef]
105. Galante, J.O.; Lemons, J.; Spector, M.; Wilson, P.D.; Wright, T.M. The biologic effects of implant materials. *J. Orthop. Res.* **1991**, *9*, 760–775. [CrossRef] [PubMed]
106. Hallab, N.J.; Mikecz, K.; Vermes, C.; Skipor, A.; Jacobs, J.J. Orthopaedic implant related metal toxicity in terms of human lymphocyte reactivity to metal-protein complexes produced from cobalt-base and titanium-base implant alloy degradation. *Mol. Cell. Biochem.* **2001**, *222*, 127–136. [CrossRef] [PubMed]
107. Hallab, N.J.; Jacobs, J.J.; Skipor, A.; Black, J.; Mikecz, K.; Galante, J.O. Systemic metal-protein binding associated with total joint replacement arthroplasty. *J. Biomed. Mater. Res.* **2000**, *49*, 353–361. [CrossRef]
108. Correa, C.B.; Pires, J.R.; Fernandes-Filho, R.B.; Sartori, R.; Vaz, L.G. Fatigue and fluoride corrosion on Streptococcus mutansadherence to titanium-based implant/component surfaces. *J. Prosthodont.* **2009**, *18*, 382–387. [CrossRef] [PubMed]
109. Fathi, M.H.; Salehi, M.; Saatchi, A.; Mortazavi, V.; Moosavi, S.B. In vitro corrosion behavior of bioceramic, metallic, and bioceramic–metallic coated stainless steel dental implants. *Dent. Mater.* **2003**, *19*, 188–198. [CrossRef]
110. Yu, F.; Addison, O.; Baker, S.J.; Davenport, A.J. Lipopolysaccharide inhibits or accelerates biomedical titanium corrosion depending on environmental acidity. *Int. J. Oral Sci.* **2015**, *7*, 179–186. [CrossRef] [PubMed]
111. Simon, B.I.; Goldman, H.M.; Ruben, M.P.; Baker, E. The role of endotoxin in periodontal disease. I. A reproducible, quantitative method for determining the amount of endotoxin in human gingival exudate. *J. Periodontol.* **1969**, *40*, 695–701. [CrossRef] [PubMed]
112. Casarin, R.C.V.; Ribeiro, É.P.; Mariano, F.S.; Nociti, F.H., Jr.; Casati, M.Z.; Gonçalves, R.B. Levels of aggregatibacter actinomycetemcomitans, Porphyromonas gingivalis, inflammatory cytokines and species-specific immunoglobulin G in generalized aggressive and chronic periodontitis. *J. Periodontal. Res.* **2010**, *45*, 635–642. [CrossRef] [PubMed]

113. Mathew, M.T.; Barão, V.A.; Yuan, J.C.-C.; Assunção, W.G.; Sukotjo, C.; Wimmer, M.A. What is the role of lipopolysaccharide on the tribocorrosive behavior of titanium? *J. Mech. Behav. Biomed. Mater.* **2012**, *8*, 71–85. [CrossRef] [PubMed]
114. Gil, F.J.; Canedo, R.; Padros, A.; Baneres, M.V.; Arano, J.M. Fretting corrosion behaviour of ball-and-socket joint on dental implants with different prosthodontic alloys. *Biomed. Mater. Eng.* **2003**, *13*, 27–34. [PubMed]
115. Hjalmarsson, L.; Smedberg, J.I.; Wennerberg, A. Material degradation in implant-retained cobalt-chrome and titanium frameworks. *J. Oral Rehabil.* **2010**, *38*, 61–71. [CrossRef] [PubMed]
116. Wataha, J.C. Biocompatibility of dental casting alloys: A review. *J. Prosthet. Dent.* **2000**, *83*, 223–234. [CrossRef]
117. Brune, D. Metal release from dental biomaterials. *Biomaterials* **1986**, *7*, 163–175. [CrossRef]
118. Wataha, J.C.; Malcolm, C.T.; Hanks, C.T. Correlation between cytotoxicity and the elements released by dental casting alloys. *Int. J. Prosthodont.* **1995**, *8*, 9–14. [PubMed]
119. Sarkar, N.; Fuys, R.; Stanford, J.W. Applications of electrochemical techniques to characterize the corrosion of dental alloys. In *Corrosion and Degradation of Implant Materials*; Syrett, B.C., Acharya, A., Eds.; ASTM: West Conshohocken, PA, USA, 1979; pp. 277–294.
120. Revathi, A.; Borrás, A.D.; Muñoz, A.I.; Richard, C.; Manivasagam, G. Degradation mechanisms and future challenges of titanium and its alloys for dental implant applications in oral environment. *Mater. Sci. Eng. C* **2017**, *76*, 1354–1368. [CrossRef] [PubMed]
121. Mathew, M.T.; Pai, P.S.; Pourzal, R.; Fischer, A.; Wimmer, M.A. Significance of tribocorrosion in biomedical applications: Overview and current status. *Adv. Tribol.* **2009**, *2009*, 1–12. [CrossRef]
122. Landolt, D.; Mischler, S.; Stemp, M. Electrochemical methods in tribocorrosion: A critical appraisal. *Electrochim. Acta* **2001**, *46*, 3913–3929. [CrossRef]
123. Licausi, M.P.; Muñoz, A.I.; Borrás, V.A. Influence of the fabrication process and fluoride content on the tribocorrosion behaviour of Ti6Al4V biomedical alloy in artificial saliva. *J. Mech. Behav. Biomed. Mater.* **2013**, *20*, 137–148. [CrossRef] [PubMed]
124. Dodds, M.W.J.; Johnson, D.A.; Yeh, C.-K. Health benefits of saliva: A review. *J. Dent.* **2005**, *33*, 223–233. [CrossRef] [PubMed]
125. Cruz, H.V.; Souza, J.C.M.; Henriques, M.; Rocha, L.A. Tribocorrosion and bio-tribocorrosion in the oral environment: The case of dental implants. In *Biomedical Tribology*; Davim, J.P., Ed.; Nova Science Publishers: New York, NY, USA, 2011; pp. 1–33.
126. Souza, J.C.M.; Henriques, M.; Teughels, W.; Ponthiaux, P.; Celis, J.-P.; Rocha, L.A. Wear and corrosion interactions on titanium in oral environment: Literature review. *J. Bio-Tribo-Corros.* **2015**, *1*, 13. [CrossRef]
127. Mischler, S. Triboelectrochemical techniques and interpretation methods in tribocorrosion: A comparative evaluation. *Tribol. Int.* **2008**, *41*, 573–583. [CrossRef]
128. Schiff, N.; Grosgogeat, B.; Lissac, M.; Dalard, F. Influence of fluoride content and pH on the corrosion resistance of titanium and its alloys. *Biomaterials* **2002**, *23*, 1995–2002. [CrossRef]
129. Huang, H.-H. Effects of fluoride concentration and elastic tensile strain on the corrosion resistance of commercially pure titanium. *Biomaterials* **2002**, *23*, 59–63. [CrossRef]
130. Kaneko, K.; Yokoyama, K.I.; Moriyama, K.; Asaoka, K.; Sakai, J.I.; Nagumo, M. Delayed fracture of beta titanium orthodontic wire in fluoride aqueous solutions. *Biomaterials* **2003**, *24*, 2113–2120. [CrossRef]
131. Ratner, B.D.; Johnston, A.B.; Lenk, T.J. Biomaterial surfaces. *J. Biomed. Mater. Res.* **1987**, *21*, 59–89. [CrossRef] [PubMed]
132. Smith, D.C.; Pilliar, R.M.; Chernecky, R. Dental implant materials. I. Some effects of preparative procedures on surface topography. *J. Biomed. Mater. Res.* **1991**, *25*, 1045–1068. [CrossRef] [PubMed]
133. Christersson, C.E.; Dunford, R.G.; Glantz, P.O.; Baier, R.E. Effect of critical surface tension on retention of oral microorganisms. *Scand. J. Dent. Res.* **1989**, *97*, 247–256. [CrossRef] [PubMed]
134. Teughels, W.; van Assche, N.; Sliepen, I.; Quirynen, M. Effect of material characteristics and/or surface topography on biofilm development. *Clin. Oral Implant. Res.* **2006**, *17*, 68–81. [CrossRef] [PubMed]
135. Ma, Y.; Lassiter, M.O.; Banas, J.A.; Galperín, M.Y.; Taylor, K.G.; Doyle, R.J. Multiple glucan-binding proteins of Streptococcus sobrinus. *J. Bacteriol.* **1996**, *178*, 1572–1577. [CrossRef] [PubMed]
136. Steinberg, D.; Eyal, S. Early formation of Streptococcus sobrinus biofilm on various dental restorative materials. *J. Dent.* **2001**, *30*, 47–51. [CrossRef]

137. Zitzmann, N.U.; Abrahamsson, I.; Berglundh, T.; Lindhe, J. Soft tissue reactions to plaque formation at implant abutments with different surface topography. An experimental study in dogs. *J. Clin. Periodontol.* **2002**, *29*, 456–461. [CrossRef] [PubMed]
138. Dmytiyk, J.J.; Fox, S.C.; Moriarty, J.D. The effects of scaling titanium implant surfaces with metal and plastic instruments on cell attachment. *J. Periodontol.* **1990**, *61*, 491–496. [CrossRef] [PubMed]
139. Fox, S.C.; Moriarty, J.D.; Kusy, R.P. The effects of scaling a titanium implant surface with metal and plastic instruments: An in vitro study. *J. Periodontol.* **1990**, *61*, 485–490. [CrossRef] [PubMed]
140. Schou, S.; Berglundh, T.; Lang, N.P. Surgical treatment of peri-implantitis. *Int. J. Oral Maxillofac. Implant.* **2004**, *19*, 140–149.
141. Claffey, N.; Clarke, E.; Polyzois, I.; Renvert, S. Surgical treatment of peri-implantitis. *J. Clin. Periodontol.* **2008**, *35*, 316–332. [CrossRef] [PubMed]
142. Augthun, M.; Tinschert, J.; Huber, A. In vitro studies on the effect of cleaning methods on different implant surfaces. *J. Periodontol.* **1998**, *69*, 857–864. [CrossRef] [PubMed]
143. Brookshire, F.V.G.; Nagy, W.W.; Dhuru, V.B.; Ziebert, G.J.; Chada, S. The qualitative effects of various types of hygiene instrumentation on commercially pure titanium and titanium alloy implant abutments: An in vitro and scanning electron microscope study. *J. Prosthet. Dent.* **1997**, *78*, 286–294. [CrossRef]
144. Meier, R.M.; Pfammatter, C.; Zitzmann, N.U.; Filippi, A.; Kuhl, S. Surface quality after implantoplasty. *Schweiz. Monatsschr. Zahnmed.* **2012**, *122*, 714–724. [PubMed]
145. Ramaglia, L.; di Lauro, A.E.; Morgese, F.; Squillace, A. Profilometric and standard error of the mean analysis of rough implant surfaces treated with different instrumentations. *Implant Dent.* **2006**, *15*, 77–82. [CrossRef] [PubMed]
146. Barbour, M.E.; O'Sullivan, D.J.; Jenkinson, H.F.; Jagger, D.C. The effects of polishing methods on surface morphology, roughness and bacterial colonisation of titanium abutments. *J. Mater. Sci. Mater. Med.* **2007**, *18*, 1439–1447. [CrossRef] [PubMed]
147. Schwarz, F.; Sahm, N.; Iglhaut, G.; Becker, J. Impact of the method of surface debridement and decontamination on the clinical outcome following combined surgical therapy of peri-implantitis: A randomized controlled clinical study. *J. Clin. Periodontol.* **2011**, *38*, 276–284. [CrossRef] [PubMed]
148. Rimondini, L.; Simoncini, F.C.; Carrassi, A. Micro-morphometric assessment of titanium plasma-sprayed coating removal using burs for the treatment of peri-implant disease. *Clin. Oral Implant. Res.* **2000**, *11*, 129–138. [CrossRef]
149. Valderrama, P.; Wilson, T.G., Jr. Detoxification of implant surfaces affected by peri-implant disease: An overview of surgical methods. *Int. J. Dent.* **2013**, *2013*, 740680. [CrossRef] [PubMed]
150. Ramel, C.F.; Lüssi, A.; Özcan, M.; Jung, R.E.; Hämmerle, C.H.F.; Thoma, D.S. Surface roughness of dental implants and treatment time using six different implantoplasty procedures. *Clin. Oral Implant. Res.* **2016**, *27*, 776–781. [CrossRef] [PubMed]
151. Rosen, P.; Qari, M.; Froum, S.; Dibart, S.; Chou, L. A pilot study on the efficacy of a treatment algorithm to detoxify dental implant surfaces affected by peri-implantitis. *Int. J. Periodontics Restor. Dent.* **2018**, *38*, 261–267. [CrossRef] [PubMed]
152. Moëne, R.; Décaillet, F.; Andersen, E.; Mombelli, A. Subgingival plaque removal using a new air-polishing device. *J. Periodontol.* **2010**, *81*, 79–88. [CrossRef] [PubMed]
153. Kozlovsky, A.; Soldinger, M.; Sperling, I. The effectiveness of the air-powder abrasive device on the tooth and periodontium: An overview. *Clin. Prev. Dent.* **1989**, *11*, 7–11. [PubMed]
154. Atkinson, D.R.; Cobb, C.M.; Killoy, W.J. The effect of an air-powder abrasive system on in vitro root surfaces. *J. Periodontol.* **1984**, *55*, 13–18. [CrossRef] [PubMed]
155. Petersilka, G.J.; Bell, M.; Haberlein, I.; Mehl, A.; Hickel, R.; Flemmig, T.F. In vitro evaluation of novel low abrasive air polishing powders. *J. Clin. Periodontol.* **2003**, *30*, 9–13. [CrossRef] [PubMed]
156. Petersilka, G.J.; Tunkel, J.; Barakos, K.; Heinecke, A.; Häberlein, I.; Flemmig, T.F. Subgingival plaque removal at interdental sites using a low-abrasive air polishing powder. *J. Periodontol.* **2003**, *74*, 307–311. [CrossRef] [PubMed]
157. Petersilka, G.J.; Steinmann, D.; Haberlein, I.; Heinecke, A.; Flemmig, T.F. Subgingival plaque removal in buccal and lingual sites using a novel low abrasive air-polishing powder. *J. Clin. Periodontol.* **2003**, *30*, 328–333. [CrossRef] [PubMed]

158. Tastepe, C.S.; Lin, X.; Donnet, M.; Wismeijer, D.; Liu, Y. Parameters that improve cleaning efficiency of subgingival air polishing on titanium implant surfaces: An in vitro study. *J. Periodontol.* **2017**, *88*, 407–414. [CrossRef] [PubMed]
159. Ronay, V.; Merlini, A.; Attin, T.; Schmidlin, P.R.; Sahrmann, P. In vitro cleaning potential of three implant debridement methods. Simulation of the non-surgical approach. *Clin. Oral Implant. Res.* **2017**, *28*, 151–155. [CrossRef] [PubMed]
160. Duarte, P.M.; Reis, A.F.; de Freitas, P.M.; Ota-Tsuzuki, C. Bacterial adhesion on smooth and rough titanium surfaces after treatment with different instruments. *J. Periodontol.* **2009**, *80*, 1824–1832. [CrossRef] [PubMed]
161. Kreisler, M.; Al Haj, H.; d'Hoedt, B. Clinical efficacy of semiconductor laser application as an adjunct to conventional scaling and root planing. *Lasers Surg. Med.* **2005**, *37*, 350–355. [CrossRef] [PubMed]
162. Kreisler, M.; Kohnen, W.; Christoffers, A.-B.; Götz, H.; Jansen, B.; Duschner, H.; D'Hoedt, B. In vitro evaluation of the biocompatibility of contaminated implant surfaces treated with an Er:YAG laser and an air powder system. *Clin. Oral Implant. Res.* **2005**, *16*, 36–43. [CrossRef] [PubMed]
163. Razzoog, M.E.; Koka, S. In vitro analysis of the effects of two air-abrasive prophylaxis systems and inlet air pressure on the surface of titanium abutment cylinders. *J. Prosthodont.* **1994**, *3*, 103–107. [CrossRef] [PubMed]
164. Finnegan, M.; Linley, E.; Denyer, S.P.; McDonnell, G.; Simons, C.; Maillard, J.Y. Mode of action of hydrogen peroxide and other oxidizing agents: Differences between liquid and gas forms. *J. Antimicrob. Chemother.* **2010**, *65*, 2108–2115. [CrossRef] [PubMed]
165. Gosau, M.; Hahnel, S.; Schwarz, F.; Gerlach, T.; Reichert, T.E.; Burgers, R. Effect of six different peri-implantitis disinfection methods on in vivo human oral biofilm. *Clin. Oral Implant. Res.* **2010**, *21*, 866–872. [CrossRef]
166. Waal, Y.C.M.; Raghoebar, G.M.; Meijer, H.J.A.; Winkel, E.G.; van Winkelhoff, A.J. Implant decontamination with 2% chlorhexidine during surgical peri-implantitis treatment: A randomized, double-blind, controlled trial. *Clin. Oral Implant. Res.* **2014**, *26*, 1015–1023. [CrossRef] [PubMed]
167. Valderrama, P.; Gonzalez, M.G.; Cantu, M.G.; Wilson, T.G. Detoxification of implant surfaces affected by peri-implant disease: An overview of non-surgical methods. *Open Dent. J.* **2014**, *8*, 77–84. [CrossRef] [PubMed]
168. Wheelis, S.E.; Gindri, I.M.; Valderrama, P.; Wilson, T.G.; Huang, J.; Rodrigues, D.C. Effects of decontamination solutions on the surface of titanium: Investigation of surface morphology, composition, and roughness. *Clin. Oral Implant. Res.* **2016**, *27*, 329–340. [CrossRef] [PubMed]
169. Könönen, M.H.O.; Lavonius, E.T.; Kivilahti, J.K. SEM observations on stress corrosion cracking of commercially pure titanium in a topical fluoride solution. *Dent. Mater.* **1995**, *11*, 269–272. [CrossRef]
170. Rodrigues, D.C.; Urban, R.M.; Jacobs, J.J.; Gilbert, J.L. In vivo severe corrosion and hydrogen embrittlement of retrieved modular body titanium alloy hip-implants. *J. Biomed. Mater. Res. Part B Appl. Biomater.* **2009**, *88*, 206–219. [CrossRef] [PubMed]
171. Sartori, R.; Correa, C.B.; Marcantonio, E., Jr.; Vaz, L.G. Influence of a fluoridated medium with different pHs on commercially pure titanium-based implants. *J. Prosthodont.* **2009**, *18*, 130–134. [CrossRef] [PubMed]
172. Muguruma, T.; Iijima, M.; Brantley, W.A.; Yuasa, T.; Kyung, H.-M.; Mizoguchi, I. Effects of sodium fluoride mouth rinses on the torsional properties of miniscrew implants. *Am. J. Orthod. Dentofac. Orthop.* **2011**, *139*, 588–593. [CrossRef] [PubMed]
173. Toniollo, M.B.; Galo, R.; Macedo, A.P.; Rodrigues, R.C.S.; Ribeiro, R.F.; Mattos, M.D.G. Effect of fluoride sodium mouthwash solutions on cpTI: Evaluation of physicochemical properties. *Braz. Dent. J.* **2012**, *23*, 496–501. [CrossRef] [PubMed]
174. Wiedmer, D.; Petersen, F.C.; Lönn-Stensrud, J.; Tiainen, H. Antibacterial effect of hydrogen peroxide-titanium dioxide suspensions in the decontamination of rough titanium surfaces. *Biofouling* **2017**, *33*, 451–459. [CrossRef] [PubMed]
175. Ntrouka, V.I.; Slot, D.E.; Louropoulou, A.; van der Weijden, F. The effect of chemotherapeutic agents on contaminated titanium surfaces: A systematic review. *Clin. Oral Implant. Res.* **2011**, *22*, 681–690. [CrossRef] [PubMed]
176. Ntrouka, V.; Hoogenkamp, M.; Zaura, E.; van der Weijden, F. The effect of chemotherapeutic agents on titanium-adherent biofilms. *Clin. Oral Implants Res.* **2011**, *22*, 1227–1234. [CrossRef] [PubMed]
177. Ramesh, D.; Sridhar, S.; Siddiqui, D.A.; Valderrama, P.; Rodrigues, D.C. Detoxification of titanium implant surfaces: Evaluation of surface morphology and bone-forming cell compatibility. *J. Bio-Tribo-Corros.* **2017**, *3*, 1–13. [CrossRef]

178. Oliveira, M.N.; Schunemann, W.V.H.; Mathew, M.T.; Henriques, B.; Magini, R.S.; Teughels, W.; Souza, J.C.M. Can degradation products released from dental implants affect peri-implant tissues? *J. Periodontal Res.* **2018**, *53*, 1–11. [CrossRef] [PubMed]
179. Souza, J.G.S.; Cordeiro, J.M.; Lima, C.V.; Barão, V.A.R. Citric acid reduces oral biofilm and influences the electrochemical behavior of titanium: An in situ and in vitro study. *J. Periodontol.* **2018**. [CrossRef] [PubMed]
180. Ungvari, K.; Pelsoczi, I.K.; Kormos, B.; Oszko, A.; Rakonczay, Z.; Kemeny, L.; Radnai, M.; Nagy, K.; Fazekas, A.; Turzo, K. Effects on titanium implant surfaces of chemical agents used for the treatment of peri-implantitis. *J. Biomed. Mater. Res. B Appl. Biomater.* **2010**, *94*, 222–229. [CrossRef] [PubMed]
181. Soukos, N.S.; Mulholland, S.E.; Socransky, S.S.; Doukas, A.G. Photodestruction of human dental plaque bacteria: Enhancement of the photodynamic effect by photomechanical waves in an oral biofilm model. *Lasers Surg. Med.* **2003**, *33*, 161–168. [CrossRef] [PubMed]
182. Kreisler, M.; Kohnen, W.; Marinello, C.; Schoof, J.; Langnau, E.; Jansen, B.; d'Hoedt, B. Antimicrobial efficacy of semiconductor laser irradiation on implant surfaces. *Int. J. Oral Maxillofac. Implant.* **2003**, *18*, 706–711.
183. Cohen, M.I.; Epperson, J.P. Application of lasers to microelectronic fabrication. In *Electron Beam and Laser Beam Technology*; Marton, L., El-Kareh, A.B., Eds.; Academic Press: New York, NY, USA, 1968; pp. 139–186.
184. Wilson, M. Photolysis of oral bacteria and its potential use in the treatment of caries and periodontal disease. *J. Appl. Bacteriol.* **1993**, *75*, 299–306. [CrossRef] [PubMed]
185. Hall, R.R. The healing of tissues incised by a carbon-dioxide laser. *Br. J. Surg.* **1971**, *58*, 222–225. [CrossRef] [PubMed]
186. Kreisler, M.; Gotz, H.; Duschner, H. Effect of Nd:YAG, Ho:YAG, Er: YAG, CO2, and GaAIAs laser irradiation on surface properties of endosseous dental implants. *Int. J. Oral Maxillofac. Implant.* **2002**, *17*, 202–211.
187. Shibli, J.A.; Theodoro, L.H.; Haypek, P.; Garcia, V.G.; Marcantonio, E. The effect of CO2 laser irradiation on failed implant surfaces. *Implant. Dent.* **2004**, *13*, 342–351. [CrossRef] [PubMed]
188. Park, C.-Y.; Kim, S.-G.; Kim, M.-D.; Eom, T.-G.; Yoon, J.-H.; Ahn, S.-G. Surface properties of endosseous dental implants after NdYAG and CO2 laser treatment at various energies. *J. Oral Maxillofac. Surg.* **2005**, *63*, 1522–1527. [CrossRef] [PubMed]
189. Park, J.H.; Heo, S.J.; Koak, J.Y.; Kim, S.K.; Han, C.H.; Lee, J.H. Effects of laser irradiation on machined and anodized titanium disks. *Int. J. Oral Maxillofac. Implant.* **2012**, *27*, 265–272.
190. Romanos, G.; Ko, H.-H.; Froum, S.; Tarnow, D. The use of CO2 laser in the treatment of peri-implantitis. *Photomed. Laser Surg.* **2009**, *27*, 381–386. [CrossRef] [PubMed]
191. Stuebinger, S.; Etter, C.; Miskiewicz, M.; Homann, F.; Saldamli, B.; Wieland, M.; Sader, R. Surface alterations of polished and sandblasted and acid-etched titanium implants after Er:YAG, carbon dioxide, and diode laser irradiation. *Int. J. Oral Maxillofac. Implant.* **2010**, *25*, 104–111.
192. Aoki, A.; Ando, Y.; Watanabe, H.; Ishikawa, I. In vitro studies on laser scaling of subgingival calculus with an erbium:YAG laser. *J. Periodontol.* **1994**, *65*, 1097–1106. [CrossRef] [PubMed]
193. Ishikawa, I.; Aoki, A.; Takasaki, A.A. Clinical application of erbium: YAG laser in periodontology. *J. Int. Acad. Periodontol.* **2008**, *10*, 22–30. [PubMed]
194. Schmage, P.; Thielemann, J.; Nergiz, I.; Scorziello, T.M.; Pfeiffer, P. Effects of 10 cleaning instruments on four different implant surfaces. *Int. J. Oral Maxillofac. Implant.* **2012**, *27*, 308–317.
195. Shin, S.-I.; Min, H.-K.; Park, B.-H.; Kwon, Y.-H.; Park, J.-B.; Herr, Y.; Heo, S.-J.; Chung, J.-H. The effect of Er:YAG laser irradiation on the scanning electron microscopic structure and surface roughness of various implant surfaces: An in vitro study. *Lasers Med. Sci.* **2011**, *26*, 767–776. [CrossRef] [PubMed]
196. Galli, C.; Macaluso, G.M.; Elezi, E.; Ravanetti, F.; Cacchioli, A.; Gualini, G.; Passeri, G. The effects of Er:YAG laser treatment on titanium surface profile and osteoblastic cell activity: An in vitro study. *J. Periodontol.* **2011**, *82*, 1169–1177. [CrossRef] [PubMed]
197. Taniguchi, Y.; Aoki, A.; Mizutani, K.; Takeuchi, Y.; Ichinose, S.; Takasaki, A.A.; Schwarz, F.; Izumi, Y. Optimal Er:YAG laser irradiation parameters for debridement of microstructured fixture surfaces of titanium dental implants. *Lasers Med. Sci.* **2013**, *28*, 1057–1068. [CrossRef] [PubMed]
198. Shin, S.-I.; Lee, E.-K.; Kim, J.-H.; Lee, J.-H.; Kim, S.-H.; Kwon, Y.-H.; Herr, Y.; Chung, J.-H. The effect of Er:YAG laser irradiation on hydroxyapatite-coated implants and fluoride-modified TiO_2-blasted implant surfaces: A microstructural analysis. *Lasers Med. Sci.* **2013**, *28*, 823–831. [CrossRef] [PubMed]

199. Ayobian-Markazi, N.; Karimi, M.; Safar-Hajhosseini, A. Effects of Er:YAG laser irradiation on wettability, surface roughness, and biocompatibility of SLA titanium surfaces: An in vitro study. *Lasers Med. Sci.* **2015**, *30*, 561–566. [CrossRef] [PubMed]
200. Takagi, T.; Aoki, A.; Ichinose, S.; Taniguchi, Y.; Tachikawa, N.; Shinoki, T.; Meinzer, W.; Sculean, A.; Izumi, Y. Effective removal of calcified deposits on microstructured titanium fixture surfaces of dental implants with erbium lasers. *J. Periodontol.* **2018**, *89*, 680–690. [CrossRef] [PubMed]
201. Romanos, G.E.; Everts, H.; Nentwig, G.H. Effects of diode and Nd:YAG laser irradiation on titanium discs: A scanning electron microscope examination. *J. Periodontol.* **2000**, *71*, 810–815. [CrossRef] [PubMed]
202. Castro, G.L.; Gallas, M.; Núñez, I.R.; Borrajo, J.L.L.; Álvarez, J.C.; Varela, L.G. Scanning electron microscopic analysis of diode laser-treated titanium implant surfaces. *Photomed. Laser Surg.* **2007**, *25*, 124–128. [CrossRef] [PubMed]
203. Gonçalves, F.; Zanetti, A.L.; Zanetti, R.V.; Martelli, F.S.; Avila-Campos, M.J.; Tomazinho, L.F.; Granjeiro, J.M. Effectiveness of 980-mm diode and 1064-nm extra-long-pulse neodymium-doped yttrium aluminum garnet lasers in implant disinfection. *Photomed. Laser Surg.* **2010**, *28*, 273–280. [CrossRef] [PubMed]
204. Leja, C.; Geminiani, A.; Caton, J.; Romanos, G.E. Thermodynamic effects of laser irradiation of implants placed in bone: An in vitro study. *Lasers Med. Sci.* **2013**, *28*, 1435–1440. [CrossRef] [PubMed]
205. Geminiani, A.; Caton, J.G.; Romanos, G.E. Temperature change during non-contact diode laser irradiation of implant surfaces. *Lasers Med. Sci.* **2012**, *27*, 339–342. [CrossRef] [PubMed]
206. Giannelli, M.; Lasagni, M.; Bani, D. Thermal effects of λ = 808 nm GaAlAs diode laser irradiation on different titanium surfaces. *Lasers Med. Sci.* **2015**, *30*, 2341–2352. [CrossRef] [PubMed]

© 2018 by the authors. Licensee MDPI, Basel, Switzerland. This article is an open access article distributed under the terms and conditions of the Creative Commons Attribution (CC BY) license (http://creativecommons.org/licenses/by/4.0/).

Review

Influence of Dental Prosthesis and Restorative Materials Interface on Oral Biofilms

Yu Hao [1,2,3,†], **Xiaoyu Huang** [1,2,3,†], **Xuedong Zhou** [1,2,3], **Mingyun Li** [1,3], **Biao Ren** [1,3], **Xian Peng** [1,3,*] **and Lei Cheng** [1,2,3,*]

1. State Key Laboratory of Oral Diseases, Sichuan University, Chengdu 610041, China; haoyu_dentist@sina.com (Y.H.); 2013181641007@stu.scu.edu.cn (X.H.); zhouxd@scu.edu.cn (X.Z.); limingyun@scu.edu.cn (M.L.); renbiao@scu.edu.cn (B.R.)
2. Department of Cariology and Endodontics, West China School of Stomatology, Sichuan University, Chengdu 610041, China
3. National Clinical Research Center for Oral Diseases, Sichuan University, Chengdu 610041, China
* Correspondence: pengx07@hotmail.com (X.P.); chenglei@scu.edu.cn (L.C.); Tel.: +86-158-8240-6893 (X.P.); +86-138-8228-4602 (L.C.)
† These authors contributed equally to this work.

Received: 27 August 2018; Accepted: 10 October 2018; Published: 14 October 2018

Abstract: Oral biofilms attach onto both teeth surfaces and dental material surfaces in oral cavities. In the meantime, oral biofilms are not only the pathogenesis of dental caries and periodontitis, but also secondary caries and peri-implantitis, which would lead to the failure of clinical treatments. The material surfaces exposed to oral conditions can influence pellicle coating, initial bacterial adhesion, and biofilm formation, due to their specific physical and chemical characteristics. To define the effect of physical and chemical characteristics of dental prosthesis and restorative material on oral biofilms, we discuss resin-based composites, glass ionomer cements, amalgams, dental alloys, ceramic, and dental implant material surface properties. In conclusion, each particular chemical composition (organic matrix, inorganic filler, fluoride, and various metallic ions) can enhance or inhibit biofilm formation. Irregular topography and rough surfaces provide favorable interface for bacterial colonization, protecting bacteria against shear forces during their initial reversible binding and biofilm formation. Moreover, the surface free energy, hydrophobicity, and surface-coating techniques, also have a significant influence on oral biofilms. However, controversies still exist in the current research for the different methods and models applied. In addition, more in situ studies are needed to clarify the role and mechanism of each surface parameter on oral biofilm development.

Keywords: biofilm; dental restorative material; surface characteristics; surface roughness; resin-based composite

---

1. Introduction

From the widely applied dental amalgams [1] to esthetic resin-based composites [2,3] and ion-release glass ionomer cements [4], direct restorative materials are generally used to reconstruct the tooth when its structure is compromised by trauma or dental caries. Besides, indirect crown restorations and dental implants have been applied to tooth and dentition defect restorations for decades [5,6]. Although these restorative materials had significant evolvement in the past few decades, the failure rates of restorations are still problems to the dentists and investigators.

As it stands, direct restorations showed an annual failure rate up to 7.9% with the main reasons of secondary caries and bulk fracture [3,7,8]. It was reported that the 5-year failure rate of fixed dental prostheses was more than 10%, due to the common complications of caries and endodontic diseases [9–11]. Although the implant survival reached 92.8–97.1% over a follow-up period of

10 years, we cannot ignore peri-implantitis, which is mainly caused by biofilm accumulation [12,13]. The prevalence of peri-implantitis varies from 11% to 47%, because of the different threshold of bone loss [14]. However, Schwendicke's study showed that an implant might cost more than 300 Euro when it comes to peri-implantitis, by comparison with a healthy implant [15]. Dental restorative materials placed in oral cavity are subjected to aggressive attack by bacteria. Components in materials will be biodegraded by the dental plaque, which will probably compromise the marginal integrity and induce the development and progression of secondary caries and peri-implantitis [12,16–18].

The oral cavity is a complex environment, where high humidity, moderate temperature, and abundance of nutrients promote the formation of differentiated microorganisms and microbial biofilms [19–21]. Biofilm formation in the oral cavity is a gradated process consisting of four stages (Figure 1) [22]:

1. acquired pellicle formation;
2. primary colonization;
3. coaggregation;
4. mature biofilm establishment.

To generate a biofilm, all surfaces exposed to the oral environment are steadily covered by a pellicle derived from the adsorption of organic and inorganic molecules in saliva. The receptors of salivary pellicle offer binding sites for floating initial bacteria cells to attach to these surfaces and form microcolonies. As time goes by, the bacteria cells aggregate, proliferate, and grow into a mushroom-shaped mature biofilm, firmly attaching to these surfaces [23,24]. Therefore, bacterial cells within the biofilm do not exist as independent entities but, rather, as a coordinated, metabolically integrated microbial community [22].

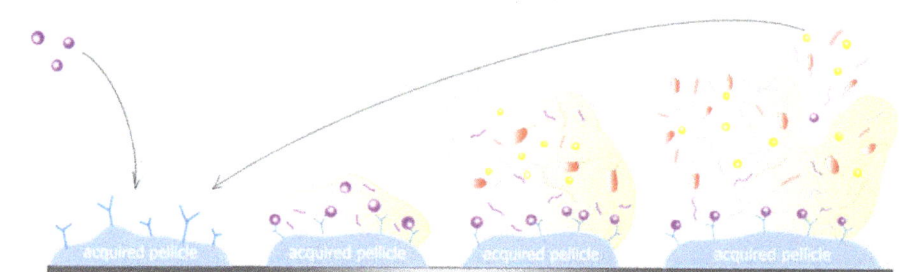

Figure 1. The process of biofilm formation in the oral cavity is divided into four stages: 1. acquired pellicle formation; 2. initial adhesion; 3. coaggregation; 4. maturation and diffusion.

Since adhesion is the crucial step of biofilm formation, understanding bacteria–surface interaction is essential for biofilm control and survival rate of restorations. The physical and chemical characteristics of dental prosthesis and restorative materials can influence pellicle coating, initial bacterial adhesion, and biofilm formation. The growing application of dental materials has presented an ever-increasing need to better understand the interactions between biofilm and material surfaces in the oral cavity. Thus, in this review, we discuss the effects of physical and chemical characteristics of different dental prosthesis and restorative material surfaces on oral biofilms.

2. Physical Characteristics of Dental Materials

2.1. Surface Roughness

Nowadays, some clinical procedures, polishing and finishing, are usually applied for smoother surfaces. Among these polishing and finishing techniques, the lowest surface roughness (SR) values could be achieved by Mylar, and followed by Al_2O_3 discs, one-step rubber points, diamond bur, and multi-blade carbide bur [25].

Many researches have demonstrated that unpolished materials surfaces could accumulate more dental biofilm than polished ones, including resin-based composites, ceramics, implant abutments, and denture bases [22,26,27]. Kim [28] investigated the surface ultrastructure, roughness of four ceramic materials (Vita Enamic, Lava Ultimate, Vitablocs Mark II, and Wieland Reflex), and assessed their promotion of biofilm development following adjustments and simulated intraoral polishing methods. It was proved that surface roughness values (Ra) were greater in all materials following these methods, resulting in more biofilm accumulation, which implied the main cause of biofilm accumulation was surface roughness. A previous study evaluated the surface roughness of 20 commercial dental composite resins after abrasive wear, with the average roughness ranging from 0.49 to 0.79 μm [29]. According to Bollen's study, the surface roughness above the threshold roughness (Ra = 0.2 μm) results in a simultaneous increase in biofilm accumulation, and no further reduction in bacterial adhesion could be observed under the threshold value [30]. In the same way, Yuan et al. demonstrated that the area of adherent bacteria was a highly linear correlation coefficient ($r = 0.893$, $P < 0.01$) when Ra < 0.80 μm, and weakly correlated with SR when Ra $\leq$ 0.20 μm ($r = 0.643$, $P < 0.01$) [31]. It indicated that factors other than SR influence biofilm formation when Ra $\leq$ 0.20 μm.

According to Ionescu et al., surface topography, the 3D characteristics of a surface with peaks and valleys distribution, could explain the crucial role of SR in biofilm formation [26]. The deeper and larger depressions may increase the contact area and provide more favorable interfaces for bacterial colonization and biofilm formation, protecting bacteria against shear forces (rinsing and brushing) during their initial reversible binding, leading to irreversible and stronger attachment [17,32]. Hence, it is difficult to eliminate microcolonies on the rough surfaces, resulting in the formation of mature biofilm [33].

The studies mentioned above were mostly done in vitro. All surfaces in the oral cavity are covered by the salivary pellicle, and the SR, one of the physical characteristics of material surfaces, is, in part, counterbalanced by the presence of the salivary pellicle [33]. Besides, as the biofilm maturing, the effect of SR on biofilm development is reduced, with the new bacteria adhering to the initial formed biofilm but not to the tested material surface [34–36]. Hence, the roughness of material surface mainly influences the initial bacterial colonization. Lorenzo [37] revealed that biofilms developed by single bacterial species or simple microbial associations are more readily influenced by surface roughness and topography than biofilm formed by complex communities. The different outcomes of the above research could be related to different methods and the development of the genetic technology.

Although improving implant osseointegration, the surface roughness has been proposed as the main feature inducing biofilm development [38]. Many studies showed that the increase of the SR could cause an exponential growth in bacterial cells [39] and facilitate biofilm formation [22]. However, compared to the resin-based composites (RBCs) and ceramic, SR is not always detrimental to the treatment because it is benefit for the attachment of osteoblast. Thus, the role of SR seems a little contradictory in this field.

With respect to the biofilm composition, Marcos [40] affirmed that there seems to be no reason to believe that implants with rough surfaces are more susceptive to fail, and his results are in accordance with the study [41] that showed a similar microbiota composition on titanium of different SR. The controversial views, above, may result from different kinds of biofilms, different incubation times and, most importantly, the different kind of titanium discs used. Some of these studies employed the commercially available ones provided by companies, and some of them employed titanium discs only for labs, which differed in more than just roughness. Marcos [40] also found no significant difference of succession kinetics of 23 microorganism species on titanium, with different Ra values, in 1, 3, 7, 14, and 21 days. Regardless, in manufacturing advanced implants, new surface treatment technology should be applied to establish a balance between osteoblast and oral bacterial attachment.

2.2. Other Physical Characteristics

The surface free energy (SFE) is related to the wettability of the material surface as an equivalent to the surface of a fluid. To determine the SFE, the contact angles (θ) are measured by three liquids differing in hydrophobicity on a specific surface [26]. A smaller contact angle implies higher SFE and higher surface hydrophilia of the material [33]. It has been reported that less biofilm formation occurred on RBC surfaces with low SFE, probably because of the similar hydrophilic properties between salivary pellicle and substratum surfaces [26,30]. The effects of SFE on biofilm formation may be inaccurate when Ra > 0.1 μm, and SR plays the primary role in biofilm accumulation, indeed. In other words, SFE influenced early adhesion of *Streptococcus mutans* (*S. mutans*) on super smooth surfaces (Ra ≤ 0.06 μm) [31].

The parameters of clinical dental materials influence on oral biofilms are intricate and co-occurring. Higher surface hydrophilia implies higher SFE, which induces more microorganism accumulation [33]. Also, the SR partially depends on inorganic filler size. Nanofilled RBCs wear by breaking out of individual primary particles. However, for microhybrid RBCs, the relatively soft matrix is worn before the fillers plucked out [29]. It is essential to control variables to study single parameters of clinical dental materials influencing oral biofilms.

3. Chemical Characteristics of Dental Materials

3.1. Resin-Based Composite

Resin-based dental materials have substantially evolved since they have been brought into the market, more than 60 years ago [2,42]. Resin-based composites, as a kind of versatile direct restorative materials, are widely used due to their excellent esthetic properties, improved mechanical characteristics, and ease of clinical handling [2,3]. Conventional RBCs are composed of four major components: a polymeric matrix, inorganic fillers, a silane coupling agent to produce a strong interface between the two phases mentioned, and initiators that induce or modulate the polymerization reaction [2]. Although the composite resin has been widely used in recent years, it has more biofilm accumulation, more frequent replacement, and shorter longevity, when compared with amalgam [43,44]. The failure of RBCs, mainly on account of secondary caries along the tooth–composite interfaces, is frequently related to biofilm formation on dental restorations [3,8].

Different kinds of components imply that the surface of a RBC is not a homogeneous interface, because of the distribution of physical-chemical phases with different chemical properties. The main base monomers of polymeric matrix used in commercial dental composites are Bis-GMA (bis-phenyl glycidyl dimethacrylate), Bis-EMA (bisphenol A ethoxylated dimethacrylate), PEGDMA (polyethylene glycol dimethacrylate), and UDMA (urethane dimethacrylate) with high viscosity, mixed with TEGDMA (triethylene glycol dimethacrylate) for dilution [2,31]. By tailoring RBC surfaces with either high carbon (matrix-rich) or high silicon (filler-rich) content from several commercially available RBCs without any antimicrobial agent, Ionescu et al. suggested that minimization of resin-matrix exposure might reduce biofilm formation on RBC surfaces because of the correlation between RBC surface carbon content and viable *S. mutans* biomass [26]. Recently, it was discovered that RBCs with a UDMA/aliphatic dimethacrylate matrix blend showed significantly higher biofilm formation on the surfaces than specimens with a Bis-GMA/TEGDMA matrix blend and analogous filler fraction, except for nanosized filler particles [45]. Another matrix, the silorane-based composite, was demonstrated to be less prone to *S. mutans* biofilm development compared with a generally used methacrylate-based composite, due to the increased hydrophobicity by silorane [46]. It was investigated that a reduced light-curing time can significantly increase the amount of unpolymerized monomers on the material surface, which might be responsible for increasing in vitro colonization on resin composite surfaces by *S. mutans* [47]. Kawai et al. reported that the specific resin components, a diglycidyl methacrylate and TEGDMA, significantly promoted glucosyltransferase (GTF) enzymes activity [48]. The GTF enzymes involved in the synthesis of water-insoluble glucan in situ entail an extracellular slime layer that promotes adhesion and the formation of dental plaque biofilms [49,50]. Consistently, the biodegradation byproduct (BBP) triethylene glycol (TEG), derived from methacrylate monomers, promotes the growth of *S. mutans* via upregulating the expression of glucosyltransferase B (*gtfB*) (involved in biofilm formation) and *yfiV* (a putative transcription regulator) in *S. mutans* [49]. Meanwhile, another BBP bishydroxypropoxyphenyl propane (BisHPPP) of Bis-GMA can also enhance the GTF enzyme activity of *S. mutans* biofilms, and modulate genes and proteins involved in biofilm formation, carbohydrate transport, and acid tolerance [51]. In conclusion, further studies are needed to explore the appropriate proportions of resin matrix and filler particles on the surface of RBCs, as well as to explore better ways to prevent resin biodegradation.

There are different sized inorganic fillers of the resin composites, including macrofill, microfill, nanofill, and hybrids. The RBC's strength and polishing ability mostly depend on the size and proportion of inorganic fillers [2]. Pereira et al. demonstrated the least biofilm formation on a nanofilled RBC (Filtek Z350™) compared with nanohybrid, microhybrid, and bulk-filled RBCs. The nanosized inorganic fillers could obtain the extensive distribution of the fillers and smoother composite surfaces after the same finishing and polishing procedures, consequently decreasing *S. mutans* adhesion [25,29,52]. Resin composites containing surface pre-reacted glass ionomer (S-PRG) filler have been reported to show less biofilm accumulation and reduced bacterial attachment. The pre-reacted glass-ionomer bioactive fillers have been fabricated by the acid–base reaction between a fluoroaluminosilicate glass and polyalkenoic acid in the presence of water. The antibacterial effects of S-PRG filler-containing resin composite is mainly attributed to release of BO_3^{3-} and F^-, and fluoride-recharging abilities [53,54]. Yoshihara et al. investigated that bioactive glass filler may promote bacterial adhesion because of the unstable surface integrity, releasing ions and dissolving, which results in rougher restoration surfaces [55].

Up to now, there is still a high secondary caries rate, probably because of relatively few commercially antibacterial resins materials applied in clinic. However, more and more experimental antibacterial components and materials have been produced in the lab [44,56–58], among which, 12-methacryloyloxydodecylpyridinium bromide (MDPB), fluoride, and nanoparticles, have been translated into clinical materials. Both experimental antibacterial materials and new commercial antibacterial materials will soon pioneer a new materials field [54,59,60].

These experimental findings (Table 1) suggest that biofilm formation is influenced by the surface chemical composition of the material, including filler size, shape, and distribution, as well as matrix composition.

Table 1. The effect of the resin-based composites on biofilm formation.

Author, Year	Resin-Based Composite	Brief	Ref.
Ionescu et al., 2012	Filtek Supreme XT; Filtek Silorane™; Grandio	The proportions of resin matrix and filler particles on the surface of resin-based composite strongly influence biofilm formation in vitro.	[26]
Brambilla et al., 2016	Filtek Silorane™; Filtek Z250™	Silorane-based composite is less prone to *S. mutans* biofilm development.	[46]
Brambilla et al., 2009	Filtek Z250™	Unpolymerized monomers on the material surface are responsible for increasing in vitro colonization by *S. mutans*.	[47]

Table 1. *Cont.*

Author, Year	Resin-Based Composite	Brief	Ref.
Kawai et al., 2000	Clearfil F II; Silux	The diglycidyl methacrylate and TEGDMA significantly promoted GTF enzymes activity	[48]
Pereira et al., 2011	Filtek Z 350™; Esthet X™; Vit-l-escence™	The least biofilm forms on a nanofilled RBC compared with nanohybrid, microhybrid, and bulk-filled RBCs.	[52]
Hahnel et al., 2014	Beautifil II	The inclusion of S-PRG fillers may reduce biofilm formation on resin composite.	[53]
Yoshihara et al., 2017	Beautifil ll; Herculite XRV Ultra	Bioactive glass filler may promote bacterial adhesion because of unstable surface integrity, releasing ions and dissolving.	[53]

3.2. Glass Ionomer Cements

Glass ionomer cements (GICs), applied as direct restorative materials and cements, feature some desirable characters, such as a chemical adhesion to enamel and dentin, and the ability to release fluoride over time [4]. It is well known that conventional GICs have biological effects and caries-inhibiting properties because of the release of surface fluoride ions [61].

Recently, many studies have reported that the fluoride of GICs can affect the acid production, acid tolerance, and extracellular polymetric substance (EPS) formation of dental plaques, especially cariogenic biofilms, such as *S. mutans* biofilms. The fluoride can reduce the proportion of *S. mutans* but increase *S. oralis* (*Streptococcus oralis*) in the dual-species biofilm, subsequently inhibiting the formation of cariogenic bacteria-dominant biofilms [62]. This phenomenon lasts during both the initial rapid and second slow release phases, which is called the biphasic pattern of fluoride release of GICs [63–66]. The release of fluoride showed a significant dependence on the experimental conditions applied, such as sterile broth, bacteria, and acid. The bacterial condition leads to the highest decrease in the release of fluoride, which can be explained by the extracellular matrix of biofilm serving as a layer that modulates the release of fluoride from the substratum materials. Furthermore, the acidic conditions can enhance the constant release of fluoride, due to its high bioavailability at low pH [67]. The result agreed with Jennifer's study, that more fluoride is released at pH 4, the acidic and cariogenic pH, than at pH 5.5 or pH 7, when these ions are most needed to inhibit caries [68]. It can be concluded that the efficiency of fluoride ions depends not only on their amount, but also on the pH value of the material during setting.

Acidic conditions promote the free fluoride ions to be released and form a weak electrolyte, hydrogen fluoride (HF, unionized fluoride) [66], and the combination of F-/HF and enzymes can modulate bacterium metabolism [69]. Adjacent to GIC restorations, an anti-caries environment is established by the fluoride, which may inhibit acidic pH efficiently due to the relatively high pKa value, 3.15, of hydrogen fluoride (HF) in vivo [70], by affecting bacterial metabolism (Figure 2), both directly (e.g., inhibition of enolase and ATPase) and indirectly (e.g., intracellular acidification) [71]. In addition, the aluminum released from Vitremer plays a vital role in inhibiting bacterial metabolism and has a synergistic effect with fluoride [71,72].

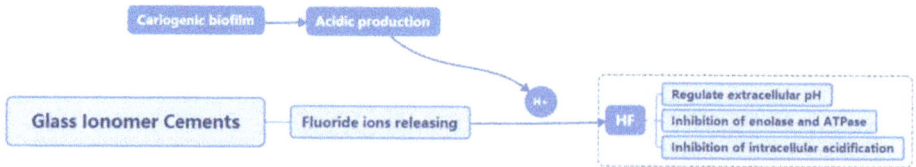

Figure 2. The relationship between fluoride of glass ionomer cements and bacterial metabolism.

Although the acid conditions of biofilms can promote fluoride release to inhibit biofilm formation, the microbial environment changes the morphology of GICs, and accelerates material aging (Figure 3). Meanwhile, the changed morphology and increased roughness can enhance the initial bacterial attachment and oral biofilm formation [17]. To demonstrate the actual effects on biofilm formation of dental material in a pragmatic way, the study models evolved consistently, from the previous water aging model to a biological aging model, from a primary caries animal model to a secondary caries animal model, and from in vitro to the in situ model used nowadays [73–75].

Figure 3. Representative SEM images of glass ionomer cement (GIC) surfaces before and after aging treatments. **A**: without any aging treatments; **B**: the GICs were immersed in water; **C**: *S. mutans* suspensions; **D**: salivary microbes' suspensions.

3.3. Amalgams

Over its long clinical history, dental amalgams have evolved and served the profession successfully and at low cost. Amalgam restorations are being phased out because of the environmental pollution and inferior esthetic appearance [1]. However, they cannot be replaced by other restoratives because of their perfect mechanical properties, longevity, and low cost [15]. The longevity of amalgam is inseparable from the lower incidence of secondary caries caused by oral biofilms.

After clinical placement, amalgam restorations undergo a series of corrosion to release a variety of metallic ions in oral cavities. It was discovered that the mercury of amalgams could deposit in the dental plaque for up to 2 µg in 24 h, whereas the aged amalgams released little mercury because of the presence of the formed passive tarnish layer on the surface of amalgams [76]. In the 1980s, the amalgam was proved to have bacteriostatic and bactericidal properties due to the metallic ions being released from the surface of the materials, such as Ag, Cu, Sn, and Hg [77]. The low biomass of oral biofilms on amalgam surfaces is probably a result of the release of toxic ions from amalgam, which mainly consists of Hg and Ag [69]. Specifically, amalgam showed lasting inhibition of both *S. mutans* and *Actinomyces viscosus* (*A. viscosus*) which played crucial roles in biofilm formation [78]. Morrier et al. investigated that the order of antimicrobial potential of elements in amalgams would be Hg > Cu > Zn, by testing a suspension of *S. mutans* and *A. viscosus* [79]. Among those metallic ions, Cu^{2+} and Zn^{2+} showed synergistic effects on the reduction in acid production in dental biofilm [80]. Amalgam also showed a robust acid-buffering ability, which can neutralize bacteria-produced acids of oral biofilm by increasing the start pH of all solutions to around 7 to 8. This should be attributed to the release of corrosion products on the amalgams surface. The tin and copper oxides are amphoteric compounds that react as a base in acidic conditions [81]. This can be related to the fact that biofilms accumulated more on composites than amalgams in the clinic. Even in the in situ study, the amalgam showed, visually, a prevalence of non-viable cells forming small clusters distributed by the biofilm compared to other materials [69]. However, no research has yet explored the mechanisms of bacteriostatic and bactericidal properties of amalgam clearly.

3.4. Dental Alloys of Indirect Restoration

After 1975, the alloys for full-cast restorations, porcelain-fused-to-metal restorations, and removable partial denture frameworks, can be divided into three kinds, high-noble alloys (Au–Pt,

Au–Pd, Au–Cu–Ag–Pd), noble alloys (Au–Cu–Ag–Pd, Pd–Cu, Pd–Ag), and base-metal alloys (Ni–Cr, Co–Cr, Ti) [5]. Oral microbial metabolites, such as acids, sulfide, and ammonia, can induce the microbial corrosion of metallic materials [82]. Dental alloys corrode and release metal irons in the oral environment which may compromise material biocompatibility and mechanical properties, and lead to the esthetic loss of dental restorations, and influence health [83].

Among the noble alloys, a high gold content alloy (88% by weight), Captek™, showed a 71% reduction in total bacterial numbers when compared to natural tooth surfaces [84]. This could be attributed to the low porosity of high nobility gold inherent in the manufacturing process and the unique electrochemical corrosion resistance [85]. Besides, metallic copper and copper-containing alloys possess a strong and rapid bactericidal effect, named "contact killing". This was induced by successive membrane damage, oxidative damage, cell death, and DNA degradation [20,86]. The surface-released free copper ions are toxic to bacteria because of their soft ionic character and their thiophilicity [86,87]. As for the base-metal alloys, a higher amount of viable microbial cells and biofilm density on prosthetic structures based on cobalt–chromium (Co–Cr) alloys was demonstrated, when compared to those based on titanium [21,88]. Mystkowska found that there were more corrosion pits on cobalt alloys than on titanium alloys [88], and that these corrosion pits increase the surfaces roughness of dental alloys, which may facilitate the subsequent accumulation of biofilm [82]. However, there was a significant increase in biofilm density and number of microbial cells of biofilm growing on both titanium and Co–Cr alloy, from 24 up to 48 h [21]. The acid produced by microorganisms induces the corrosion of Cr_2O_3 and TiO_2, the passive films, which are responsible for corrosion resistance and biocompatibility of the alloys [82,89–91] (Table 2).

Table 2. The influence of different dental alloys on the biofilm formation.

Author, Year	Resin-Based Composite	Brief	Ref.
Zappala et al., 1996	Gold alloy	High-noble alloys showed a significant reduction in biofilm because of the low porosity and unique electrochemical corrosion resistance.	[85]
Grass et al., 2011	Metallic copper	Metallic copper processes strong and rapid bactericidal effect, named "contact killing".	[20]
Mystkowska et al., 2016	Co–Cr-based alloy	Co–Cr alloys developed more pits and viable microbial cells than titanium alloys after degradation.	[88]
McGinley et al., 2013	Ni-based alloy	Ni-based dental casting alloys induced elevated levels of cellular toxicity compared with S. mutans-treated Co–Cr-based dental casting alloys.	[91]
Souza et al., 2013	Titanium	The presence of S. mutans colonies on the titanium negatively affected its corrosion resistance due to the titanium-passive film.	[21]

Zhang et al. discovered that corroded alloy surfaces could upregulate gene expression of the glucosyltransferase BCD, glucan-binding proteins B, fructosyltransferase, and lactate dehydrogenase in S. mutans, which play critical roles in bacteria adherence and biofilm accumulation [82]. Microorganisms of biofilm decrease the pH by producing acidic substances and dissolve the surface oxides of the dental alloys to reduce the corrosion resistance of the metal [92]. In turn, the changed surfaces of the dental alloys can accelerate the virulence gene expression and biofilm formation [82]. Therefore, this bacteria-adhesion and corrosion cycle can accelerate the corrosion process and, finally, induce failure of the dental alloys' restoration.

3.5. Ceramic

In recent years, adhesively cemented ceramic restorations, such as inlays/onlays, veneers, and crowns, have been used as the main approach for minimally invasive esthetic restorations in anterior and posterior teeth [93]. However, its clinical failure is related to a lot of factors, such as marginal misfit, surface irregularities, and cement excess, which may favor the accumulation of microorganisms, compromising clinical restoration longevity [94].

Both surface roughness and surface free energy have been found to influence initial microbial adherence decisively [40], due to compositional and microstructural differences, and bacterial colonization was thought to differ from one ceramic material to another. Sebastian [95] employed different kind of ceramics, glass/lithium disilicate glass/glass-infiltrated zirconia/partially sintered zirconia/hipped zirconia ceramic as the specimens, and the glass plates were used as a control. He found that the lithium disilicate glass ceramic had the highest values for Ra, whereas the lowest values were found for the glass ceramic, the partially sintered zirconia, and the hipped zirconia ceramic. Furthermore, salivary protein coating caused a significant increase in surface free energy and the polarity of these ceramics, except for the control material. However, after salivary protein coating, only the control material showed higher values for streptococcal adhesion than all ceramic materials. The same study [96], which was performed in vivo, demonstrated significant differences in biofilm formation with various types of dental ceramics. In particular, zirconia exhibited low biofilm accumulation. Thus, except for its high intensity, low biofilm accumulation makes zirconia a promising material for various indications. The different results of the two studies [95,96] may be related to the different models (in vitro, in vivo) they applied (Table 3).

Table 3. The influence of different ceramic on the biofilm formation.

Author, Year	Ceramic	Brief	Ref.
Hahnel et al., 2009	Glass, lithium disilicate glass, glass-infiltrated zirconia, partially sintered zirconia, hipped zirconia ceramic	Only slight and random differences in streptococcal adhesion were found between the various ceramic materials, and control material showed higher values for streptococcal adhesion than all ceramic materials.	[95]
Bremer et al., 2011	Veneering glass-ceramic, lithium disilicate glass-ceramic, yttrium-stabilized zirconia (Y-TZP), hot isostatically pressed (HIP) Y-TZP ceramic, and HIP Y-TZP ceramic with 25% alumina	The study in vivo showed significant difference in biofilm formation with various types of dental ceramics; especially zirconia exhibited low biofilm accumulation.	[96]
Kim et al., 2017	Commercially available ceramic materials: Vita Enamic, Lava Ultimate, Vitablocs Mark II, and Wieland Reflex	All materials, except for Vitablocs Mark II, promoted significantly greater biofilm growth.	[28]

3.6. Dental Implant

Over the last decades, the use of dental implants has become a common way of restoring dentition defect [6]. The implant survival rate reaches to 92.8–97.1% over a follow-up period of 10 years, but dental implants easily become infectious, due to oral pathogenic bacteria [12,13,97]. Two main etiologies of peri-implantitis are oral biofilms and occlusal overload [98], among which, oral biofilms developed on dental implants play a significant role in peri-implantitis' pathogenesis. The peri-implantitis can cause implant loss in the absence of prevention and therapy [99,100]. The implant may be attached by saliva, blood, and oral bacterial cells during and after the implant surgery, and bacterial cells attached to the abutment harm the surrounding gingiva. All the above-mentioned points would affect the healing and restoration following surgery [101].

We begin with the abutment, since pathogenic bacteria usually attach on it first, causing peri-implant mucositis [102]. Hence, peri-implant tissue inflammation, as a consequence of biofilms on abutments in the subgingival region, is currently considered as a major contributor to implant loss [103]. Avila [103] found that, in the case of saliva-derived biofilm, the number of cells and the density of the biofilm on ZrO_2 were lower than on titanium materials. Zirconia abutments have a lower possibility for bacterial attachment, which is similar to the study above [104,105], and some researchers thought that the surface free energy is more critical on zirconia abutment surfaces [106]. Cássio's [107] 16S rDNA sequencing results agreed with previous studies [108,109] that the titanium accumulated more biofilm and more species of microorganisms. Two studies [110,111] found that the early bacterial communities were low in genome counts at the very beginning of implant surgery for both the zirconia and titanium abutment materials. As time goes by, both materials showed similar microbial counts and diversity, the same as on teeth. The different results may be related to no criterion for these products and testing methods. Zirconia is used widely for its esthetic property nowadays, and maybe the zirconia abutment will replace the titanium abutment for the lower bacteria attachment. However, substantial evidence is needed to prove its excellent properties in microbiological field.

When the peri-implant mucositis progress to peri-implantitis, more attention should be paid to the implant surface (Figure 4). About implant surface treatment techniques, there are mainly four kinds of coating techniques: alumina coating, titanium plasma spraying (TPS), biomimetic calcium phosphate (CaP) coating and plasma sprayed hydroxyapatite (HA) coating [112]. The coating techniques contribute to critical positive effects of dental implant application. Most authors [113,114] agreed that a suitable coating technique may enhance the mechanical properties of the dental implants. However, these techniques have several limitations including poor long-term adherence of the coating to the substrate material [115], nonuniformity in thickness of the deposited layer, variations in crystallinity [116], and composition of the coating, which influence the biofilm formation on the surface [112] (Table 4). However, none of studies shows the single factor of different coating techniques so far, because different coating techniques are related to different surface characteristics, which we have discussed in other sections of this review, further studies about the coating techniques should be performed.

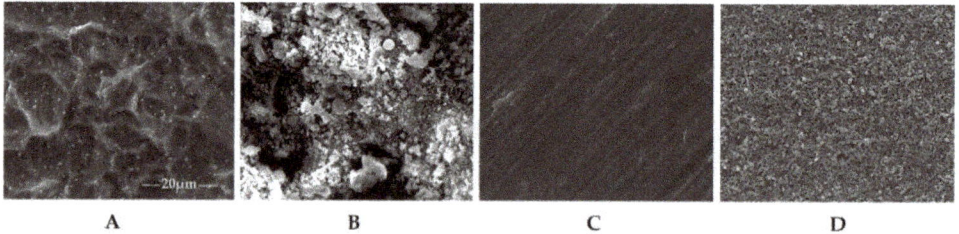

Figure 4. Four kinds of titanium implant surface treatment show different SEM imagines. **A**: Sandblasting and acid etching technique (SLA); **B**: plasma sprayed hydroxyapatite coating (HA); **C**: machined treatment (machined); **D**: microarc oxidation (MAO).

It has been found out that, except for surface roughness and surface free energy [119], the type of the biomaterial itself can also influence biofilm formation and subsequent plaque accumulation on implant surfaces [21]. Two investigations have shown less inflammatory cells in the peri-implant soft tissue of zirconia in comparison with titanium or other metals [104,105]. Additionally, Zhao's [106] study showed that neither roughness nor hydrophobicity had a decisive influence on the biofilm formation that occurred on three different implant materials, comprising titanium (Ti, cold-worked, grade 4), titanium–zirconium alloy (TiZr, 15% (wt) Zr) and zirconium oxide (ZrO_2, Y-TZP). Same as Zhao's result, in the 3-species biofilm (*Streptococcus sanguinis*, *Fusobacterium nucleatum*, and *Porphyromonas gingivalis*), the analysis showed that there were no significant differences between titanium and zirconia in terms of total biofilm mass and metabolism. However, zirconia revealed significantly reduced plaque thickness. Regarding human plaque biofilms, microbiological techniques showed statistically significant reduction in biofilm formation for zirconia compared to titanium. The result suggested that not only surface roughness or surface hydrophilicity might be important factors for biofilm formation, but also material composition—metals compared to ceramics—suggesting a reduced disposition for peri-implant plaque and subsequent potential peri-implant infections on zirconia compared to titanium implant surfaces [32,120]. Nowadays, topography, surface charge, roughness, hydrophobicity, and chemistry have been investigated for many years. Besides, some new techniques have been studied, like nanoscale surface roughness, negatively charged surfaces, super hydrophilic surfaces and super hydrophobic surfaces, and they have all been demonstrated to reduce bacterial adhesion [32].

Table 4. The influence of different titanium surface treatments on the biofilm formation.

Author, Year	Different Titanium Surfaces	Brief	Ref.
Patrick et al., 2013	Machined, stained, acid-etched, or sandblasted/acid-etched (SLA)	After the colonization for 2, 4, and 8 h, there seems no difference between these titanium discs. Up to 16.5 h, the SLA surface showed the highest trend for the bacterial colonization	[117]
Matos et al., 2011	Micro-arc oxidation (MAO), glow discharge plasma (GDP), machined, and sandblasted surfaces	The counts of *F. nucleatum* were lower for MAO treatment at early biofilm phase (16.5 h), while the plasma treatment did not affect the viable microorganism counts. Biofilm extracellular matrix was similar among these groups, except for GDP, with the lowest protein content.	[118]
Al-Ahmad et al., 2010	Machined titanium (Tim), modified titanium (TiUnite)	No significant differences in biofilm composition on the implant surfaces. Besides, the influence of roughness and material on biofilm formation was compensated by biofilm maturation	[35]
de Freitas et al., 2011	Machined, blasted, HA-coated	The titanium discs were put into volunteers' oral cavity and were tested after 1, 3, 7, 14, and 21 days. There was no statistically significant difference between the kinetics of bacterial species succession and the different surfaces.	[40]
Bevilacqua et al., 2018	Machined surface(M), laser-treated surface (LT), sandblasted surface (SB)	The biofilm developed in vivo for 1 day and 4 days showed no statistical difference between 3 kinds of discs. In vitro, when the biofilm was formed by *P. aeruginosa*, M showed less biomass and biofilm average thickness. As for the biofilm developed by mixed salivary bacteria, SB showed less biomass and average biofilm thickness.	[37]

4. Conclusions

As discussed in this review, bacterial adhesion and biofilm formation can be strongly influenced by surface characteristics of dental materials, which include chemical compositions, surface roughness, surface free energy, surface topography, ions release, and others. In conclusion, every possible particular chemical composition (organic matrix, inorganic filler, fluoride, and various metallic ions) can enhance or inhibit biofilm formation. Irregular topography and rough surfaces provide favorable interfaces for bacterial colonization, protecting bacteria against shear forces during their initial reversible binding and biofilm formation. Besides, the surface free energy, hydrophobicity, surfaces coating techniques also have a significant influence on oral biofilm.

However, the "ideal" surface characteristics have not been identified yet, and results have varied from different methods and models. One of the major drawbacks of current research is the limitation of the in vitro study. In vitro studies are not always able to completely simulate the complicated conditions presented in the oral environment. Thus, further in situ studies are much needed to clarify the role and mechanism of each surface parameter on oral biofilm formation. Finally, the goal is to produce robust, long-lasting dental materials which will reduce costly replacements and significantly ameliorate oral health.

Author Contributions: Y.H. and X.H. drafted the manuscript. X.Z., M.L., B.R., X.P., L.C. edited and added valuable insights into the manuscript. All authors read and approved the final manuscript.

Acknowledgments: This research was supported by The National Key Research Program of China 2017YFC0840100 and 2017YFC0840107 (L.C.), National Natural Science Foundation of China (81870759), the Youth Grant of the Science and Technology Department of Sichuan Province, China 2017JQ0028 (L.C.), Innovative Research Team Program of Sichuan Province (L.C.)

Conflicts of Interest: The authors declare no conflict of interest.

References

1. Shenoy, A. Is it the end of the road for dental amalgam? A critical review. *J. Conserv. Dent.* **2008**, *11*, 99–107. [CrossRef] [PubMed]
2. Ferracane, J.L. Resin composite—State of the art. *Dent. Mater.* **2011**, *27*, 29–38. [CrossRef] [PubMed]
3. Demarco, F.F.; Correa, M.B.; Cenci, M.S. Longevity of posterior composite restorations: Not only a matter of materials. *Dent. Mater.* **2012**, *28*, 87–101. [CrossRef] [PubMed]

4. Joel, H. Berg, Glass ionomer cements. *Pediatr. Dent.* **2002**, *24*, 430–438.
5. Wataha, J.C. Alloys for prosthodontic restorations. *J. Prosthet. Dent.* **2002**, *87*, 351–363. [CrossRef] [PubMed]
6. Renvert, S.; Quirynen, M. Risk indicators for peri-implantitis. A narrative review. *Clin. Oral Implants Res.* **2015**, *26*, 15–44. [CrossRef] [PubMed]
7. Laske, M.; Opdam, N.J.; Bronkhorst, E.M. Longevity of direct restorations in Dutch dental practices. Descriptive study out of a practice based research network. *J. Dent.* **2016**, *46*, 12–17. [CrossRef] [PubMed]
8. Delaviz, Y.; Finer, Y.; Santerre, J.P. Biodegradation of resin composites and adhesives by oral bacteria and saliva: A rationale for new material designs that consider the clinical environment and treatment challenges. *Dent. Mater.* **2014**, *30*, 16–32. [CrossRef] [PubMed]
9. Goodacre, C.J. Bernal, Guillermo, Rungcharassaeng, Kitichai,, Clinical complications in fixed prosthodontics. *J. Prosthet. Dent.* **2003**, *90*, 31–41. [CrossRef]
10. Toman, M.; Toksavul, S. Clinical evaluation of 121 lithium disilicate all-ceramic crowns up to 9 years. *Quintessence Int.* **2015**, *46*, 189–197. [CrossRef] [PubMed]
11. Layton, D. A critical appraisal of the survival and complication rates of tooth-supported all-ceramic and metal-ceramic fixed dental prostheses the application of evidence-based dentistry. *Int. J. Prosthodont.* **2011**, *24*, 417–427. [PubMed]
12. Albrektsson, T.; Donos, N. Implant survival and complications. In Proceedings of the Third EAO Consensus Conference, Pfäffikon, Schwyz, Switzerland, 15–18 February 2012; pp. 63–65.
13. Srinivasan, M.; Vazquez, L.; Rieder, P.; Moraguez, O. Survival rates of short (6 mm) micro-rough surface implants: A review of literature and meta-analysis. *Clin. Oral Implants Res.* **2014**, *25*, 539–545. [CrossRef] [PubMed]
14. Robertson, K.; Shahbazian, T.; MacLeod, S. Treatment of peri-implantitis and the failing implant. *Dent. Clin. N. Am.* **2015**, *59*, 329–343. [CrossRef] [PubMed]
15. Schwendicke, F.; Tu, Y.K.; Stolpe, M. Preventing and Treating Peri-Implantitis: A Cost-Effectiveness Analysis. *J. Periodontol.* **2015**, *86*, 1020–1029. [CrossRef] [PubMed]
16. Li, Y.; Carrera, C.; Chen, R.; Li, J. Degradation in the dentin-composite interface subjected to multi-species biofilm challenges. *Acta Biomater.* **2014**, *10*, 375–383. [CrossRef] [PubMed]
17. Park, J.W.; Song, C.W.; Jung, J.H. The effects of surface roughness of composite resin on biofilm formation of Streptococcus mutans in the presence of saliva. *Oper. Dent.* **2012**, *37*, 532–539. [CrossRef] [PubMed]
18. Cheng, L.; Zhang, K.; Zhang, N.; Melo, M.A.S.; Weir, M.D.; Zhou, X.D.; Bai, Y.X.; Reynolds, M.A.; Xu, H.H.K. Developing a New Generation of Antimicrobial and Bioactive Dental Resins. *J. Dent. Res.* **2017**, *96*, 855–863. [CrossRef] [PubMed]
19. Dewhirst, F.E.; Chen, T.; Izard, J. The human oral microbiome. *J. Bacteriol.* **2010**, *192*, 5002–5017. [CrossRef] [PubMed]
20. Grass, G.; Rensing, C.; Solioz, M. Metallic copper as an antimicrobial surface. *Appl. Environ. Microbiol.* **2011**, *77*, 1541–1547. [CrossRef] [PubMed]
21. Souza, J.; Mota, R.R.; Sordi, M.B.; Passoni, B.B. Biofilm Formation on Different Materials Used in Oral Rehabilitation. *Braz. Dent. J.* **2016**, *27*, 141–147. [CrossRef] [PubMed]
22. Teughels, W.; Van Assche, N.; Sliepen, I.; Quirynen, M. Effect of material characteristics and/or surface topography on biofilm development. *Clin. Oral Implants Res.* **2006**, *17*, 68–81. [CrossRef] [PubMed]
23. Wang, Z.; Shen, Y.; Haapasalo, M. Dental materials with antibiofilm properties. *Dent. Mater.* **2014**, *30*, e1–e16. [CrossRef] [PubMed]
24. Teranaka, A.; Tomiyama, K.; Ohashi, K. Relevance of surface characteristics in the adhesiveness of polymicrobial biofilms to crown restoration materials. *J. Oral Sci.* **2017**, *60*, 129–136. [CrossRef] [PubMed]
25. Cazzaniga, G.; Ottobelli, M.; Ionescu, A.C. In vitro biofilm formation on resin-based composites after different finishing and polishing procedures. *J. Dent.* **2017**, *67*, 43–52. [CrossRef] [PubMed]
26. Ionescu, A.; Wutscher, E.; Brambilla, E. Influence of surface properties of resin-based composites on in vitro Streptococcus mutans biofilm development. *Eur. J. Oral Sci.* **2012**, *120*, 458–465. [CrossRef] [PubMed]
27. Haralur, S.B. Evaluation of efficiency of manual polishing over autoglazed and overglazed porcelain and its effect on plaque accumulation. *J. Adv. Prosthodont.* **2012**, *4*, 179–186. [CrossRef] [PubMed]
28. Kim, K.H.; Loch, C.; Waddell, J.N.; Tompkins, G. Surface Characteristics and Biofilm Development on Selected Dental Ceramic Materials. *Int. J. Dent.* **2017**, *2017*, 7627945. [CrossRef] [PubMed]

29. Han, J.M.; Zhang, H.; Choe, H.S.; Lin, H.; Zheng, G.; Hong, G. Abrasive wear and surface roughness of contemporary dental composite resin. *Dent. Mater. J.* **2014**, *33*, 725–732. [CrossRef] [PubMed]
30. Curd, M.L.; Bollen, P.L. Marc Quirynen Comparison of surface roughness of oral hard materials to the threshold surface roughness for bacterial plaque retention: A review of the literature. *Dent. Mater.* **1997**, *13*, 258. [CrossRef]
31. Yuan, C.; Wang, X.; Gao, X. Effects of surface properties of polymer-based restorative materials on early adhesion of Streptococcus mutans in vitro. *J. Dent.* **2016**, *54*, 33–40. [CrossRef] [PubMed]
32. Song, F.; Koo, H.; Ren, D. Effects of Material Properties on Bacterial Adhesion and Biofilm Formation. *J. Dent. Res.* **2015**, *94*, 1027–1034. [CrossRef] [PubMed]
33. Cazzaniga, G.; Ottobelli, M.; Ionescu, A. Surface properties of resin-based composite materials and biofilm formation A review of the current literature. *Am. J. Dent.* **2015**, *28*, 311–320. [PubMed]
34. Frojd, V.; Chavez de Paz, L. In situ analysis of multispecies biofilm formation on customized titanium surfaces. *Mol. Oral Microbiol.* **2011**, *26*, 241–252. [CrossRef] [PubMed]
35. Al-Ahmad, A.; Wiedmann-Al-Ahmad, M.; Faust, J.; Bachle, M.; Follo, M.; Wolkewitz, M.; Hannig, C.; Hellwig, E.; Carvalho, C.; Kohal, R. Biofilm formation and composition on different implant materials in vivo. *J. Biomed. Mater. Res. B Appl. Biomater.* **2010**, *95*, 101–109. [CrossRef] [PubMed]
36. Dezelic, T.G.B.; Schmidlin, P.R. Multi-species Biofilm Formation on Dental Materials and an Adhesive Patch. *Oral Health Prev. Dent.* **2009**, *7*, 47–53. [CrossRef] [PubMed]
37. Bevilacqua, L.; Milan, A.; Del Lupo, V.; Maglione, M.; Dolzani, L. Biofilms Developed on Dental Implant Titanium Surfaces with Different Roughness: Comparison Between In Vitro and In Vivo Studies. *Curr. Microbiol.* **2018**, *75*, 766–772. [CrossRef] [PubMed]
38. Wennerberg, A.; Albrektsson, T. Effects of titanium surface topography on bone integration: A systematic review. *Clin. Oral Implants Res.* **2009**, *20* (Suppl. S4), 172–184. [CrossRef] [PubMed]
39. Da Silva, C.H.F.P.; Vidigal, G.M., Jr.; de Uzeda, M.; de Almeida Soares, G. Influence of Titanium Surface Roughness on Attachment of Streptococcus Sanguis: An in vitro study. *Implant Dent.* **2005**, *14*, 88–93. [CrossRef]
40. De Freitas, M.M.; da Silva, C.H.; Groisman, M.; Vidigal, G.M., Jr. Comparative analysis of microorganism species succession on three implant surfaces with different roughness: An in vivo study. *Implant Dent.* **2011**, *20*, e14–e23. [CrossRef] [PubMed]
41. Größner-Schreiber, B.; Teichmann, J.; Hannig, M.; Dörfer, C.; Wenderoth, D.F.; Ott, S.J. Modified implant surfaces show different biofilm compositions under in vivo conditions. *Clin. Oral Implants Res.* **2009**, *20*, 817–826. [CrossRef] [PubMed]
42. Pfeifer, C.S. Polymer-Based Direct Filling Materials. *Dent. Clin. N. Am.* **2017**, *61*, 733–750. [CrossRef] [PubMed]
43. Spencer, P.; Ye, Q.; Misra, A. Proteins, pathogens, and failure at the composite-tooth interface. *J. Dent. Res.* **2014**, *93*, 1243–1249. [CrossRef] [PubMed]
44. Zhang, N.; Melo, M.A.S.; Weir, M.D. Do Dental Resin Composites Accumulate More Oral Biofilms and Plaque than Amalgam and Glass Ionomer Materials? *Materials* **2016**, *9*, 888. [CrossRef] [PubMed]
45. Ionescu, A.; Brambilla, E.; Wastl, D.S. Influence of matrix and filler fraction on biofilm formation on the surface of experimental resin-based composites. *J. Mater. Sci. Mater. Med.* **2015**, *26*, 1–7. [CrossRef] [PubMed]
46. Brambilla, E.; Ionescu, A.; Cazzaniga, G.; Ottobelli, M. Influence of Light-curing Parameters on Biofilm Development and Flexural Strength of a Silorane-based Composite. *Oper. Dent.* **2016**, *41*, 219–227. [CrossRef] [PubMed]
47. Brambilla, E.; Gagliani, M.; Ionescu, A. The influence of light-curing time on the bacterial colonization of resin composite surfaces. *Dent. Mater.* **2009**, *25*, 1067–1072. [CrossRef] [PubMed]
48. Kawai, K.; Tsuchitani, Y. Effects of resin composite components on glucosyltransferase of cariogenic bacterium. *J. Biomed. Mater. Res.* **2000**, *51*, 123–127. [CrossRef]
49. Khalichi, P.; Singh, J.; Cvitkovitch, D.G. The influence of triethylene glycol derived from dental composite resins on the regulation of Streptococcus mutans gene expression. *Biomaterials* **2009**, *30*, 452–459. [CrossRef] [PubMed]
50. Bowen, W.H.; Koo, H. Biology of Streptococcus mutans-derived glucosyltransferases: Role in extracellular matrix formation of cariogenic biofilms. *Caries Res.* **2011**, *45*, 69–86. [CrossRef] [PubMed]

51. Sadeghinejad, L.; Cvitkovitch, D.G.; Siqueira, W.L. Mechanistic, genomic and proteomic study on the effects of BisGMA-derived biodegradation product on cariogenic bacteria. *Dent. Mater.* **2017**, *33*, 175–190. [CrossRef] [PubMed]
52. Pereira, C.A.; Eskelson, E.; Cavalli, V. Streptococcus mutansBiofilm Adhesion on Composite Resin Surfaces After Different Finishing and Polishing Techniques. *Oper. Dent.* **2011**, *36*, 311–317. [CrossRef] [PubMed]
53. Hahnel, S.; Wastl, D.S. Streptococcus mutans biofilm formation and release of fluoride from experimental resin-based composites depending on surface treatment and S-PRG filler particle fraction. *J. Adhes. Dent.* **2014**, *16*, 313–321. [CrossRef] [PubMed]
54. Miki, S.; Kitagawa, H.; Kitagawa, R. Antibacterial activity of resin composites containing surface pre-reacted glass-ionomer (S-PRG) filler. *Dent. Mater.* **2016**, *32*, 1095–1102. [CrossRef] [PubMed]
55. Yoshihara, K.; Nagaoka, N.; Maruo, Y. Bacterial adhesion not inhibited by ion-releasing bioactive glass filler. *Dent. Mater.* **2017**, *33*, 723–734. [CrossRef] [PubMed]
56. Liang, J.; Li, M.; Ren, B.; Wu, T. The anti-caries effects of dental adhesive resin influenced by the position of functional groups in quaternary ammonium monomers. *Dent. Mater.* **2018**, *34*, 400–411. [CrossRef] [PubMed]
57. Khurshid, Z.; Naseem, M.; Sheikh, Z.; Najeeb, S. Oral antimicrobial peptides: Types and role in the oral cavity. *Saudi Pharm. J.* **2016**, *24*, 515–524. [CrossRef] [PubMed]
58. Ge, Y.; Wang, S.; Zhou, X.; Wang, H.; Xu, H.H.; Cheng, L. The Use of Quaternary Ammonium to Combat Dental Caries. *Materials* **2015**, *8*, 3532–3549. [CrossRef] [PubMed]
59. Carlo, H.L.; Bonan, P.R.F.; Franklin, G.G. In vitro effect of S. mutans biofilm on fluoride/MDPB-containing adhesive system bonded to caries-affected primary dentin. *Am. J. Dent.* **2014**, *37*, 227–232.
60. Khurshid, Z.; Zafar, M.; Qasim, S.; Shahab, S. Advances in Nanotechnology for Restorative Dentistry. *Materials* **2015**, *8*, 717–731. [CrossRef] [PubMed]
61. Kramer, N.; Schmidt, M.; Lücker, S. Glass ionomer cement inhibits secondary caries in an in vitro biofilm model. *Clin. Oral Investig.* **2018**, *22*, 1019–1031. [CrossRef] [PubMed]
62. Jung, J.E.; Cai, J.N.; Cho, S.D. Influence of fluoride on the bacterial composition of a dual-species biofilm composed of Streptococcus mutans and Streptococcus oralis. *Biofouling* **2016**, *32*, 1079–1087. [CrossRef] [PubMed]
63. Neilands, J.; Troedsson, U.; Sjodin, T.; Davies, J.R. The effect of delmopinol and fluoride on acid adaptation and acid production in dental plaque biofilms. *Arch. Oral Biol.* **2014**, *59*, 318–323. [CrossRef] [PubMed]
64. Pandit, S.; Kim, H.J.; Song, K.Y.; Jeon, J.G. Relationship between fluoride concentration and activity against virulence factors and viability of a cariogenic biofilm: In vitro study. *Caries Res.* **2013**, *47*, 539–547. [CrossRef] [PubMed]
65. Chau, N.P.; Pandit, S.; Jung, J.-E. Long-term anti-cariogenic biofilm activity of glass ionomers related to fluoride release. *J. Dent.* **2016**, *47*, 34–40. [CrossRef] [PubMed]
66. Mayanagi, G.; Igarashi, K.; Washio, J.; Domon-Tawaraya, H.; Takahashi, N. Effect of fluoride-releasing restorative materials on bacteria-induced pH fall at the bacteria–material interface: An in vitro model study. *J. Dent.* **2014**, *42*, 15–20. [CrossRef] [PubMed]
67. Hahnel, S.; Ionescu, A.C.; Cazzaniga, G.; Ottobelli, M.; Brambilla, E. Biofilm formation and release of fluoride from dental restorative materials in relation to their surface properties. *J. Dent.* **2017**, *60*, 14–24. [CrossRef] [PubMed]
68. Moreau, J.L.; Xu, H.H. Fluoride releasing restorative materials: Effects of pH on mechanical properties and ion release. *Dent. Mater.* **2010**, *26*, e227–e235. [CrossRef] [PubMed]
69. Padovani, G.C.; Fucio, S.B.; Ambrosano, G.M.; Correr-Sobrinho, L.; Puppin-Rontani, R.M. In situ bacterial accumulation on dental restorative materials. CLSMCOMSTAT analysis. *Am. J. Dent.* **2016**, *28*, 3–8.
70. Nakajo, K.; Takahashi, Y.; Kiba, W.; Imazato, S.; Takahashi, N. Fluoride ion released from glass-ionomer cement is responsible to inhibit the acid production of caries-related oral streptococci. *Interface Oral Health Sci.* **2007**, *25*, 263–264. [CrossRef]
71. Hayacibara, M.F.; Rosa, O.P.; Koo, H. Effects of fluoride and aluminum from ionomeric materials on S. mutans biofilm. *J. Dent. Res.* **2003**, *82*, 267–271. [CrossRef] [PubMed]
72. Fucio, S.B.; Paula, A.B.; Sardi, J.C.O. Streptococcus Mutans Biofilm Influences on the Antimicrobial Properties of Glass Ionomer Cements. *Braz. Dent. J.* **2016**, *27*, 681–687. [CrossRef] [PubMed]

73. Zhou, X.; Wang, S.; Peng, X.; Hu, Y.; Ren, B. Effects of water and microbial-based aging on the performance of three dental restorative materials. *J. Mech. Behav. Biomed. Mater.* **2018**, *80*, 42–50. [CrossRef] [PubMed]
74. Wu, T.; Li, B.; Zhou, X.; Hu, Y.; Zhang, H. Evaluation of Novel Anticaries Adhesive in a Secondary Caries Animal Model. *Caries Res.* **2018**, *52*, 14–21. [CrossRef] [PubMed]
75. Xue, Y.; Lu, Q.; Tian, Y.; Zhou, X. Effect of toothpaste containing arginine on dental plaque-A randomized controlled in situ study. *J. Dent.* **2017**, *67*, 88–93. [CrossRef] [PubMed]
76. Lyttle, H.A.; Bowden, G.H. The level of mercury in human dental plaque and interaction in vitro between biofilms of Streptococcus mutans and dental amalgam. *J. Dent. Res.* **1993**, *72*, 1320–1324. [CrossRef] [PubMed]
77. Morrier, J.J.; Barsotti, O.; Blanc-Benon, J.; Rocca, J.P.; Dumont, J. Antibacterial properties of five dental amalgams an in vitro study. *Dent. Mater.* **1989**, *5*, 310–313. [CrossRef]
78. Beyth, N.; Domb, A.J.; Weiss, E.I. An in vitro quantitative antibacterial analysis of amalgam and composite resins. *J. Dent.* **2007**, *35*, 201–206. [CrossRef] [PubMed]
79. Morrier, J.J.; Suchett-Kaye, G.; Nguyen, D.; Rocca, J.P.; Blanc-Benon, J.; Barsotti, O. Antimicrobial activity of amalgams, alloys and their elements and phases. *Dent. Mater.* **1998**, *5*, 310–313. [CrossRef]
80. Afseth, J.; Oppermann, R.V.; Rolla, G. Thein vivoeffect of glucose solutions containing Cu^{++} and Zn^{++} on the acidogenicity of dental plaque. *Acta Odontol. Scand.* **2009**, *38*, 229–233. [CrossRef]
81. Nedeljkovic, I.; De Munck, J.; Slomka, V. Lack of Buffering by Composites Promotes Shift to More Cariogenic Bacteria. *J. Dent. Res.* **2016**, *95*, 875–881. [CrossRef] [PubMed]
82. Zhang, S.; Qiu, J.; Ren, Y. Reciprocal interaction between dental alloy biocorrosion and Streptococcus mutans virulent gene expression. *J. Mater. Sci. Mater. Med.* **2016**, *27*, 78. [CrossRef] [PubMed]
83. Lu, C.; Zheng, Y.; Zhong, Q. Corrosion of dental alloys in artificial saliva with Streptococcus mutans. *PLoS ONE* **2017**, *12*, e0174440. [CrossRef] [PubMed]
84. Goodson, J.M.; Shoher, I.; Imber, S.; Som, S.; Nathanson, D. Reduced dental plaque accumulation on composite gold alloy margins. *J. Periodontal Res.* **2001**, *36*, 252–259. [CrossRef] [PubMed]
85. Zappala, C.; Shoher, I.; Battaini, P. Microstructural Aspects of the Captek™ Alloy for PorcelainFused-to-Metal Restorations. *J. Esthet. Dent.* **1996**, *8*, 151. [CrossRef] [PubMed]
86. HansSalima, M. Physicochemical properties of copper important for its antibacterial activity and development of a unified model. *Biointerphases* **2016**, *11*, 018902. [CrossRef]
87. Molteni, C.; Abicht, H.K.; Solioz, M. Killing of bacteria by copper surfaces involves dissolved copper. *Appl. Environ. Microbiol.* **2010**, *76*, 4099–4101. [CrossRef] [PubMed]
88. Mystkowska, J. Biocorrosion of dental alloys due to Desulfotomaculum nigrificans bacteria. *Acta Bioeng. Biomech.* **2016**, *18*, 87–96. [CrossRef] [PubMed]
89. Ward, B.C.; Webster, T.J. The effect of nanotopography on calcium and phosphorus deposition on metallic materials in vitro. *Clin. Oral Implants Res.* **2006**, *27*, 3064–3074. [CrossRef] [PubMed]
90. Souza, J.C.; Ponthiaux, P. Corrosion behaviour of titanium in the presence of Streptococcus mutans. *J. Dent.* **2013**, *41*, 528–534. [CrossRef] [PubMed]
91. McGinley, E.L.; Dowling, A.H.; Moran, G.P. Influence of S. mutans on base-metal dental casting alloy toxicity. *J. Dent. Res.* **2013**, *92*, 92–97. [CrossRef] [PubMed]
92. Lucchetti, M.C.; Fratto, G.; Valeriani, F. Cobalt-chromium alloys in dentistry: An evaluation of metal ion release. *J. Prosthet. Dent.* **2015**, *114*, 602–608. [CrossRef] [PubMed]
93. Pereira, S.; Anami, L.C.; Pereira, C.A. Bacterial Colonization in the Marginal Region of Ceramic Restorations: Effects of Different Cement Removal Methods and Polishing. *Oper. Dent.* **2016**, *41*, 642–654. [CrossRef] [PubMed]
94. Anami, L.C.; Pereira, C.A.; Guerra, E. Morphology and bacterial colonisation of tooth/ceramic restoration interface after different cement excess removal techniques. *J. Dent.* **2012**, *40*, 742–749. [CrossRef] [PubMed]
95. Hahnel, S.; Rosentritt, M.; Handel, G. Surface characterization of dental ceramics and initial streptococcal adhesion in vitro. *Dent. Mater.* **2009**, *25*, 969–975. [CrossRef] [PubMed]
96. Bremer, F.; Grade, S.; Kohorst, P.; Stiesch, M. In vivo biofilm formation on different dental ceramics. *Quintessence Int.* **2011**, *42*, 565. [PubMed]
97. Klinge, B.; Meyle, J. Peri-implant tissue destruction. The Third EAO Consensus Conference 2012. *Clin. Oral Implants Res.* **2012**, *23* (Suppl. S6), 108–110. [CrossRef] [PubMed]

98. Serino, G.; Strom, C. Peri-implantitis in partially edentulous patients: Association with inadequate plaque control. *Clin. Oral Implants Res.* **2009**, *20*, 169–174. [CrossRef] [PubMed]
99. Lindhe, J.; Meyle, J.; Group, D. Peri-implant diseases: Consensus Report of the Sixth European Workshop on Periodontology. *J. Clin. Periodontol.* **2008**, *35*, 282–285. [CrossRef] [PubMed]
100. Zitzmann, N.U.; Berglundh, T. Definition and prevalence of peri-implant diseases. *J. Clin. Periodontol.* **2008**, *35*, 286–291. [CrossRef] [PubMed]
101. Subramani, K.; Jung, R.E.; Molenberg, A.; Hämmerle, C.H. Biofilm on dental implants: A review of the literature. *Int. J. Oral Maxillofac. Implants* **2009**. [CrossRef]
102. De Avila, E.D.; Avila-Campos, M.J.; Vergani, C.E.; Spolidorio, D.M.; Mollo Fde, A., Jr. Structural and quantitative analysis of a mature anaerobic biofilm on different implant abutment surfaces. *J. Prosthet. Dent.* **2016**, *115*, 428–436. [CrossRef] [PubMed]
103. Elter, C.; Heuer, W.; Demling, A.; Hannig, M.; Heidenblut, T.; Bach, F.W.; Stiesch-Scholz, M. Supra- and subgingival biofilm formation on implant abutments with different surface characteristics. *Int. J. Oral Maxillofac. Implants* **2008**, *23*, 327–334. [PubMed]
104. Degidi, M.; Artese, L.; Scarano, A.; Perrotti, V.; Gehrke, P.; Piattelli, A. Inflammatory Infiltrate, Microvessel Density, Nitric Oxide Synthase Expression, Vascular Endothelial Growth Factor Expression, and Proliferative Activity in Peri-Implant Soft Tissues Around Titanium and Zirconium Oxide Healing Caps. *J. Periodontol.* **2006**, *77*, 73–80. [CrossRef] [PubMed]
105. Welander, M.; Abrahamsson, I.; Berglundh, T. The mucosal barrier at implant abutments of different materials. *Clin. Oral Implants Res.* **2008**, *19*, 635–641. [CrossRef] [PubMed]
106. Zhao, B.; van der Mei, H.C.; Subbiahdoss, G.; de Vries, J.; Rustema-Abbing, M.; Kuijer, R.; Busscher, H.J.; Ren, Y. Soft tissue integration versus early biofilm formation on different dental implant materials. *Dent. Mater.* **2014**, *30*, 716–727. [CrossRef] [PubMed]
107. Nascimento, C.; Pita, M.S.; Santos Ede, S.; Monesi, N.; Pedrazzi, V.; Albuquerque Junior, R.F.; Ribeiro, R.F. Microbiome of titanium and zirconia dental implants abutments. *Dent. Mater.* **2016**, *32*, 93–101. [CrossRef] [PubMed]
108. Do Nascimento, C.; Pita, M.S.; Pedrazzi, V. In vivo evaluation of Candida spp. adhesion on titanium or zirconia abutment surfaces. *Arch. Oral Biol.* **2013**, *58*, 853–861. [CrossRef] [PubMed]
109. Nascimento, C.D.; Pita, M.S.; Fernandes, F.H.N.C.; Pedrazzi, V.; de Albuquerque Junior, R.F.; Ribeiro, R.F. Bacterial adhesion on the titanium and zirconia abutment surfaces. *Clin. Oral Implants Res.* **2014**, *25*, 337–343. [CrossRef] [PubMed]
110. De Freitas, A.R.; Silva, T.S.O.; Ribeiro, R.F.; de Albuquerque Junior, R.F.; Pedrazzi, V.; do Nascimento, C. Oral bacterial colonization on dental implants restored with titanium or zirconia abutments: 6-month follow-up. *Clin. Oral Investig.* **2018**. [CrossRef] [PubMed]
111. Raffaini, F.C.; Freitas, A.R.; Silva, T.S.O.; Cavagioni, T.; Oliveira, J.F.; Albuquerque Junior, R.F.; Pedrazzi, V.; Ribeiro, R.F.; do Nascimento, C. Genome analysis and clinical implications of the bacterial communities in early biofilm formation on dental implants restored with titanium or zirconia abutments. *Biofouling* **2018**, *34*, 173–182. [CrossRef] [PubMed]
112. Jemat, A.; Ghazali, M.J.; Razali, M.; Otsuka, Y. Surface Modifications and Their Effects on Titanium Dental Implants. *Biomed. Res. Int.* **2015**, *2015*, 791725. [CrossRef] [PubMed]
113. Aparicio, C.; Rodriguez, D.; Gil, F.J. Variation of roughness and adhesion strength of deposited apatite layers on titanium dental implants. *Mater. Sci. Eng. C Mater. Biol. Appl.* **2011**, *31*, 320–324. [CrossRef]
114. San Thian, E.; Huang, J.; Barber, Z.H.; Best, S.M.; Bonfield, W. Surface modification of magnetron-sputtered hydroxyapatite thin films via silicon substitution for orthopaedic and dental applications. *Surf. Coat. Technol.* **2011**, *205*, 3472–3477. [CrossRef]
115. He, F.M.; Yang, G.L.; Li, Y.N. Early bone response to sandblasted, dual acid-etched and H2O2/HCl treated titanium implants: An experimental study in the rabbit. *Int. J. Oral Maxillofac. Surg.* **2009**, *38*, 677–681. [CrossRef] [PubMed]
116. Yang, C.Y.; Lee, T.M.; Lu, Y.Z.; Yang, C.W.; Lui, T.S.; Kuo, A.; Huang, B.W. The influence of plasma spraying parameters on the characteristics of fluorapatite coatings. *J. Med. Biol. Eng.* **2010**, *30*, 91–98.
117. Schmidlin, P.R.; Mueller, P.; Attin, T.; Wieland, M.; Hofer, D.; Guggenheim, B. Polyspecies biofilm formation on implant surfaces with different surface characteristics. *J. Appl. Oral Sci.* **2013**, *21*, 48–55. [CrossRef] [PubMed]

118. Matos, A.O.; Ricomini-Filho, A.P.; Beline, T.; Ogawa, E.S.; Costa-Oliveira, B.E.; de Almeida, A.B.; Nociti Junior, F.H.; Rangel, E.C.; da Cruz, N.C.; Sukotjo, C.; et al. Three-species biofilm model onto plasma-treated titanium implant surface. *Colloids Surf. B Biointerfaces* **2017**, *152*, 354–366. [CrossRef] [PubMed]
119. Al-Ahmad, A.; Wiedmann-Al-Ahmad, M.; Fackler, A.; Follo, M.; Hellwig, E.; Bachle, M.; Hannig, C.; Han, J.S.; Wolkewitz, M.; Kohal, R. In vivo study of the initial bacterial adhesion on different implant materials. *Arch. Oral Biol.* **2013**, *58*, 1139–1147. [CrossRef] [PubMed]
120. Roehling, S.; Astasov-Frauenhoffer, M.; Hauser-Gerspach, I.; Braissant, O.; Woelfler, H.; Waltimo, T.; Kniha, H.; Gahlert, M. In Vitro Biofilm Formation on Titanium and Zirconia Implant Surfaces. *J. Periodontol.* **2017**, *88*, 298–307. [CrossRef] [PubMed]

© 2018 by the authors. Licensee MDPI, Basel, Switzerland. This article is an open access article distributed under the terms and conditions of the Creative Commons Attribution (CC BY) license (http://creativecommons.org/licenses/by/4.0/).

MDPI
St. Alban-Anlage 66
4052 Basel
Switzerland
Tel. +41 61 683 77 34
Fax +41 61 302 89 18
www.mdpi.com

International Journal of Molecular Sciences Editorial Office
E-mail: ijms@mdpi.com
www.mdpi.com/journal/ijms

www.ingramcontent.com/pod-product-compliance
Lightning Source LLC
LaVergne TN
LVHW070049120526
838202LV00101B/1846